Sustainability in Plant and Crop Protection

Series editor

Aurelio Ciancio, Sezione di Bari, Consiglio Nazionale delle Ricerche Istituto per la Protezione delle Piante, Bari, Italy

More information about this series at http://www.springer.com/series/13031

Md. Aslam Khan • Wasim Ahmad
Editors

Termites and Sustainable Management

Volume 2 - Economic Losses and Management

Springer

Editors
Md. Aslam Khan
Department of Biology, Faculty of Science
Jazan University
Jazan, Saudi Arabia

Wasim Ahmad
Department of Zoology, Section of Nematology
Aligarh Muslim University
Aligarh, Uttar Pradesh, India

Sustainability in Plant and Crop Protection
ISBN 978-3-319-68725-4 ISBN 978-3-319-68726-1 (eBook)
https://doi.org/10.1007/978-3-319-68726-1

Library of Congress Control Number: 2017957654

Printed on acid-free paper

This Springer imprint is published by Springer Nature
The registered company is Springer International Publishing AG
The registered company address is: Gewerbestrasse 11, 6330 Cham, Switzerland

Foreword

Termites are the most important component of the living world accounting for about ten per cent of animal biomass in tropics and subtropics. Besides the essential roles played by termites in organic matter decomposition, nutrient cycling, soil airing, and draining, they are an important pest of temperate and tropical regions. They are responsible for massive loss to agricultural crops, forests, commercial plantations, and wooden structures in buildings, and billions of dollars are spent on its control.

Traditional termite control methods involve injecting hundreds of liters of synthetic insecticides to the soil or the use of termite bait products containing insect growth regulators. The former poses risk to both man and environment and the latter is relatively expensive. There is increasing general recognition of the need to improve termite management practices and reducing the harmful environmental effects. This led to a paradigm shift in the termite management strategies being practiced. In this juncture, it is praiseworthy to have a book comprising of different methods for sustainable management of this notorious pest.

The current volume should prove a very timely action in this direction. Different chapters in this book provide valuable information in this regard. The editors of this volume together with the authors of the individual chapters have made a remarkable contribution in collating the up-to-date information on sustainable and ecofriendly management against termite pest.

This information could be useful for researchers, educators, students, and industry persons for understanding and developing ecofriendly and sustainable termite management strategies. This book comprehensively addresses various methods related to sustainable management of termite pest through the expertise of leading authors worldwide. Finally, this book in the series Sustainability in Plant and Crop Protection is highly innovative in covering both basic information and effective management of termite pest.

I congratulate the editors and the various authors of this volume for such a splendid contribution on this notorious pest.

Bengaluru, India

S. Ayyappan
NABARD Chair Professor &
Former Director General (ICAR)

Preface

This is the second volume on termites presented in the series *Sustainability in Plant and Crop Protection*. The previous one dealt with biology, social behaviour and economic importance of these pests. In the present volume, the editors and authors focus on many applied aspects of this important group of insects, including biological factors underpinning the economic losses they induce and related management issues. The volume unravels the many facets of this fundamental and often hidden side of the forest and soil ecology, with particular attention to the tropics. The role of termites in environment construction and maintenance is also examined, considering not only the soil and crop ecology but also the many issues related to anthropic space and activities.

The volume considers many aspects, starting with termites biology, fungus-growing species and the damages observed on tropical crops. Management issues are afforded and reviewed in subsequent chapters, dealing with forest and agricultural systems. The following contributions include reviews or experimental data, with methods and applied technologies dealing with indoor and outdoor pest control.

Today, biodiversity conservation and environment protection represent unavoidable strategies in any sustainable management plan. In this view, several types of eco-friendly approaches in termite control are examined in chapters spanning from organic agriculture and biological control to the use of botanicals and other products.

As for many other pests, human-induced changes (first of all agriculture and deforestation) that permanently modified the environment are indeed part of the problem. In this view, termite problems should be considered the result of a mismatch between the original, pristine environment (with all its builders, food webs and organisms) and the growing human needs, such as land required for food production and space needed for lodging. This situation may be ultimately viewed as a conflict between two different types of social species, that is termites and humankind.

This aspect led to the identification and adoption of solutions to the termite problems, such as the use of physical barriers or chemical compounds. Being external to the microcosm in which termites coevolved with their food sources and environment, external factors produce a significant, immediate pressure on the pests' biology and dynamics, as well as on the surrounding environment. These issues are revised in the final chapters that also offer the reader a clear view about the impact of termites on human activities and actual possible solutions.

Being complex in their nature, studies of this kind are, therefore, welcome, and their results should be added to the progress and knowledge already achieved. The scientists who edited and contributed to the volume have a long-term experience in the field, for which they represent a leading edge, producing innovative and updated research efforts. The reader may look with interest at the informations provided, which are supported by rich bibliographies.

We hope that these data will provide us a clearer insight on the complexity of the ecology and role of these insects, in order to coexist with them with minimal damage to the environment, on human activities and our future social needs.

Aurelio Ciancio
IPSP CNR Bari, Italy
SUPP Series Editor

Preface

Termites are among the most successful group of insects on earth and contribute significantly to most of the world's ecosystems. Their economic importance is twofold, both extremely beneficial as well as extremely injurious to man. Termite species gain pest status when they damage building materials or agronomic and forestry commodities. As pests, they have been associated with crops since immemorial times. They can damage right from sowing till the harvest of the crops. Billions of dollars are spent annually throughout the world to control and prevent termite infestation. The cryptic nature and social organization of termites explain why infestations can be difficult to study and control. In the past, attention was mainly directed towards their biology, systematics, sociality and symbioses. It is therefore imperative to study applied termite biology and control.

Chemical control is the most popular and effective method. About two thirds of treatments by pest control companies rely on the use of liquid insecticides in soil. However, the deleterious effect of continuous usage of chemical termiticides is a serious concern, and many of the most effective chemical agents are now banned under environmental regulations. So, there is a dire need to develop termite management strategies with the least possible cost and harm to non-target organisms, present in the environment.

This volume comprises 13 chapters in an attempt to bring all available information on sustainable and eco-friendly termite management. The previous volume considered the biology, social behaviour and economic importance of these insects. Chapters in this book dealing with damage and specific management of fungus-growing termites provide a review on most recent methodologies used for management. Termites damage crops from sowing till harvest. As it is difficult to detect damages in the field, usually it is too late when the symptoms are noticed. A separate chapter on issues related to Indian agriculture and the contemporary practices being followed by majority of the Indian farmers is quite informative. Similarly, a case study for termites infesting Malaysian forests constitutes an important contribution. Various issues related to integrated and eco-friendly termite management in tropical conditions have been addressed comprehensively. Potential role of microbes has also been discussed in detail in other chapters. The information contained under these chapters should help termite management in a way that natural resources can be used and maintained for the generations to come. Similarly, the chapter on physical barriers contributes a wealth of information that can be useful all over the world where termite is a problem. Emphasis has been laid on reviewing contribution of synthetic chemical insecticides in termite management. A separate chapter dealing with standard norms in wood protection constitutes a significant step in this direction. A further chapter throws light on the potential of biotechnology as a tool in management.

Present volume covers several facets of termites with a focus given on eco-friendly means of management. We hope that it will be helpful to students, teachers, researchers and industry technicians. We are highly grateful to all the authors for providing their expertise in the form of stimulating contributions. Thanks are due to the Head of the Biology Department and Dean of the Faculty of Science, Jazan University, Jazan, for their constant support. We are grateful to Dr. Aurelio Ciancio, Bari, Italy, for including these two volumes dealing with termites and sustainable management in the Springer series "Sustainability in Plant and Crop Protection". We extend our thanks to the Springer International team for their generous cooperation at every stage of the book production. Md. Aslam Khan acknowledges the Research Centre for Environmental Studies, Jazan University, KSA, for the financial assistance CERS 7/2013.

Professor Wasim Ahmad is thankful to the Dean of Faculty of Life Sciences and the Vice Chancellor of Aligarh Muslim University for the encouragement and support.

Jazan, Saudi Arabia Md. Aslam Khan
Aligarh, India Wasim Ahmad

Contents

Chapter 1
The Fungus-Growing Termites: Biology, Damage on Tropical Crops and Specific Management

M. Diouf and C. Rouland-Lefevre

Contents

M. Diouf (✉)
Département ECOEVO, équipe Biologie des termites et fonctionnement des écosystèmes tropicaux (EcoTerm). Faculté des Sciences et Technologie, Institut d'Ecologie et des Sciences de l'Environnement de Paris (IEES, Paris), 61 Avenue du Général de Gaulle, 94010 Créteil, France
e-mail: michel.diouf@u-pec.fr

C. Rouland-Lefevre
Département ECOEVO, équipe Biologie des termites et fonctionnement des écosystèmes tropicaux (EcoTerm). Centre IRD France Nord, Institut d'Ecologie et des Sciences de l'Environnement de Paris (IEES, Paris), 32 Avenue Henri Varagnat, 93143 Bondy, France

M.A. Khan, W. Ahmad (eds.), *Termites and Sustainable Management*, Sustainability in Plant and Crop Protection,
https://doi.org/10.1007/978-3-319-68726-1_1

Abstract Termites have a significant ecological role in natural, tropical ecosystems. However, some species (around 10%) have potential to become pests in agricultural systems. In particular, some species from the family Macrotermitinae, also known as fungus-growing termites, are described as significant crop pests. Widely present in Africa and Asia but absent from the other continents, these termites are featured by an original digestive symbiosis with a basidiomycete fungus belonging to the genus *Termitomyces.* This specific fungus allows a fast and efficient digestion of plant material by the host. Consequently, these termites have a great ability to adapt to various food sources, making them a potentially harmful group during the introduction of new crops. Despite the economic impact of these crop pests, there is to date no method for rapid and specific management of these insects. The existing control strategies have weak efficiency and require excessive amounts of chemicals. Generally, these techniques consist of repeated spraying of pesticides on the floor, leading to a more or less massive destruction of nontarget species and a gradual deterioration of the environment. More specific control methods have been tested as alternatives and have clearly shown better efficiency, compared to conventional methods. These are mainly based on baits treated with attractants (pheromones) or on growth inhibitors specifically targeting the fungal symbiont. This chapter provides a review of the biology of fungus-growing termites, focusing on their specific nutrition mode, and presents the most recent methodology used for their management.

Keywords Macrotermitinae • *Termitomyces* • Symbiosis • Crop pests • Control strategies

1.1 Introduction

1.1.1 Biology and Taxonomy of Fungus-Growing Termites

Fungus-growing termites constitute the subfamily of Macrotermitinae, belonging to the Termitidae family, known as flagellate-free termites (Eggleton 2006). This family represents the overwhelming majority of current termite species and, above all, of tropical species (Eggleton 2011). The loss of fermentative flagellates has been one of the most critical steps in the evolution of termites (Eggleton 2006). This loss has been compensated in different ways by the different groups of Termitidae. In most subfamilies of Termitidae, it was compensated by a higher complexification/compartmentation of the gut and by an evolution toward a hummus-feeding diet (with frequent reversions toward a wood-feeding guild). The subfamily Macrotermitinae, however, compensated this loss largely through the adoption of an extra-digestive fungus symbiont (Eggleton 2006). This hypothesis is supported by recent phylogenies confirming the monophyly of the Termitidae and nesting Macrotermitinae sometimes with the genus *Sphaerotermes* (building symbiont-free comb) as a sister group of all the remaining subfamilies of Termitidae (Inward et al. 2007; Legendre et al. 2008; Engel et al. 2009). This particularity of fungus-growing termites is also reflected by the simple structure of their gut, which is rather similar to the gut structure of lower termites (with gut flagellates), than to those of other Termitidae (Bignell and Eggleton 1995). The monophyly of the fundus-growing termites suggests a single origin of their association with the fungal symbiont. The genus *Sphaerotermes*, which is generally associated to fungus-growing termites, belongs in fact to another subfamily (Sphaerotermitinae) (Engel and Krishna 2004).

Besides the adverse effects that some fungus-growing species can have on some human activities, Macrotermitinae are acknowledged to play a crucial ecological role in their natural environment. They play a significant role in organic matter cycling and soil structuration (Wood and Sands 1978; Collins 1981; Bignell and Eggleton 2000; Holt and Lepage 2000; Jouquet et al. 2005; Schuurman 2012), particularly in African savannahs. Around 20% of the carbon mineralization is attributable to their activity in certain African savannahs, where their specific diversity is generally significant (Bignell and Eggleton 2000). In these ecosystems, with high density of fungus-growing termites, they similarly contribute to the improvement of the chemical and physical properties of soil (Holt and Lepage 2000). Their impact on the carbon fluxes is less apparent in tropical forest ecosystems (1–2% of C mineralization) (Bignell and Eggleton 2000), which yet is their native natural habitat (Aanen and Eggleton 2005).

Fungus-growing termites are exclusively present in Tropical Africa (including in Madagascar) and in Indo-Malayan and Arabic areas of Asia. The diversity of fungus-growing termites is estimated to around 330 species, affiliated to 12 genera. The number of genera could be slightly lower since phylogenetic analyses have revealed that the genus *Hypotermes*, which was thought to be endemic in Asia, nested within the genus *Odontotermes* (Aanen et al. 2002). All genera (except the putative genus *Hypotermes*) are present in Africa against only four in Asia and only one in Madagascar (Eggleton 2000; Nobre et al. 2010), though the presence of two more genera (*Odontotermes* and *Ancistrotermes*) in the Malagasy area has been reported (Eggleton and Davies 2003).

1.1.2 Nest Construction

As for other social insects living in densely populated colonies, termites have to face various constraints, including the maintenance of thermo-hygrometric conditions and gas exchanges in a range compatible with their biology. They overcome these constraints largely thanks to their nest structure. Of all the termites, and probably of all nonhuman animals, structures built by Macrotermitinae are of the most complex and the most impressive in nature (Eggleton 2011). Apart from their size and complexity, mounds of Macrotermitinae differ from those of other termites by the noninclusion of feces in the building material (Grasse 1984; Noirot and Darlington 2000). A synthetic and simplified view of their nest structure generally considers an underground part housing the brood and the fungus comb and a less populated epigeal part with highly variable height, which is mainly dedicated to construction and repair and to the regulation of internal parameters at the nest-air interface. However, this simplified view hides a higher variability in nest structure and construction modes (Korb 2011). The aspects related to ventilation, thermoregulation, and homeostasis in the nests of Macrotermitinae, responsible for suitable and stable conditions for the termite and the fungal symbiont regardless of the fluctuations of external parameters, have been reviewed by Korb (2011) and recently enriched by a new study (King et al. 2015). These aspects are not detailed in this chapter. According to the significance of the epigeal part on one hand and to the concentration of diffuse character of the underground part, two types of mounds that are worth taking into account in pest management can be distinguished in Macrotermitinae (Grasse 1984; Noirot and Darlington 2000).

Subterranean Nests These are entirely hypogeal nests consisting in many diffuse chambers containing fungus combs and disposed either parallel to the soil surface or at different depths. These chambers communicate between them and with the royal cell by a dense network of galleries. For few species, a more or less significant part of the nest can be perceptible above the ground level. This is the case of nests of genera such as *Microtermes*, *Ancistrotermes*, *Acanthotermes*, and many species of *Odontotermes*.

Epigeal Nests They emerge from a relatively common subterranean stage made of a single chamber containing the primary fungus comb and the colony population and bounded by a spherical inner wall. This chamber and the wall then gradually expand as the size of the nest increases and are surrounded by a huge empty space (without brood) which may limit the access of predators. When the nest emerges above soil surface, the continuation of the nest growth can follow two distinct developmental models: (1) either the single chamber extends laterally, which results in a suppression of the empty surrounding space and gives rise to a more diffuse internal part of the nest, as is the case of *M. subhyalinus* in West African savannahs and *M. falciger*, or (2) this chamber remains isolated and is open outside the nest by a chimney system. Mounds of species such as *M. jeanneli* and *O. transvaalensis* (Ruelle 1985; Turner 1994) have a single chimney, while those of species such as *M. bellicosus* and *M. subhyalinus* from Kenya (reviewed by Korb 2011) and *O. obesus* (King et al. 2015), *O. latericius*, and *O. transvaalensis* in South Africa (Coaton 1961; Grasse 1984) are open outside by multiple chimneys. It should also be noted that except the case of synonymies, same species may adopt different construction modes depending on the habitat, in response to the environmental constraints (Korb 2011).

1.1.3 Social Organization

As in most of the Termitidae subfamilies, a colony population of Macrotermitinae is made on three main castes, the sexual line, soldiers, and workers, supplemented with juvenile stages (eggs, larvae, nymphs, pre-soldiers). However, this simplified structure conceals further subdivisions of castes in morphs performing different roles in the colony functioning.

The Sexual Line This is represented by the royal pair, the king and the queen, originating from nymphs. Nymphs already differ from larvae by the second instar, though the presence of wing pads (Grasse 1984; Noirot 1985). As in other Termitidae, the imagoes (alates) emerge after five nymphal molts (Noirot 1985). The subfamily Macrotermitinae is restricted to tropical areas. Consequently, the production of sexual individuals is seasonal, contrary to neuters which are continuously produced. Following the swarming flight, alates of both sexes either from the same colony or from different colonies form couples, and each of them builds a copularium which will become the royal cell. This royal cell is accessible to feeding workers and soldiers, but not to the brood. In most cases, the sexual line is monogamous. However, some cases of polygyny or polyandry can occur, as, for example, in *Macrotermes michaelseni* (Darlington 1985; Brandl et al. 2001; Kaib et al. 2001; Brandl et al. 2004; Hacker et al. 2005) and *Macrotermes herus* (Darlington 1988). The coexistence in these cases mainly stems from pleometrosis.[1] The presence of multiple reproductives has also been reported in *M. subhyalinus* (Mulatu and Emana 2015). The presence of multiple reproducers can also be induced by the loss of primary reproducers. In this case, secondary reproductives are nymphoids or adultoids, since neotenic ergatoids are unknown in this subfamily (Grasse 1984; Myles 1999). A colony orphaned, for a period in which it is devoid of nymphs and alates, is therefore doomed to disappear (Grasse 1982). From a biological point of view, the nonexistence of ergatoids in Macrotermitinae contrasts with the high-energy investment made for the construction of these highly structured colonies.

The Worker Caste This caste is polymorph in fungus-growing termites, consisting in two morphs: the major workers, which are males deriving from the larvae of the third instar, and the minor workers, which are females derived from the third instar of the small larvae (Mensa-Bonsu 1978; Okot-Kotber 1981; Grasse 1982; Neoh and Lee 2009). In addition to this polymorphism, there is an age-dependent polyethism in the worker caste of fungus-growing termites, with distinction between (1) old workers featured by a darken abdomen (related to the alimentary bolus), which are responsible for collecting the raw food outside the nest and in return feed on the mature part of the fungus comb rich in mycelia and not in conidia, and (2) young workers, which stay inside the nest and are responsible for the construction and inoculation of the fungus comb. For this task, they ingest the raw food brought to the nest by old workers, and following a rapid intestinal transit without a real

[1] Pleometrosis is the co-foundation by many alates.

digestion, pellet it on the top of the existing comb. One function of this transit among others is the mixing of the ingested material with the fungal conidia which are the main nutritional resource of these young workers (Leuthold et al. 1989). Young workers are also food providers for the brood and newly moulted individuals through stomodeal trophallaxis. This polyethism of the worker caste has been described in many species not only from the genus *Macrotermes* (Okot-Kotber 1981; Badertscher et al. 1983; Rouland-Lefevre 2000; Hinze et al. 2002; Korb 2011) but also from *Odontotermes* (Li et al. 2015, 2016) and may be universal within the subfamily.

The Soldier Caste This is certainly the most variable caste with regard to the origin (including the sex) and the number of morphs, as compared to the other castes. The presence of two distinct soldier morphs, all of which are females derived from the third instar of small larvae (minor soldiers) or from the small workers of the first stage (major soldiers), is the most common case in fungus-growing termites (*Macrotermes*, *Ancistrotermes*, *Pseudacanthotermes*) (Noirot 1955; Mensa-Bonsu 1978; Grasse 1982; Eggleton 2000). However, there are some exceptions to this general model. For instance, in genera such as *Odontotermes*, *Protermes*, and *Microtermes*, the soldier castes likely consist in a single morph entirely derived from the small larvae of the third instar and are therefore females. On the contrary, the distinction of up to three soldier morphs has been reported in *Acanthotermes acanthothorax*. These morphs are all from the same sex (males) and originate, respectively, from the second instar (minor soldiers), from the third instar of small larvae (medium soldiers), or from minor workers of the first stage (major soldiers) (Noirot 1955; Grasse 1982). As for workers, the distinct morphs of the soldier caste may perform defensive functions at different colony levels. For instance, in *Macrotermes jeanneli*, minor soldiers escort workers inside and outside the nest, while major soldiers guard the colony entrances, including the galleries and chimney (Leuthold et al. 2004).

1.1.4 Colony Odors and Nestmate Recognition

For a long time, the chemical communication in termites has been considered less important as compared to Hymenoptera such as bees and ants. However, this idea may have been skewed by a smaller number of investigations regarding this aspect of the termite biology. The bulk of interactions governing the social life within a colony is controlled by chemical signals. These signals are involved in reproduction (sexual pheromones), colony defense (alarm, defense, and aggregation signals), and nestmate and intercolonial recognition and forage (trail-following and recruitment pheromones) (Bordereau and Pasteels 2011).

From a pest management perspective, among the wide range of molecules involved in the chemical communication, two main categories regulating the interactions between workers are particularly interesting: trail pheromones and cuticular hydrocarbons.

Trail-Following Pheromones These chemicals have a single source: the sternal glands present in all the termite castes (Noirot 1969; Quennedey et al. 2008; Costa-Leonardo et al. 2009; Bordereau and Pasteels 2011; Bagneres and Hanus 2015). They are released when the termite presses its abdomen on the substrate. These pheromones are involved in the orientation and the recruitment of individuals during the foraging activities. Their secretion pattern varies depending on whether workers are actively exploring food or have already found it, thereby triggering a precise rush of workers to the identified substrate (Reinhard and Kaib 2001). One limitation of the use of these volatiles for targeted control of termite pests is the very low variability within Termitidae on one hand and within lower termites on the other hand (Peppuy et al. 2001; Costa-Leonardo et al. 2009; Bordereau and Pasteels 2011). This surprising redundancy is illustrated by the description of less than ten distinct pheromones in some sixty termite species, from diverse subfamilies (Bordereau and Pasteels 2011; Bagneres and Hanus 2015).

Cuticular Hydrocarbons (CHCs) CHCs are a mixture of straight-chain, methylated, saturated, and unsaturated hydrocarbons (Lockey 1988; Estrada-Pena et al. 1994; Ye et al. 2007; Bagneres and Hanus 2015) involved in the acceptance or rejection of individuals as nestmates or non-nestmates. CHCs are synthetized by oenocytes, particularly cells either associated to the epidermal cells or located to the periphery of the fat body, and then carried by transport proteins (lipophorins) through hemolymph. However, their way of transfer from lipophorins to epidermal cells and how they pass through these cells toward epicuticle are still unresolved points (Bagneres and Blomquist 2010). Given their role in interindividual recognition, CHCs are crucial in the social behavior of termites and in the sharing of special and alimentary resources between sympatric species (Adams 1991; Bagneres et al. 1991; Takahashi and Gassa 1995; Clement and Bagneres 1998; Kaib et al. 2002; Howard and Blomquist 2005). Indeed, little or no aggressive behavior exists between individuals with similar CHC composition and vice versa (Haverty and Thorne 1989; Adams 1991; Haverty et al. 1999; Kaib et al. 2004).

Opposite to the highly conserved trail-following pheromones, CHCs are much more variable. Variations can be recorded even between conspecific individuals from different colonies, but the larger variation magnitudes are observed among distinct species (Estrada-Pena et al. 1994; Haverty and Nelson 1997; Takematsu and Yamaoka 1999; Haverty et al. 2000; Peppuy et al. 2001; Page et al. 2002). Given this high variability, CHCs were expected to be an interesting chemical for taxonomic identification and termite targeting. In this respect, some studies performed on some lower termites have highlighted correlations between variations of CHC phenotypes and mitochondrial and/or nuclear markers, in cryptic species (Jenkins et al. 2000; Kutnik et al. 2004; Copren et al. 2005; Aldrich and Kambhampati 2007). A similar trend could be expected in Termitidae, including in Macrotermitinae. From the findings reported for different species of *Odontotermes*, it has been suggested that CHCs could be useful in refining phylogenetic and evolutionary relationships among species from this genus (Kaib et al. 1991). However, the studies performed afterward on CHCs of other Macrotermitinae, especially from the genus *Macrotermes*, highlighted a significant intraspecific variability among colonies

(Bagine et al. 1994; Kaib et al. 2002, 2004; Marten et al. 2009, 2010), suggesting that distinct CHC phenotypes were genetically very close and did not correspond to cryptic species. The same was true for colonies of *M. subhyalinus* from the Comoe National Park of Ivory Coast (Kaib et al. 2004). The comparison of the CHCs from various colonies of *Macrotermes* species collected from West and East Africa confirmed that the delimitation of species based on CHC phenotypes did not match the delimitation based on genetic markers (Marten et al. 2009). Therefore, they may only be used in taxonomy as a complement to the commonly used tools, such as genetic markers.

The heritability in CHC composition is likely very weak. Due to the position of the epicuticle (where the CHCs are located) at the interface with the environment, the secreted CHCs may face a significant influence of various extrinsic factors. The inquilinism, whose effect could be further magnified in Macrotermitinae due to the longevity of their colony, and thus the cohabitation, may represent a significant source of variability.

High CHC similarities have been reported between *Macrotermes herus* and the termitophile *Termitobia herus* (Marten et al. 2010), suggesting a mutual influence on CHCs of both taxa. This influence could be amplified by the varying composition of the inquiline fauna between colonies. Another potential source of CHC variation is related to the symbionts. These consist in two categories in Macrotermitinae, the fungal symbiont and gut microbiota, known to vary slightly among colonies of the same species (Boucias et al. 2013). In *Zootermopsis nevadensis*, it has been reported that a precursor in the biosynthesis of CHCs was produced by the microbiota (Guo et al. 1991). This suggests that the gut procaryotes may influence the production of CHCs by their host. The influence of the fungal symbiont is also highly probable, since strain-specific odors have been evidenced from spores of the *Termitomyces* associated to various species of *Macrotermes* (Rouland-Lefevre et al., personal communication).

1.2 Termite-*Termitomyces* Association

Fungus-growing termites of the family Termitidae encompass all termite subfamilies having secondarily lost the protozoan gut symbionts. They are the only termite group having developed a clearly identified extra-digestive mutualism with fungi of the genus *Termitomyces*. The termite trace fossils found in the Northern Chad Basin already dated the presence of this association to the Upper Miocene-Lower Pliocene (7–10 Myrs ago) (Duringer et al. 2006; Duringer et al. 2007). More recent trace fossils from Rukwa Rift Basin in Tanzania, dated to be around 25 Myrs ago, have suggested an African Paleogene origin of this fungiculture (around 31 Myrs ago) (Roberts et al. 2016).

According to recent phylogenetic analyses, fungus-growing termites are a monophyletic subfamily. The genus *Sphaerotermes*, building combs without *Termitomyces* and which was considered as having secondarily lost this symbiont, is now established as belonging to another termite lineage (Engel and Krishna 2004; Inward et al. 2007). As for the fungus partners, they also fell within the single genus *Termitomyces*, member of Agaricales, in which they are nested within Lyophyllaceae, and their closer nonsymbiotic sister group is the genus *Lyophyllum*. Similarly to the genus, *Termitomyces* forms a monophyletic group within Lyophyllaceae (Moncalvo et al. 2000; FroSlev et al. 2003). This monophyly of both interacting taxa suggests on one hand that the symbiosis has a single origin and, on the other hand, that no reversion to the free-living state did occur, thus underlining the obligatory character of this mutualism (Aanen et al. 2002; Rouland-Lefevre et al. 2002; Aanen and Eggleton 2005).

As compared to the diversity of the termite partners (330 species), the specific diversity of the fungal symbionts is much lower, being estimated to about 40 described species. However, the morphology of the carpophore generally used to identify fungal species is not suitable enough to distinguish cryptic species (FroSlev et al. 2003), and this could have led to an underestimation of this diversity. Nevertheless, this highly unbalanced host-symbiont ratio suggests a frequent sharing of a same fungus species by several host species. As for the initial event which led to the definitive adoption of this mutualism, two main theories are advanced (Aanen and Jacobus 2006).

The "antagonistic equilibrium theory" was proposed several decades ago by Heim (1977). He suggested that the comb was a natural part of the nest serving as a suitable habitat for the termite brood and from which the termite seeks to suppress the supposedly parasitic fungus. Following this reasoning, *Sphaerotermes* has been the only genus which has successfully removed this presumptive parasitic fungus. However, the membership of *Sphaerotermes* to a termite clade distinct from the Macrotermitinae (mentioned above) and the obligatory nature of mutualism are body of facts contradicting this hypothesis.

The "consumption first" theory (Sands 1969) argues that the symbiosis resulted from the adoption of a beneficial white-rot fungus, which was first a significant food resource. The ancestor of the *Termitomyces* used the fecal material (cartoon) as a substrate for the nest construction. By fungal digestion this substrate became more palatable for the host. This new resource (including fungal biomass and predigested organic matter) was likely behind this association and led to the currently known strict mutualism. The use of rot fungi and their degradation products as food resource is common in termites (Rouland-Lefèvre 2000). The crucial step for the ancestor of Macrotermitinae has therefore been the acquisition of the aptitude to manipulate the fungal partner. Another scenario, "transmission first," according to which the termite was first used by the fungus for dispersal (as a vector) before using it as a resource, was also proposed (Korb and Aanen 2003; Mueller et al. 2005).

Though their ecological impact is more significant in savannahs (Wood and Sands 1978; Buxton 1981; Eggleton 2011; Jones and Eggleton 2011), the Macrotermitinae-*Termitomyces* association likely appeared in the rainy forest of Tropical Africa (Aanen and Eggleton 2005). The phylogenetic reconstructions made from the bulk of current genera (10 out of 11), from the main geographical areas of the group, support this hypothesis. The fungiculture then repeatedly extended to the savannahs and to Asia and Madagascar. The group of white-rot fungi (including *Termitomyces*) has requirements in temperature (with a narrow range) and moisture (close to saturation) found only in the tropical rainy forest and not in savannahs, often marked by significant variations in temperatures and generally low moisture. The conquest of savannahs by partners has likely been made possible through the creation by the host of a homeostatic environment in the inner part of the nest, with nearly optimal conditions (close to those of the forest) for *Termitomyces*, regardless of the external fluctuations (Darlington et al. 1997; Korb 2011).

These optimal conditions for the fungus rarely occur sustainably in savannahs. They enable the growth and the pre-degradation of plant matter by the fungus, thus proving a new significant resource for the host (Aanen and Eggleton 2005). This rainy forest-origin hypothesis is also supported by the genus-level diversity which is higher in the forest as compared to savannahs. Indeed, while all the genera found in savannahs are also present in forests, some genera such as *Acanthotermes*, *Synacanthotermes*, and *Protermes* are endemic to the forest habitat (Aanen and Eggleton 2005), suggesting a "filtering effect" at the genus-level in the dispersal toward savannahs (Aanen and Eggleton 2005).

The higher diversity in forest relative to savannah does not apply to the species level, presumably owing to several adaptive radiations achieved by genera represented in this habitat (Aanen and Eggleton 2005).

1.2.1 The Construction of the Fungus Comb and Mechanisms of the Symbiosis

This building of the fungus comb is performed by the worker caste, featured by a high polyethism in Macrotermitinae (Badertscher et al. 1983; Leuthold et al. 1989; Traniello and Leuthold 2000). As previously mentioned, the older adult workers forage plant material outside and carry it in the nest and feed on the mature part of the comb, predigested and rich in fungal mycelium. As for young workers inside the nest, they preferentially feed on the fungus conidia also called "mycotêtes" (Fig. 1.1). The plant matter brought to the nest by the old morph is ingested by these young workers, and following a first intestinal transit without real digestion processes, it is defecated in the form of pellets, constituent material of the fungus comb. The biological and physiological significance of this first intestinal transit is still not fully understood, but it may allow a mixing of the ingested material with conidia (inoculation) and possibly the removal of eventual unwanted microbial competitors, including spores of others *Termitomyces* (Aanen 2006; Nobre and Aanen 2012) thereby favoring a fungus monoculture per colony (Aanen et al. 2002, 2007, 2009; Katoh et al. 2002). It would also allow the inoculation of the material with gut-derived bacteria found in the fungus comb (Otani et al. 2016), whose function in the comb processing is still not elucidated.

Fig. 1.1 The fungus comb of a Macrotermitinae showing the conidia (or mycotêtes) and mass diffuse mycelia (Photo, C. Rouland-Lefèvre)

Despite the unique origin of mutualism, several variants of fungus combs exist, depending on the organic matter composition, the building mode, and the dynamics of the comb (Rouland-Lefèvre 2000). For instance, within the same genus, the material used in comb construction varies between forest and congeneric species from savannah. For *Macrotermes* forest species such as *M. muelleri* and *M. ivorensis*, the comb is made with discs of dry leaves, while species from savannah, such as *M. subhyalinus* and *M. bellicosus*, use wood pieces or grass twigs (Rouland-Lefevre 2000). For other genera of fungus-growing termites such as *Pseudacanthotermes*, *Acanthotermes*, *Odontotermes*, *Microtermes*, *Ancistrotermes*, various types of plant fragments (roots, leaves and stems) can be used. This variability found in the type of substrate also applies to the chemical composition of fungus combs, according to species. Nevertheless, apart from some rare cases, the content in major macromolecules such as lignin (10–25%) and cellulose (±20%) varies only slightly (Rouland-Lefèvre 2000).

Besides these parameters, the presence or absence of concomitance between building and consumption allows distinguishing two types of dynamics. In species of *Macrotermes* and *Odontotermes*, the fungus comb is subjected to a continuous turnover. While the upper part is continuously fed with new fecal pellets, the lower part predigested by the fungus and rich in fungal mycelium is consumed (Rouland-Lefevre 2000; Rouland-Lefevre and Bignell 2002; Nobre et al. 2011a, b). Opposite to this model, in species from *Pseudacanthotermes*, once the fungus comb construction is completed, no new material is added during ripening period. It is then entirely consumed before the construction of a new comb, in an empty chamber. Between these two extremes, the fungus combs of species from *Ancistrotermes* and *Microtermes* have intermediate dynamics. New material is added to the old comb but far less regularly than in *Macrotermes* and *Odontotermes* (Rouland-Lefevre 2000).

With regard to the respective roles of partners in this mutualism, it has been theorized that before being a direct nutritional resource for the host, the adopted white-rot fungus was first key in depolymerizing lignocellulose, the major component of the host diet (Sands 1969). Nevertheless, several aspects on the respective role of both partners remain to be discovered. The contribution of the host on the biology of the fungus can take various forms:

- The provision of a mechanically pre-pulverized substrate and facilitation of the contact (inoculation) through the first intestinal transit.
- The provision of a habitat with stable and near-optimal conditions for the fungus regardless of the fluctuations of the external parameters (Turner 1994, 2001; Darlington et al. 1997; Korb 2003). This may have been determinant in the dispersion of the couple toward areas drier than their native habitat (Aanen and Eggleton 2005).
- The dispersal of the fungus which is unable to thrive outside the termite nest. According to species, this dispersal is performed either by the sexual line (swarming alates) or by workers.
- The function of an "external rumen" is also suggested by some authors who consider that the first gut transit undoubtedly allows mixing the substrate with the fungal spores and enzymes but also the addition of the host enzymes either produced by the termite itself or by its microbiota (Collins 1983; Nobre and Aanen 2012), the synergetic action of which permitting an optimal degradation of the substrate.

As for the fungal partner, its role in the symbiosis has often been put forward probably due to an unintentional parallelism made with the role of gut protozoa and prokaryotes. The contribution of *Termitomyces* on the lignocellulose decomposition process is acknowledged by almost all studies. Nevertheless, a generalization of their functions is difficult, given the still limited number of studies on one hand and the substantial variability between species on the other hand. However, four main functions of the fungus seem to emerge:

- The degradation of lignin which has been presented as the primary role of *Termitomyces*, which makes the other associated macromolecules such as cellulose and hemicellulose available and degradable (Grasse and Noirot 1958; Grasse 1982; Hyodo et al. 2000). This function has been proven in many species from the genus *Macrotermes* (Rohrmann 1978; Hyodo et al. 2000; Hyodo et al. 2003) but does not seem to be the main function for the fungus species associated to the genus *Odontotermes-Hypotermes*, *Ancistrotermes pakistanicus* and *Pseudacanthotermes militaris* (Hyodo et al. 2003).

- The fungus is also a food resource rich in proteins, overall in its conidia form (Botha and Eicker 1992). This special diet is yet significantly reflected in the composition of the gut microbiota of Macrotermitinae, predominantly made of bacterial groups involved in amino acids and polysaccharides fermentation such as *Bacteroidetes* and *Firmicutes* (Otani et al. 2014, 2016; Poulsen et al. 2014; Poulsen 2015).
- The fungus is a significant provider of enzymes, catalyzing the depolymerization of the major structural carbohydrates (cellulases, hemi-cellulases), thereby decreasing the substrate C/N ratio. The relative contribution of the symbiont cellulases and hemi-cellulases has been assessed in several studies. It is particularly variable according to species and is likely related to the type of substrate. This issue has been extensively discussed in previous reviews (Rouland-Lefevre 2000; Nobre et al. 2011a, b; Nobre and Aanen 2012). Nevertheless, recent studies on the microbiota, which is clearly shaped by the nature of the ingested substrate, highlight a highly similar pattern within the subfamily. Their microbiota is primarily marked by a general predominance of *Bacteroidetes* and *Firmicutes* known for their involvement in cello-oligomer degradation (Liu et al. 2011; Zhang et al. 2014; Brune and Dietrich 2015), while the cellulose-degrading bacteria such *Spirochaetes* and *Fibrobacteres*-TG3 are underrepresented as compared to their abundance in the gut of wood-feeding termites from their sister subfamilies (Poulsen and Boomsma 2005; Hongoh et al. 2006; Mathew et al. 2012; Bastien et al. 2013; Dietrich et al. 2014; Poulsen 2015; Otani et al. 2016; Diouf et al. 2015). Given the established impact of the ingested diet in shaping the gut microbiota (Mikaelyan et al. 2015), this convergent structure of the microbiota suggests a similar involvement of *Termitomyces* in the processing of the comb.

1.2.2 The Interaction Macrotermitinae-Termitomyces: Specificity and Coevolution

1.2.2.1 The Global Coevolutionary Pattern

The first outstanding fact in this symbiosis is the significant disproportion in specific richness between the interacting partners (330 species for Macrotermitinae versus ~40 species for *Termitomyces*). Despite the probable underestimation of the real specific diversity of the symbiont due to the existence of cryptic species (FroSlev et al. 2003; Makonde et al. 2013), this high disproportion suggests that a single symbiont species may be shared by several host species. This has been highlighted by several studies documenting that the same species of *Termitomyces* could be adopted by termites from several species and sometimes from different genera (Rouland-Lefevre et al. 2002; Aanen et al. 2007; Osiemo et al. 2010; Nobre et al. 2011a, b; Makonde et al. 2013). The reverse is also true since a single termite species can be associated with several species of *Termitomyces* (Aanen et al. 2002; Osiemo et al. 2010; Nobre et al. 2011a, b; Makonde et al. 2013). From the data available on GenBank, Makonde and colleagues estimated that, in average, one host species was related to 1.5 to 2 symbiont species. From the bulk of studies on termite-*Termitomyces* co-speciation, it may be stated that the conceptual view relating one fungus species to a single host species can be definitely ruled out. Nevertheless, several fungus switches occur within the same genus, so that, in general, each specific host genus interacts with a single symbiont clade (Aanen et al. 2002; Rouland-Lefevre et al. 2002; Osiemo et al. 2010; Nobre et al. 2011a, b; Poulsen 2015).

Based on the internal transcribed spacers (ITS) of the ribosomal RNA gene regions, which are most frequently used as molecular markers for fungi, around 40–47% of variations were recorded between fungal clades associated to different genera (Aanen et al. 2007; Osiemo et al. 2010) versus about 18% for those related to a single genus (Aanen et al. 2007). A valid coevolutionary signal is therefore acknowledged at the genus level, evidenced by the congruence of phylogenies between fungal clades and host genera. However, few occasional switches with higher magnitude have been reported by many studies. For instance, it was the case for a fungal strain common to *Ancistrotermes*, *Odontotermes*, and *Microtermes* (Rouland-Lefevre et al. 2002); to *Ancistrotermes*, *Microtermes*, and *Synacanthotermes* (Aanen et al. 2002); to *Odontotermes*, *Protermes*, and *Microtermes*; or to *Macrotermes* and *Microtermes* (Makonde et al. 2013). One of the most spectacular switches is the case of very close fungal strains between *Microtermes*, *Ancistrotermes*, *Acanthotermes*, and *Synacanthotermes* from diverse origins (Nobre et al. 2011a, b).

This situation illustrates that recent switches followed by diversification though host genera and geographic areas might occur for certain generalist fungal species (Nobre et al. 2011a, b). However, this king of ubiquity seems to be very scarce in "specialist" fungal strains such as the *M. natalensis* symbiont (De Fine Licht et al. 2007). Despite these variations, a consistent element characterizing this symbiosis is the uniqueness of the fungal strain per colony. Indeed, in the overwhelming majority of studies, a single fungal strain could be identified in each nest (Katoh et al. 2002; Rouland-Lefèvre et al. 2002; Shinzato et al. 2005; De Fine Licht et al. 2006; Aanen et al. 2009; Makonde et al. 2013). The occasionally reported presence of two fungal strains in the same nest might stem from inquilinism (Makonde et al. 2013) and does not contradict the monoculture of single fungus per comb. This monoculture and the host-symbiont coevolution, noted at least at the genus level, raise the question of factors allowing the successful selection of a single strain, which yet in most of the cases is laterally acquired from the surrounding environment at each generation, among spores of other species.

1.2.2.2 Putative Factors Ensuring the Limitation of Symbiont Switches and De Novo Acquisitions

In a mutualistic interaction, a key factor limiting genetic variations, and thereby guaranteeing the specificity of the interaction, is the accuracy of the transmission of symbionts from parents to offsprings (Kaltenpoth et al. 2014). In fungus-growing termites, two transmission modes of the fungal symbionts have been described:

The Horizontal Transmission

This is by far the most widespread and, presumably, the ancestral mode (Korb and Aanen 2003). For the corresponding species, the primary fungus comb is inoculated with sexual spores released in the surrounding environment by the fungus fruiting bodies of adult nests. These spores are collected and brought to the incipient colony by the first cohort of workers (Johnson 1981; Sieber 1983; Wood and Thomas 1989). In this case, there is no direct connection between reproductive cycles of both partners. However, in nature, the emergence of the first workers from the new colonies coincides with the fructification (production of spores) of the right fungus from the surrounding adult colonies (Kone et al. 2011). The seasonality of fungus fructification (Heim 1977) and its close dependency on the hydro- and hygrometric parameters (Kone et al. 2011) make this transmission mode more risky, as regards the variations of the environmental conditions.

The Uniparental Vertical Transmission

The fungal conidia used to inoculate the primary comb of incipient colonies are carried in the gut of swarming alates of one sex. The conidia are carried by male alates in *Macrotermes bellicosus* (Johnson et al. 1981; Nobre et al. 2011a, b) and by females in the complex of species from the genus *Microtermes* (Johnson 1981; Johnson et al. 1981; Nobre et al. 2011a, b). This transmission mode is less common and has likely been acquired independently by the host lineages mentioned above. The fungal symbionts associated to these taxa are also characterized by the nonexistence of fructifications. Contrary to the horizontal transmission mode, the risk of failure in the initiation of a viable primary fungus comb is very low in this case, and the corresponding taxa can be successfully reared in the laboratory, without any external supply of the fungal inoculum (Johnson et al. 1981). Moreover, being independent from the existing spores in their environment, the vertical transmission mode may facilitate long-distance dispersal of the host as, for example, in case of an island environment conquest (Nobre et al. 2010).

Given the opportunity of symbiont exchanges related to collecting the spores of an unwanted strain, the association was reasonably expected to be more versatile in species adopting horizontal transmission and vice versa. Surprisingly, in view of the phylogenic reconstructions, the transmission mode is not likely the primary driver of the interaction specificity. For instance, the fungal partners of *Microtermes* is closely related to horizontally transmitted strains associated to various genera (Rouland-Lefevre et al. 2002; Nobre et al. 2010; Osiemo et al. 2010). Likewise, *M. bellicosus* can adopt fungal strains from various clades within which they are more related to horizontally acquired strains (Aanen et al. 2002; Osiemo et al. 2010; Nobre et al. 2011a, b; Poulsen 2015). This also indicates that even for supposedly uniparental vertical transmission, horizontal transfers may also occur. It is also useful to recall that the vertical transmission is not uniform in all non-fruiting fungal species. For instance, the fructification has never been reported for the symbiont of

M. natalensis which yet adopts a horizontal transmission mode through a not yet clarified mechanism (De Fine Licht et al. 2006). The loss of the fructification is therefore occasional and may have different causes (Korb and Aanen 2003): (1) the conidia carried by swarming alates are made of primary (monocaryotic) mycelia, unable to produce fructification; (2) otherwise (conidia with dicaryotic mycelia), the fructification may be suppressed by the host itself. For instance, according to laboratory observations, *M. natalensis* suppresses the fructification of the symbiont by foraging the fungal primordia (De Fine Licht et al. 2005).

Since the lateral mode of transmission is the general rule, the great enigma concerns the mechanism used by the host to restrict the genetic variability of the symbiont at each generation. Several assumptions have been made in this regard, some of which begun to be tested, providing still ungeneralizable conclusions. The diversification in the mutualism (Rouland-Lefevre 2000) and the maintenance of a more or less specific interaction may be dictated by the host nutritional needs that the fungus must meet and which vary from a host species to another (Nobre and Aanen 2012). The selection of the right fungus by the host may be (1) active, for instance, through an over-foraging of potential competitors, or (2) passive, for example, through the comb chemical composition or physical and chemical conditions to which spores are submitted during the gut transit or (3) occur via specific fungicidal compounds, secreted either by the host or its associated microbiota (Nobre et al. 2011a, b; Um et al. 2013). This assumption is supported by the existence of *Bacillus* spp. associated to *M. natalensis* (Um et al. 2013) or *Odontotermes formosanus* (Mathew et al. 2012), inhibiting the growth of the competitors of *Termitomyces*. Moreover, several other bacteria such as *Actinobacteria* of the genus *Streptomyces*, frequently isolated from the gut of workers and from the fungus comb of Macrotermitinae, may be involved in selection processes (Carr et al. 2012; Kim et al. 2014; Beemelmanns et al. 2016).

A frequency-dependent selection of the symbiont has also been proposed by Aanen and colleagues at the issue of an investigation performed with *M. natalensis* (Aanen et al. 2009). According to these authors, the strain selected, among the diverse bacterial spores brought to the nest by workers, is the one with the highest spore density and thus with the highest rate of mycelia fusion. This study was, however, performed with spores of different genotypes but from the same species and therefore deserves to be validated with other less related spores. However, it is in accordance with the assumption that the selection of the right fungus strain is favored by the time lag between fructification periods of sympatric fungal species and by the synchronization between the fructifications of the right fungus and the emergence of the firs cohort of workers in new colonies (Grasse 1982; Kone et al. 2011). Moreover, this concomitance may be enabled by the drastic biomass loss resulting from the swarming (averaging 40% loss of biomass in some species) (Wood and Sands 1978) and the resulting biomass of conidia which are not foraged that can produce fruiting bodies, whose dehiscence coincides with the emergence of first workers, in incipient nests.

1.3 Macrotermitinae as Crop Pest

Thanks to their very efficient digestive symbiosis with fungus, members of Macrotermitinae are the major termite pests of tropical crops. This group is particularly active in Africa, Southeastern Arabia, India, and Southeast Asia (Wood 1996). The main genera considered as pests (of arable crops, trees, and wood in service) are *Odontotermes*, *Microtermes*, *Macrotermes*, *Pseudacanthotermes*, and *Ancistrotermes* (Harris 1969; Wood et al. 1987; Wood and Cowie 1988; Logan 1992; Rouland et al. 1993; Mora et al. 1996). The major pest species in Africa and Asia are summarized in Table 1.1. Due to the problem of taxonomy within *Odontotermes* and *Microtermes* genera, the corresponding species were very often unknown.

1.4 Sustainable Management

In the tropics, control of termite pests has usually consisted in using powerful and persistent organochlorine insecticides such as dieldrin and aldrin (Harris 1969; Wood and Pearce 1991). However, the use of these compounds has been restricted in the USA since the 1970s of the last century. Afterward, the evidence of their toxicity to mammals and of their harmful environmental effects resulted in a general prohibition in 1985 (Wood and Pearce 1991). Organophosphates (malathion, chlorpyrifos, dichlorvos), carbamates (aldicarb, carbosulfan), and chlorinated phenols (pentachlorophenol) have been brought forward to the market to replace them, with some success (Wardell 1987). Other persistent insecticides such as the phenylpyrazole fipronil (Bobe et al. 1998; Sharma et al. 2008) or thiamethoxam (Maienfisch et al. 2001) were proposed during the 1990s. After several years of use, it has been accepted that the control strategies based on chemicals had weak efficiency against fungus-growing termite species, required excessive quantities of chemicals (with an estimated cost around 20 billion € worldwide), and were environmentally unfriendly (Singh et al. 2014). Alternative control methods, more specific to Macrotermitinae pest species, were then tested and have clearly shown a better efficiency, compared to purely chemical methods.

1.4.1 Cultural Practices Without Chemical Means

Healthy plants can sometimes be damaged by termites, but infected or stressed plants are generally more exposed (Cowie and Wood 1989). For instance, in semiarid areas, water stress increases the mortality of young plants through an increased susceptibility to termite attacks (Logan and El Bakri 1990; Rao et al. 2000). Furthermore, cropping and tree plantation increasingly extend into marginal zones,

Table 1.1 Inventory of the species of fungus-growing termites reported in the literature as causing damage

Genera	Africa	Crops	References	Asia	Crops	References
Macrotermes	*M. subhyalinus*	Sugar crops	Collins (1984)	*M. gilvus*	*Eucalyptus*	Kranz et al. (1981)
		Maize crops	Cowie and Wood (1989), Sekamatte and Okwakol (2007)			
	M. bellicosus	Groundnut	El Amin et al.(1983), and Wood and Pearce (1991)	*M.* sp.	Grasslands	Mugerwa (2015)
		Eucalyptus	Sands (1960), Brown (1965), and Chilima (1991)			
		Coconut and palm	Aisgbonhi (1985)	*M. annandalei*	*Eucalyptus*	Li-Ying and Waterhouse (1997)
	M. natalensis	*Eucalyptus*	Sands (1960), Brown (1965), and Chilima (1991)	*M. barneyi*	Sugar cane	Li-Ying and Waterhouse (1997)
		Groundnut	El Amin et al. (1983)			
	M. falciger	*Eucalyptus*	Sands (1960), Brown (1965), and Chilima (1991)			
		Maize	Sekamatte and Okwakol (2007)			
	M. michaelseni	*Eucalyptus*	Sands (1960), Brown (1965), and Chilima (1991)			
		Forage plants				
		Cereal crops	Logan (1991)			
Odontotermes	*O. smeathmani*	Sugar crops	Collins (1984)	*O. obesus*	Sugar crops	Parihar (1985)
		Coconut and palm	Logan and El Bakri (1990)		*Eucalyptus*	Rajagopal (1982)
					Coffee	Kranz et al. (1981)
	O. badius	Cotton			Cereal crops	Hashmi et al. (1983), and Singh et al. (2014)
		Maize	Sekamatte and Okwakol (2007)		Cotton	Kranz et al. (1981)
					Coconut and palm	Kranz et al. (1978)

	O. sp.	*Eucalyptus*	Logan and El Bakri (1990)		Fruit trees	Parihar (1981)
		Maize	Sekamatte and Okwakol (2007)			
		Coconut and palm	Han et al. (1998)	*O. formosanus*	Sugar crops	
		Mango trees	Sane et al. (2016)		Coffee	Kranz et al. (1981)
					Fruit trees	Kranz et al. (1981)
	O. nilensis	Market gardens	Han and Ndiaye (1998)		Coconut and palm	Li et al. (2010)
				O. redemanni	Sugar crops	Khumasinghe and Ranasinghe (1988)
				O. bellahunensis	Coconut and palm	Reddy (1983)
				O. gurdaspurens	*Eucalyptus*	Parihar (1981), and Rajagopal (1982)
				O. horni	Fruit trees	Parihar (1981)
Microtermes	*M. subhyalinus*	Sugar crops	Renoux et al. (1991), and Mora et al. (1996)			
				M. tragerdhi	Coconut and palm	Abushama and Kambal (1977)
	M. lepidus	Groundnut			Sugar crops	
	M. albopartitus	Cereal crops	Nkukika (1989, 1994), and Cowie and Wood (1989)	*M. obesi*	Cereal crops	Hashmi et al. (1983), and Singh et al. (2014)
	M. sp.	*Eucalyptus*	Mitchell (1989)	*M. thoracalis*	Groundnut	Kranz et al. (1981)
		Coffee			Cereal crops	
		Cereal	Nkukika (1989, 1994), and Cowie and Wood (1989)		Cotton	Tiben et al. (1990)
		Fruit trees	Sane (2016)		Market gardens	

(continued)

Table 1.1 (continued)

Genera	Africa	Crops	References	Asia	Crops	References
Ancistrotermes	*A. guineensis*	Sugar crops	Rouland et al. (1993)			
				A. dimorphus	Fruit trees	Li et al. (2010)
	A. cavithorax	Mango trees	Han and Ndiaye (1996, 1998)		Coconut and palms	Tang et al. (2006)
	A. latinotus	Cereal crops	Sands (1960), Brown (1965), and Chilima (1991)			
		Cotton				
	A. amphidon	*Eucalyptus*	Wardell (1987), and Wood and Pearce (1991)			
Pseudacanthotermes	*P. spiniger*	Cereal crops	Nkunika (1989, 1994), and Cowie and Wood (1989)			
	P. militaris	*Eucalyptus*	Sands (1960), Brown (1965), and Chilima (1991)			
		Maize	Sekamatte and Okwakol (2007)			

with less suitable soil (Black and Okwakol 1997). Yet, as a general rule, the maintenance of healthy plants is crucial: water stress, nutrient deficiency, the competition for light enhanced by the high planting density, as well as the wounds resulting from pruning operations increase the risks of termite attacks (UNEP-FAO 2000). Farming practices, especially those related to plant care, are therefore crucial in preventing damages related to termites. Cultural control comprises any means and actions contributing to improve or maintain the vigor of the plant but can also include means reducing the number of termites or modifying their behavior.

1.4.1.1 Practices Improving Plant Vigor

The choice of a location with low risks of termite attacks must be a priority in any agricultural holding project. As a general rule, termites have indeed their own habitat. They often cause damage on crops because these were planted in their natural habitat (Bignell 2006; Zaremski et al. 2009). Using termite-resistant plant varieties, or plant species which are well adapted to pedo-climatic conditions, thereby avoiding stress and crop vulnerability, must be considered (Logan and El Bakri 1990). Intensive cultivation, and in particular monoculture, reduces fertility and modifies soil structure (Black and Okwakol 1997). It has been reported that, under these conditions, plants are less vigorous and more susceptible to termite attacks (Cowie and Wood 1989; Sileshi et al. 2005). Crop rotations, especially with food legumes or fallowing, can remedy changes of soil structure (Sekamatte et al. 2003; Sileshi et al. 2008) and are therefore recommended to mitigate termite attacks.

Rates of plant attacks by Macrotermitinae very often depend on hydrological conditions. This factor has been evidenced in cotton, groundnut, and *Eucalyptus* plantations, suggesting an increased vulnerability to termites in dry zones, where drought conditions prevail (Cowie and Wood 1989). Insufficient irrigation has also been suggested as a cause of new termite attacks on date palms in Sudan (Logan and El Bakri 1990), on sugarcane in Sri Lanka (Khumasinghe and Ranasinghe 1988), and on mango trees in Senegal (Sane et al. 2016). In low-rainfall regions, irrigation can reduce water stress and thereby lessen termite attacks (Sharma et al. 2004; Sane 2016).

Enhancing the diversity of non-harmful termite groups, thereby limiting the diversity of pest species through competitive interactions, is another method for mitigating termite damage (Sane 2016). This can be achieved by the restriction of pesticide use, by mulching with dead or green plant material, or by intercroppings (Sileshi et al. 2008). For mulching treatments, the pest termites may use the mulch as an alternative food resource. As a result, crops are less exposed to damage (Rao et al. 2000). However, one side effect of mulching is the increased termite activity, resulting in more damage when the mulch has been entirely consumed. Intercroppings using resistant plants may also limit damage by termites, by making target plants less accessible.

1.4.1.2 Practices to Reduce Termite Density

The simplest method used when termite mounds are epigeous is digging them and destroying the queen (Rust and Saran 2006; Sileshi et al. 2008). For subterranean nests, visual cues (ventilation holes, excreta) are used for localization. The arboreal nests (built inside tree hollows) are also removed. To get access to the nest, the bark of the plant is first removed and then placed back, after colony removal (Gui-Xiang et al. 1994; UNEP-FAO 2000; Zhong and Liu 2002). Some cover crops, such as *Pueraria*, seem to reduce the occurrence of termites when used as a fallow (Renoux et al. 1991). Plant extracts or mineral preparations may also be useful: adding chopped leaves of *Euphorbia tirucalli*, *Aloe graminicola*, *Melia azedarach*, or *Lippia javanica* to the planting holes and mulching with *Cassia siamea* or *Azadirachta indica* are recommended practices (Wardell 1987; Verma et al. 2009). In Malawi and Mozambique, planting cuttings of *Euphorbia tirucalli* was the most commonly used method for mitigating termite damage on maize crops. In termite-infested fields, applying wood ash, crushed fruits of *Swartzia madagascariensis*, and leaf concoctions of neem (*Azadirachta indica*) or *Tephrosia vogelii* in planting holes is generally suggested (Sileshi et al. 2008).

1.4.2 Biological Control

This approach comprises the manipulation of predators, pathogens, or parasites or alternatively the use of natural products primarily from plants with the view of reducing pest populations to an economically acceptable level. The main predators of termites are ants. Increasing ant populations using sugar and meat baits resulted in the effective reduction of both termite populations and subsequent damage on maize crops in Uganda (Sekamatte et al. 2001). This method has been also used in Malawi where predatory ants are attracted by farmers in termite-infested crops by means of leftover pork or beef meat (Sileshi et al. 2008). Nevertheless, despite this diversity of predators, most of the work on biological control has focused on microbial pathogens: viruses, bacteria, fungi, protists, and nematodes. The main microbial groups acknowledged as biological control agents against Macrotermitinae pest species are entomopathogenic fungi, bacteria, and nematodes.

1.4.2.1 Fungi

Metarhizium anisopliae and *Beauveria bassiana* appear quite effective against termites. Their spores can be introduced into nests associated with baits or applied as a powder, either on the nests or directly on termites (Starnes et al. 1993; Sun et al. 2003). The existing studies have shown that conidia of *M. anisopliae* var. *dcjhyium* used to an average concentration of $3 \cdot 10^8$ conidia/ml were highly virulent for *O. formosanus*, causing approximately 100% mortality within 3 days post-inoculation (Dong et al. 2009).

1.4.2.2 Bacteria

Three different species of HCN-producing rhizobacteria, *Rhizobacterium radiobacter*, *Alcaligenes latus*, and *Aeromonas caviae*, were effective in controlling *Odontotermes obesus* under in vitro conditions (Devi et al. 2007).

1.4.2.3 Nematodes

Entomopathogenic nematodes have also been used for biological control of Macrotermitinae (Qasim et al. 2015). The nematode strains *Steinernema carpocapsae* K27 and *S. kushidai* E2 have demonstrated control capacity against reproductives of several species of Macrotermitinae: *Ancistrotermes guineensis*, *Pseudacanthotermes spiniger*, *Odontotermes* sp., and *Macrotermes bellicosus* (Rouland et al. 1996). All castes are susceptible to infection, but the life cycle of the nematode can only be completed in the alates (Benmoussa-Haichour et al. 1998).

1.4.2.4 The Repellency of Particular Plant Species on Termites

A number of studies have been conducted on the issue of using some natural products such as essential oils and extracts of plants, leaves, roots, fruits, or seeds or of wood or plant-derived resins, for control of termites (Verma et al. 2009). Previous studies documented the efficiency of fresh lantana leaf mulch as a repellent barrier to termites (Ding and Hu 2010). Likewise, leaf hexane extract of *Aristolochia bracteolata*; ethyl acetate extract of *A. paniculata*, *Datura metel*, and *Euphorbia prostrata*; or methanol extract of *Acacia lineata* and *D. metel* were capable of triggering higher termite mortality within 24 h after application (Liu et al. 2015). Monoterpenes extracted from aromatic plants have also a potential to be used as new natural termiticides (Xie et al. 2014).

1.4.3 The Integrated Pest Management

As stated above, the mutualistic relationship between Macrotermitinae and *Termitomyces* is obligatory and very specific since a given termite species is often associated with a single *Termitomyces* species, never existing in a free-living state (i.e., outside the termite nest). The absolute necessity of the fungal symbiont can be exploited for controlling populations of fungus-growing termites using fungicides. The effectiveness of fungicides specifically targeted toward the symbionts of Macrotermitinae has ever been tested (Wardell 1987; El Bakri et al. 1989). The death of the symbiont blocks the assimilation of the foraged food, thus leading to the death of the whole termite colony. Such a fungicide-based control using Erpacide® has been used in the sugarcane plantations of SONASUT (Societe

Nationale Sucrière du Tchad) against the same termite group and resulted in the destruction of about 20% of colonies (Rouland et al. 1993; Rouland-Lefevre and Mora 2002). Though these fungicides are used below the toxic dose and therefore do cause little or no harm to the environment, their spreading at the field scale is often less effective and more costly than a bait-based control. For this purpose, a promising anti-termite strategy would consist in combining isolated and characterized specific trail pheromones as attractants, added to the baits containing the poison. As foraging workers collecting the food outside are the only food providers for the whole colony (reproductives, soldiers, young workers), the toxic compounds contained in the collected food will be transmitted to all the members of the colony through food exchanges, thus leading to the eradication of the whole colony. The main limitation of using baits containing chemical biocides alone is the ability of termites to detect and avoid the poisoned baits, even at very low concentrations. Placing traps containing poisoned food supplemented with synthetic pheromones as attractant around infested orchards would avoid this detection. Termite trail pheromones have been investigated for a long time, and Sillam-Dussès et al. (2010) showed that (3Z,6Z,8E)-dodeca-3,6,8-trien-1-ol and (1E,5E,9E,12R)-1,5,9-trimethyl- 12-(1-methylethenyl)-1,5,9-cyclotetradecatriene are trail pheromones common to many species of termite pests. The low variability of trail pheromones between termites from different groups suggests the feasibility of synthesizing a compound or a blend of compounds as attractants for a wide range of pest species. Associating these attractants with the biocidal molecules described above would significantly improve the selective nature and effectiveness of termite control using baits. Furthermore, spreading these synthetic attractants over wide areas could contribute misleading and further dispersing individuals, thus avoiding high local density of termite pests (Igwe and Eze 2015).

1.5 Conclusion

Thanks to their special capacity of degradation resulting at least partially to their association with a fungus partner, Macrotermitinae are key actors in the functioning of natural tropical ecosystems, especially in Africa and some Asian areas. As a result of their peculiar capacity of degradation of diverse types of organic matter, they may cause significant damages when their natural habitat has been modified and often replaced with farming or orchard planting. Fungus-growing termites are thus one of the most prominent crop pests in Africa, Southeastern Arabia, India, and Southeast Asia, and their control is a major concern. The first control strategies, based on the use of excessive amounts of chemicals such as organochlorine and organophosphate insecticides and other biocide molecules, showed limitations against fungus-growing termites, in addition to being environmentally unfriendly. Natural repellents, provided either by planting repellent species or adding in the form of mulch or plant extracts, as well as biological control agents have also been tested, with proven local successes. One of the most prominent strategies for a more targeted control of termite pests consists in using poisoned baits. However, the effectiveness of this approach is limited by the capability of termites to recognize and avoid these baits. One promising alternative of this classical biocide-treated bait drawback is the use of trail pheromones as attractants. These molecules naturally secreted by termites for the orientation and the recruitment of workers are getting increasingly identified. The demonstrated link between the stress state of plants and the magnitude of termite attacks has also led farmers to improve their cultural practices to limit the stress and thereby plant vulnerability and attacks.

References

Aanen, D. K. (2006). As you reap, so shall you sow: Coupling of harvesting and inoculating stabilizes the mutualism between termites and fungi. *Biology Letters, 2*, 209–212.

Aanen, D. K., & Eggleton, P. (2005). Fungus-growing termites originated in African rain forest. *Current Biology, 15*, 851–855.

Aanen, D. K., & Jacobus, J. B. (2006). The evolutionary origin and maintenance of the mutualistic symbiosis between termites and fungi. In K. Bourtzis & T. A. Miller (Eds.), *Insect symbiosis* (Vol. 2, pp. 79–95). London: CRC Press.

Aanen, D. K., Eggleton, P., Rouland-Lefèvre, C., Guldberg-Frøslev, T., Rosendahl, S., & Boomsma, J. J. (2002). The evolution of fungus-growing termites and their mutualistic fungal symbionts. *PNAS, 99*, 14887–14892.

Aanen, D. K., Ros, V. I., de Fine Licht, H. H., Mitchell, J., de Beer, Z. W., Slippers, B., Rouland-Lefèvre, C., & Boomsma, J. J. (2007). Patterns of interaction specificity of fungus-growing termites and *Termitomyces* symbionts in South Africa. *BMC Evolutionary Biology, 7*, 115.

Aanen, D. K., de Fine Licht, H. H., Debets, A. J. M., Kerstes, N. A. G., Hoekstra, R. F., & Boomsma, J. J. (2009). High symbiont relatedness stabilizes mutualistic cooperation in fungus-growing termites. *Science, 326*, 1103–1106.

Abushama, F. T., & Kambal, M. A. (1977). The role of sugars in the food-selection of the termite *Microtermes traegardhi* (Sjost.) *Zeitschrift für Angewandte Entomologie, 84*, 250–255.

Adams, E. S. (1991). Nest-mate recognition based on heritable odors in the termite *Microcerotermes arboreus*. *PNAS, 88*, 2031–2034.

Aisagbonhi, C. I. (1985). A survey of the destructive effect of *Macrotermes bellicosus* Smeathman (Isoptera: Termitidae – Macrotermitinae) on coconut seednuts at NIFOR, Benin, Nigeria. *International Journal of Pest Management, 35*, 380–381.

Aldrich, B. T., & Kambhampati, S. (2007). Population structure and colony composition of two *Zootermopsis nevadensis* subspecies. *Heredity, 99*, 443–451.

Badertscher, S., Gerber, C., & Leuthold, R. H. (1983). Polyethism in food supply and processing in termite colonies of *Macrotermes subhyalinus* (Isoptera). *Behavioral Ecology and Sociobiology, 12*, 115–119.

Bagine, R. K. N., Brandl, R., & Kaib, M. (1994). Species delimitation in *Macrotermes* (Isoptera: Macrotermitidae): Evidence from *Epicuticular Hydrocarbons*, morphology, and ecology. *Annals of the Entomological Society of America, 87*, 498–506.

Bagnères, A., & Blomquist, G. (2010). Site of synthesis, mechanism of transport and selective deposition of hydrocarbons. In G. J. Blomquist & A. G. Bagnères (Eds.), *Insect hydrocarbons* (pp. 75–99). Cambridge, UK: Cambridge University Press.

Bagnères, A. G., & Hanus, R. (2015). Communication and social regulation in termites. In L. Aquiloni & E. Tricarico (Eds.), *Social recognition in invertebrates* (pp. 193–248). Cham: Springer.

Bagneres, A. G., Killian, A., Clement, J. L., & Lange, C. (1991). Interspecific recognition among termites of the genus *Reticulitermes*: Evidence for a role for the cuticular hydrocarbons. *Journal of Chemical Ecology, 17*, 2397–2420.
Bastien, G., Arnal, G., Bozonnet, S., Laguerre, S., Ferreira, F., Fauré, R., Henrissat, B., Lefèvre, F., Robe, P., Bouchez, O., Noirot, C., Dumon, C., & O'Donohue, M. (2013). Mining for hemicellulases in the fungus-growing termite *Pseudacanthotermes militaris* using functional metagenomics. *Biotechnology for Biofuels, 6*, 1–15.
Beemelmanns, C., Guo, H., Rischer, M., & Poulsen, M. (2016). Natural products from microbes associated with insects. *Beilstein Journal of Organic Chemistry, 12*, 314–327.
Benmoussa-Haîchour, D., Reversat, G., & Rouland, C. (1998). Sensitivity of the different castes of higher termites to an infection with entomopathogenic nematodes – role of the lipids composition. *Actes Coll Insect Soc, 11*, 45–52.
Bignell, D. E. (2006). Termites as soil engineers and soil processors. In H. König & A. Varma (Eds.), *Intestinal microorganisms of termites and other invertebrates* (pp. 183–220). Berlin/Heidelberg: Springer.
Bignell, D. E., & Eggleton, P. (1995). On the elevated intestinal pH of higher termites (Isoptera: Termitidae). *Insectes Sociaux, 42*, 57–69.
Bignell, D. E., & Eggleton, P. (2000). Termites in ecosystems. In T. Abe, D. E. Bignell, & M. Higashi (Eds.), *Termites: Evolution, sociality, symbioses, ecology* (pp. 363–387). Dordrecht: Springer.
Black, H. I. J., & Okwakol, M. J. N. (1997). Soil biodiversity, agricultural intensification and agroecosystem function in the tropics: The role of termites. *Applied Soil Ecology, 6*, 37–53.
Bobé, A., Cooper, J. F., Coste, C. M., & Muller, M. A. (1998). Behaviour of fipronil in soil under Sahelian plain field conditions. *Pesticide Science, 52*, 275–281.
Bordereau, C., & Pasteels, J. M. (2011). Pheromones and chemical ecology of dispersal and foraging in termites. In E. D. Bignell, Y. Roisin, & N. Lo (Eds.), *Biology of termites: A modern synthesis* (pp. 279–320). Dordrecht: Springer.
Botha, W. J., & Eicker, A. (1992). Nutritional value of *Termitomyces* mycelial protein and growth of mycelium on natural substrates. *Mycological Research, 96*, 350–354.
Boucias, D. G., Cai, Y., Sun, Y., Lietze, V. U., Sen, R., Raychoudhury, R., & Scharf, M. E. (2013). The hindgut lumen prokaryotic microbiota of the termite *Reticulitermes flavipes* and its responses to dietary lignocellulose composition. *Molecular Ecology, 22*, 1836–1853.
Brandl, R., Hacker, M., Bagine, N. R. K., & Kaib, M. (2001). Geographic variation of polygyny in the termite *Macrotermes michaelseni* (Sjöstedt). *Insectes Sociaux, 48*, 134–137.
Brandl, R., Hacker, M., Bagine, R. K. N., & Kaib, M. (2004). Yearly variation in polygyny in the termite *Macrotermes michaelseni* (Sjöstedt). *Insectes Sociaux, 51*, 294–298.
Brown, K. W. (1965). Termite control research in Uganda (with special reference to the control of attacks in *Eucalyptus* plantations). *East Afr Agr Forest J, 31*, 218–223.
Brune, A., & Dietrich, C. (2015). The gut microbiota of termites: Digesting the diversity in the light of ecology and evolution. *Annual Review of Microbiology, 69*, 145–166.
Buxton, R. D. (1981). Changes in the composition and activities of termite communities in relation to changing rainfall. *Oecologia, 51*, 371–378.
Carr, G., Poulsen, M., Klassen, J. L., Hou, Y., Wyche, T. P., Bugni, T. S., Currie, C. R., & Clardy, J. (2012). Microtermolides A and B from termite-associated *Streptomyces* sp. and structural revision of Vinylamycin. *Organic Letters, 14*, 2822–2825.
Chilima, C. Z. (1991). Termite control in young *Eucalyptus* plantation in Malawi using controlled release insecticides. *Common Forest Review, 70*, 237–247.
Clément, J. L., & Bagnères, A. G. (1998). Nestmate recognition in termites. In R. K. Vander Meer, M. D. Breed, K. Espelie, & M. L. Winston (Eds.), *Pheromone communication in social insects: Ants, wasps, bees and termites* (pp. 125–155). Boulder: Westview Inc Col.
Coaton, W. (1961). Association of termites and fungi. *African Wild life, 15*, 39–54.
Collins, N. M. (1981). Consumption of wood by artificially isolated colonies of the fungus-growing termite *Macrotermes bellicosus*. *Entomologia Experimentalis et Applicata, 29*, 313–320.

Collins, N. M. (1983). The utilization of nitrogen resources by termites (Isoptera). In J. Lee, S. McNeill, & I. Rorison (Eds.), *Nitrogen as an ecological factor* (pp. 381–412). Oxford: Blackwell Scientific Publications.

Collins, N. M. (1984). Termite damage and crop loss studies in Niger- assessment of damage to upland sugarcane. *Tropical Pest Management, 30*, 26–28.

Copren, K. A., Nelson, L. J., Vargo, E. L., & Haverty, M. I. (2005). Phylogenetic analyses of mtDNA sequences corroborate taxonomic designations based on cuticular hydrocarbons in subterranean termites. *Molecular Phylogenetics and Evolution, 35*, 689–700.

Costa-Leonardo, A. M., Casarin, F. E., & Lima, J. T. (2009). Chemical communication in isoptera. *Neotropical Entomology, 38*, 1–6.

Cowie, R., & Wood, T. (1989). Damage to crops, forestry and rangeland by fungus-growing termites (Termitidae: Macrotermitinae) in Ethiopia. *Sociobiology, 15*, 139–153.

Darlington, J. (1985). Multiple primary reproductives in the termite *Macrotermes michaleseni* (Sjöstedt). In J. Watson, B. Okot-Kotber, & C. Noirot (Eds.), *Caste differentiation in social insects* (pp. 187–200). Oxford: Pergamon Press.

Darlington, J. P. E. C. (1988). Multiple reproductives in nests of *Macrotermes herus* (Isoptera: Termitidae). *Sociobiology, 14*, 347–351.

Darlington, J., Zimmerman, P., & Greenberg, J. (1997). Production of metabolic gases by nestsof the termite *Macrotermes jeanneli* in Kenya. *Journal of Tropical Ecology, 13*, 491–510.

DeFine Licht, H. H., Andersen, A., & Aanen, D. K. (2005). *Termitomyces* sp. associated with the termite *Macrotermes natalensis* has a heterothallic mating system and multinucleate cells. *Mycological Research, 109*, 314–318.

DeFine Licht, H. H., Boomsma, J. J., & Aanen, D. K. (2006). Presumptive horizontal symbiont transmission in the fungus-growing termite *Macrotermes natalensis*. *Molecular Ecology, 15*, 3131–3138.

DeFine Licht, H. H., Boomsma, J. J., & Aanen, D. K. (2007). Asymmetric interaction specificity between two sympatric termites and their fungal symbionts. *Ecological Entomology, 32*, 76–81.

Devi, K. K., Seth, N., Kothamasi, S., & Kothamasi, D. (2007). Hydrogen cyanide-producing Rhizobacteria kill subterranean termite *Odontotermes obesus* (Rambur) by cyanide poisoning under *In Vitro* conditions. *Current Microbiology, 54*, 74–78.

Dietrich, C., Köhler, T., & Brune, A. (2014). The cockroach origin of the termite gut microbiota: Patterns in bacterial community structure reflect major evolutionary events. *AEM, 80*, 2261–2269.

Ding, W., & Hu, X. P. (2010). Antitermitic effect of the *Lantana camara* plant on subterranean termites (Isoptera: Rhinotermitidae). *Insect Science, 17*, 427–433.

Diouf, M., Roy, V., Mora, P., Frechault, S., Lefebvre, T., Hervé, V., Rouland-Lefèvre, C., & Miambi, E. (2015). Profiling the succession of bacterial communities throughout the life stages of a higher termite Nasutitermes arborum (Termitidae, Nasutitermitinae) using 16S rRNA gene pyrosequencing. *PLOS ONE, 10*(10), e0140014.

Dong, C., Zhang, J., Huang, H., Chen, W., & Hu, Y. (2009). Pathogenicity of a new China variety of *Metarhizium anisopliae* (*M. Anisopliae* var. Dcjhyium) to subterranean termite *Odontotermes formosanus*. *Microbiological Research, 164*, 27–35.

Duringer, P., Schuster, M., Genise, J. F., Likius, A., Mackaye, H. T., Vignaud, P., & Brunet, M. (2006). The first fossil fungus gardens of Isoptera: Oldest evidence of symbiotic termite fungiculture (Miocene, Chad basin). *Naturwissenschaften, 93*, 610–615.

Duringer, P., Schuster, M., Genise, J. F., Mackaye, H. T., Vignaud, P., & Brunet, M. (2007). New termite trace fossils: Galleries, nests and fungus combs from the Chad basin of Africa (Upper Miocene–Lower Pliocene). *Palaeogeography Palaeoclimatology, 251*, 323–353.

Eggleton, P. (2000). Global patterns of termite diversity. In T. Abe, D. E. Bignell, & M. Higashi (Eds.), *Termites: Evolution, sociality, symbioses, ecology* (pp. 25–51). Dordrecht: Springer.

Eggleton, P. (2006). The termite gut habitat: Its evolution and co-evolution. In H. Konig & A. Varma (Eds.), *Intestinal microorganisms of termites and other invertebrates* (pp. 373–404). Berlin/Heidelberg: Springer.

Eggleton, P. (2011). An introduction to termites: Biology, taxonomy and functional morphology. In E. D. Bignell, Y. Roisin, & N. Lo (Eds.), *Biology of termites: A modern synthesis* (pp. 1–26). Dordrecht: Springer.
Eggleton, P., & Davies, R. G. (2003). Isoptera, termites. In S. M. Goodman & J. P. Benstead (Eds.), *The natural history of Madagascar* (pp. 654–660). Chicago: University of Chicago Press.
El Amin, E. M., Ishag, H. M., & Burhan, H. O. (1983). Important factors affecting the yield of groundnuts (*Arachis hypogaea* L) in the Sudan. *Zeitschrift für Angewandte Zoologie, 70*, 29–55.
El Bakri, A., Eldein, N., Kambal, M., Thomas, R., & Wood, T. (1989). Effect of fungicide impregnated food on the viability of fungus combs and colonies of *Microtermes* sp. nr. albopartitus (Isoptera: Macrotermitinae). *Sociobiology, 15*, 175–180.
Engel, M. S., & Krishna, K. (2004). Family-group names for termites (Isoptera). *American Museum Novitates, 3432*, 1–9.
Engel, M. S., Grimaldi, D. A., & Krishna, K. (2009). Termites (Isoptera): Their phylogeny, classification, and rise to ecological dominance. *American Museum Novitates, 3650*, 1–27.
Estrada-Pena, A., Castella, J., & Moreno, J. A. (1994). Using cuticular hydrocarbon composition to elucidate phylogenies in tick populations (Acari: Ixodidae). *Acta Tropica, 58*, 51–71.
FroSlev, T. G., Aanen, D. K., Laessoe, T., & Rosendahl, S. (2003). Phylogenetic relationships of *Termitomyces* and related taxa. *Mycological Research, 107*, 1277–1286.
Grasse, P. P. (1982). *Termitologia. Tome I: Anatomie, Physiologie et Reproduction de Termites* (p. 676). Paris: Ed Masson.
Grasse, P. P. (1984). *Termitologia: Tome II Fondation des societes, constructions* (p. 613). Paris: Ed Masson.
Grasse, P. P., & Noirot, C. (1958). La meule des termites champignonnistes et sa signification symbiotique. *Annales des Sciences Naturelles – Zoologie et Biologie Animale, 11*, 113–128.
Gui-Xiang, L., Zi-Rong, D., & Biao, Y. (1994). Introduction to termite research in China. *Journal of Applied Entomology, 117*, 360–369.
Guo, L., Quilici, D. R., Chase, J., & Blomquist, G. J. (1991). Gut tract microorganisms supply the precursors for methyl-branched hydrocarbon biosynthesis in the termite, *Zootermopsis nevadensis*. *Insect Biochemistry, 21*, 327–333.
Hacker, M., Kaib, M., Bagine, R. K. N., Epplen, J. T., & Brandl, R. (2005). Unrelated queens coexist in colonies of the termite *Macrotermes michaelseni*. *Molecular Ecology, 14*, 1527–1532.
Han, S. H., & Ndiaye, A. B. (1996). Dégâts causés par les termites (Isoptera) sur les arbres fruitiers dans la region de Dakar (Sénégal). Actes Coll. *Insectes Sociaux, 10*, 111–117.
Han, S. H., & Ndiaye, A. B. (1998). L'attaque des cultures maraîchères par les termites (Isoptera) dans la région de Dakar (Sénégal). *Actes Coll Insectes Sociaux, 11*, 37–43.
Han, S. H., Tokro, G. P., Tano, Y., & Lepage, M. (1998). Dégâts des termites dans les plantations de palmiers à huile en Côte d'Ivoire: évaluation et méthode de lutte. *Plantations Rech Dévelop, 5*, 119–126.
Harris, W. (1969). *Termites as pests of crops and trees* (p. 41). London: Common Wealth Agricultural Bureau.
Hashmi, A. A., Hussain, M. M., & Itaf, M. (1983). Insect pest complex of wheat crop. *Pakistan Journal of Zoology, 15*, 169–176.
Haverty, M. I., & Nelson, L. J. (1997). Cuticular hydrocarbons of *Reticulitermes* (Isoptera: Rhinotermitidae) from northern California indicate undescribed species. *Comparative Biochemistry and Physiology. B, 118*, 869–880.
Haverty, M. I., & Thorne, B. L. (1989). Agonistic behavior correlated with hydrocarbon phenotypes in dampwood termites, *Zootermopsis* (Isoptera: Termopsidae). *Journal of Insect Behavior, 2*, 523–543.
Haverty, M. I., Copren, K. A., Getty, G. M., & Lewis, V. R. (1999). Agonistic behavior and cuticular hydrocarbon phenotypes of colonies of *Reticulitermes* (Isoptera: Rhinotermitidae) from northern California. *Annals of the Entomological Society of America, 92*, 269–277.
Haverty, M. I., Woodrow, R. J., Nelson, L. J., & Grace, J. K. (2000). Cuticular hydrocarbons of termites of the Hawaiian islands. *Journal of Chemical Ecology, 26*, 1167–1191.

Heim, R. (1977). *Termites et champignons* (p. 177). Paris: Ed Société Nouvelle des éditions Boubée.

Hinze, B., Crailsheim, K., & Leuthold, H. R. (2002). Polyethism in food processing and social organisation in the nest of *Macrotermes bellicosus* (Isoptera, Termitidae). *Insectes Sociaux, 49*, 31–37.

Holt, J. A., & Lepage, M. (2000). Termites and soil properties. In T. Abe, D. E. Bignell, & M. Higashi (Eds.), *Termites: Evolution, sociality, symbioses, ecology* (pp. 389–407). Dordrecht: Springer Netherlands.

Hongoh, Y., Ekpornprasit, L., Inoue, T., Moriya, S., Trakulnaleamsai, S., Ohkuma, M., Noparatnaraporn, N., & Kudo, T. (2006). Intracolony variation of bacterial gut microbiota among castes and ages in the fungus-growing termite Macrotermes gilvus. *Molecular Ecology, 15*, 505–516.

Howard, R., & Blomquist, G. (2005). Ecological, behavioral, and biochemical aspects of insect hydrocarbons. *Annual Review of Entomology, 50*, 371–393.

Hyodo, F., Inoue, T., Azuma, J. I., Tayasu, I., & Abe, T. (2000). Role of the mutualistic fungus in lignin degradation in the fungus-growing termite *Macrotermes gilvus* (Isoptera; Macrotermitinae). *Soil Biology and Biochemistry, 32*, 653–658.

Hyodo, F., Tayasu, I., Inoue, T., Azuma, J. I., Kudo, T., & Abe, T. (2003). Differential role of symbiotic fungi in lignin degradation and food provision for fungus-growing termites (Macrotermitinae: Isoptera). *Functional Ecology, 17*, 186–193.

Igwe, O., & Eze, P. (2015). Chemistry of trail pheromones from *Cubitermes* termites (Amitermes Dentatus): An innovation in Pest management. *American Chemical Science Journal, 6*, 16–24.

Inward, D. J. G., Vogler, A. P., & Eggleton, P. (2007). A comprehensive phylogenetic analysis of termites (Isoptera) illuminates key aspects of their evolutionary biology. *Molecular Phylogenetics and Evolution, 44*, 953–967.

Jenkins, T. M., Haverty, M. I., Basten, C. J., Nelson, L. J., Page, M., & Forschler, B. T. (2000). Correlation of mitochondrial haplotypes with cuticular hydrocarbon phenotypes of sympatric *Reticulitermes* species from the south eastern United States. *Journal of Chemical Ecology, 26*, 1525–1542.

Johnson, R. A. (1981). Colony development and establishment of the fungus comb in *Microtermes* sp. nr. *usambaricus* (Sjostedt) (Isoptera: Macrotermitinae) from Nigeria. *Insectes Sociaux, 28*, 3–12.

Johnson, R. A., Thomas, R. J., Wood, T. G., & Swift, M. J. (1981). The inoculation of the fungus comb in newly founded colonies of some species of the *Macrotermitinae* (Isoptera) from Nigeria. *Journal of Natural History, 15*, 751–756.

Jones, D. T., & Eggleton, P. (2011). Global biogeography of termites: A compilation of sources. In E. D. Bignell, Y. Roisin, & N. Lo (Eds.), *Biology of termites: A modern synthesis* (pp. 477–498). Dordrecht: Springer.

Jouquet, P., Barre, P., Lepage, M., & Velde, B. (2005). Impact of subterranean fungus-growing termites (Isoptera, Macrotermitiane) on chosen soil properties in a West African savanna. *Biology and Fertility of Soils, 41*, 365–370.

Kaib, M., Brandl, R., & Bagine, R. K. N. (1991). Cuticular hydrocarbon profiles: A valuable tool in termite taxonomy. *Naturwissenschaften, 78*, 176–179.

Kaib, M., Hacker, M., & Brandl, R. (2001). Egg-laying in monogynous and polygynous colonies of the termite *Macrotermes michaelseni* (Isoptera, Macrotermitidae). *Insectes Sociaux, 48*, 231–237.

Kaib, M., Franke, S., Francke, W., & Brandl, R. (2002). Cuticular hydrocarbons in a termite: Phenotypes and a neighbour–stranger effect. *Physiological Entomology, 27*, 189–198.

Kaib, M., Jmhasly, P., Wilfert, L., Durka, W., Franke, S., Francke, W., Leuthold, R. H., & Brandl, R. (2004). Cuticular hydrocarbons and aggression in the termite *Macrotermes subhyalinus*. *Journal of Chemical Ecology, 30*, 365–385.

Kaltenpoth, M., Roeser-Mueller, K., Koehler, S., Peterson, A., Nechitaylo, T. Y., Stubblefield, J. W., Herzner, G., Seger, J., & Strohm, E. (2014). Partner choice and fidelity stabilize coevolution in a Cretaceous-age defensive symbiosis. *PNAS, 111*, 6359–6364.

Katoh, H., Miura, T., Maekawa, K., Shinzato, N., & Matsumoto, T. (2002). Genetic variation of symbiotic fungi cultivated by the macrotermitine termite *Odontotermes formosanus* (Isoptera: Termitidae) in the Ryukyu Archipelago. *Molecular Ecology, 11*, 1565–1572.

Khumasinghe, N., & Ranasinghe, M. (1988). Incidence of termite damage in sugar cane grown in Sri Lanka. *Beitr Trop Landwirt, 26*, 303–307.

Kim, K. H., Ramadhar, T. R., Beemelmanns, C., Cao, S., Poulsen, M., Currie, C. R., & Clardy, J. (2014). Natalamycin A, an ansamycin from a termite-associated *Streptomyces* sp. *Chemical Science, 5*, 4333–4338.

King, H., Ocko, S., & Mahadevan, L. (2015). Termite mounds harness diurnal temperature oscillations for ventilation. *PNAS, 112*, 11589–11593.

Koné, N. A., Dosso, K., Konaté, S., Kouadio, J. Y., & Linsenmair, K. E. (2011). Environmental and biological determinants of *Termitomyces* species seasonal fructification in central and southern Cote d'Ivoire. *Insectes Sociaux, 58*, 371–382.

Korb, J. (2003). Thermoregulation and ventilation of termite mounds. *Naturwissenschaften, 90*, 212–219.

Korb, J. (2011). Termite mound architecture, from function to construction. In E. D. Bignell, Y. Roisin, & N. Lo (Eds.), *Biology of termites: A modern synthesis* (pp. 349–373). Dordrecht: Springer.

Korb, J., & Aanen, D. K. (2003). The evolution of uniparental transmission of fungal symbionts in fungus-growing termites (Macrotermitinae). *Behavioral Ecology and Sociobiology, 53*, 65–71.

Kranz, J., Schmutterer, H., & Koch, W. (1978). *Diseases, pests and weeds in tropical crops* (p. 329). London: Wiley.

Kranz, J., Schmutterer, H., & Koch, W. (1981). *Maladies, Ravageurs et Mauvaises Herbes des Cultures Tropicales, Termites, Fourmis blanches* (pp. 304–306). Berlin et Hambourg: Verlag Paul Parey.

Kutnik, M., Uva, P., Brinkworth, L., & BagneRes, A. G. (2004). Phylogeography of two European *Reticulitermes* (Isoptera) species: The Iberian refugium. *Molecular Ecology, 13*, 3099–3113.

Legendre, F., Whiting, M. F., Bordereau, C., Cancello, E. M., Evans, T. A., & Grandcolas, P. (2008). The phylogeny of termites (Dictyoptera: Isoptera) based on mitochondrial and nuclear markers: Implications for the evolution of the worker and pseudergate castes, and foraging behaviors. *Molecular Phylogenetics and Evolution, 48*, 615–627.

Leuthold, R. H., Badertscher, S., & Imboden, H. (1989). The inoculation of newly formed fungus comb with *Termitomyces* in *Macrotermes* colonies (Isoptera, Macrotermitinae). *Insectes Sociaux, 36*, 328–338.

Leuthold, R. H., Triet, H., & Schildger, B. (2004). Husbandry and breeding of African Giant Termites (*Macrotermes jeanneli*) at Berne Animal Park. *Zool Garten N F, 74*, 26–37.

Li, H. J., ZG, X., & Deng, T. F. (2010). Species of termites attacking trees in China. *Sociobiology, 56*, 109–120.

Li, H., Yang, M., Chen, Y., Zhu, N., Lee, C. Y., Wei, J. Q., & Mo, J. (2015). Investigation of age polyethism in food processing of the fungus-growing termite *Odontotermes formosanus* (Blattodea: Termitidae) using a laboratory artificial rearing system. *Journal of Economic Entomology, 108*, 266–273.

Li, H., Dietrich, C., Zhu, N., Mikaelyan, A., Ma, B., Pi, R., Liu, Y., Yang, M., Brune, A., & Mo, J. (2016). Age polyethism drives community structure of the bacterial gut microbiota in the fungus-cultivating termite *Odontotermes formosanus*. *Environmental Microbiology, 18*, 1440–1451.

Liu, N., Yan, X., Zhang, M., Xie, L., Wang, Q., Huang, Y., Zhou, X., Wang, S., & Zhou, Z. (2011). Microbiome of fungus-growing termites: A new reservoir for mining lignocellulase genes. *Applied and Environmental Microbiology, 77*, 48–56.

Liu, G., Cornwell, W. K., Cao, K., Hu, Y., Logtestijn, R. S. P. V., Yang, S., Xie, X., Zhang, Y., Ye, D., Pan, X., Ye, X., Huang, Z., Dong, M., & Cornelissen, J. H. C. (2015). Termites amplify the effects of wood traits on decomposition rates among multiple bamboo and dicot woody species. *Journal of Ecology, 103*, 1214–1223.

Li-Ying, L., & Waterhouse, D. F. (1997). *The distribution and importance of arthropod pests and weeds of agriculture and forestry plantations in southern China*, *Report 67* (p. 73). Canberra: Australian Centre for International Agricultural Research.

Lockey, K. H. (1988). Lipids of the insect cuticle: Origin, composition and function. *Comparative Biochemistry and Physics B, 89*, 595–645.

Logan, J. M. W. (1991). Damage to sorghum by termites (Isoptera: Macrotermitinae) in the lower Shire Valley, Malawi. *Sociobiology, 19*, 305–307.

Logan, J. (1992). Termites (Isoptera): A pest or resource for small farmers in Africa? *Tropical Science, 32*, 71–79.

Logan, J., & El Bakri, A. (1990). Termite damage to date palms (*Phoenix dactylifera L*) in northern Sudan with particular reference to the Dongola District. *Tropical Science, 99*, 1363–1368.

Maienfisch, P., Angst, M., Brandl, F., Willi, F., Dieter, H., Hartmut, K., Werner, K., Alfred, R., Robert, S., Adrian, S., & Hansjurg, W. (2001). Chemistry and biology of thiamethoxam: A second generation neonicotinoid. *Pest Management Science, 57*, 906–913.

Makonde, H. M., Boga, H. I., Osiemo, Z., Mwirichia, R., Stielow, J. B., Goker, M., & Klenk, H. P. (2013). Diversity of *Termitomyces* associated with fungus-farming termites assessed by cultural and culture-independent methods. *PloS One, 8*, e56464.

Marten, A., Kaib, M., & Brandl, R. (2009). Cuticular hydrocarbon phenotypes do not indicate cryptic species in fungus-growing termites (Isoptera: Macrotermitinae). *Journal of Chemical Ecology, 35*, 572–579.

Marten, A., Kaib, M., & Brandl, R. (2010). Are cuticular hydrocarbons involved in speciation of fungus-growing termites (Isoptera: Macrotermitinae)? In M. Glaubrecht (Ed.), *Evolution in action: Case studies in adaptive radiation, speciation and the origin of biodiversity* (pp. 283–306). Berlin\Heidelberg: Springer.

Mathew, G. M., YM, J., Lai, C. Y., Mathew, D. C., & Huang, C. C. (2012). Microbial community analysis in the termite gut and fungus comb of *Odontotermes formosanus*: The implication of *Bacillus* as mutualists. *FEMS Microbiology Ecology, 79*, 504–517.

Mensa-Bonsu, A. (1978). Differentiation des castes hez les termites superieurs (Isoptera: Termitidae): Determination précoce des larves neutres et sexuées, cycle saisonnier, sexués de remplacement. Thesis, Université de Dijon.

Mikaelyan, A., Dietrich, C., Kohler, T., Poulsen, M., Sillam-Dussès, D., & Brune, A. (2015). Diet is the primary determinant of bacterial community structure in the guts of higher termites. *Molecular Ecology, 24*, 5284–5295.

Mitchell, M. R. (1989). Susceptibility to termite attack of various tree species planted in Zimbabwe. In D. J. Roland (Ed.), *Trees for the tropics*, *Monograph 10* (pp. 215–226). Canberra: Australian Centre for International Agricultural Research.

Moncalvo, J. M., Lutzoni, F. M., Rehner, S. A., Johnson, J., & Vilgalys, R. (2000). Phylogenetic relationships of agaric fungi based on nuclear large subunit ribosomal DNA sequences. *Systematic Biology, 49*, 278–305.

Mora, P., Rouland, C., & Renoux, J. (1996). Foraging, nesting and damage caused by *Microtermes subhyalinus* (Isoptera: Termitidae) in a sugarcane plantation in the Central African Republic. *Bulletin of Entomological Research, 86*, 387–395.

Mueller, U. G., Gerardo, N. M., Aanen, D. K., Six, D. L., & Schultz, T. R. (2005). The evolution of agriculture in insects. *Annual Review of Ecology, Evolution, and Systematics, 36*, 563–595.

Mugerwa, S. (2015). Magnitude of the termite problem and its potential anthropogenic causes in Nakasongola district of Uganda. Jap Soc. *Grassland Science, 61*, 75–82.

Mulatu, W., & Emana, G. (2015). Assessment of the type and number of reproductives in M. subhyalinus colonies in Ghimbi district western Ethiopia. *Advanced Research Journal of Microbiology, 2*, 74–77.

Myles, T. G. (1999). Review of secondary reproduction in termites (Insecta: Isoptera) with comments on its role in termite ecology and social evolution. *Sociobiology, 33*, 1–43.

Neoh, K. B., & Lee, C. Y. (2009). Developmental stages and castes of two sympatric subterranean termites *Macrotermes gilvus* and *Macrotermes carbonarius* (Blattodea: Termitidae). *Annals of the Entomological Society of America, 102*, 1091–1098.

Nkunika, P. O. Y. (1989). *Termites: A dilemma for the small scale farmer, Prod Farming (August)* (pp. 7–10). Lusaka: Zambia National Farmers Union.
Nkunika, P. O. Y. (1994). Control of termites in Zambia: Practical realities. *International Journal of Tropical Insect Science, 15*, 241–245.
Nobre, T., & Aanen, D. K. (2012). Fungiculture or termite husbandry? The ruminant hypothesis. *Insects, 3*, 307.
Nobre, T., Eggleton, P., & Aanen, D. K. (2010). Vertical transmission as the key to the colonization of Madagascar by fungus-growing termites? *Proceedings of the Royal Society of London B: Biological Sciences, 277*, 359–365.
Nobre, T., Kon, E. N. A., Konat, E. S., Linsenmair, K. E., & Aanen, D. K. (2011a). Dating the fungus-growing termites' mutualism shows a mixture between ancient codiversification and recent symbiont dispersal across divergent hosts. *Molecular Ecology, 20*, 2619–2627.
Nobre, T., Rouland-Lefevre, C., & Aanen, D. K. (2011b). Comparative biology of fungus cultivation in termites and ants. In E. D. Bignell, Y. Roisin, & N. Lo (Eds.), *Biology of termites: A modern synthesis* (pp. 193–210). Dordrecht: Springer.
Noirot, C. (1955). Recherches sur le polymorphisme des termites supérieurs (Termitidae). *Annales des Sciences Naturelles – Zoologie et Biologie Animale, 17*, 399–595.
Noirot, C. (1969). Glands and secretions. In K. Krishna & F. Weesner (Eds.), *Biology of termites* (Vol. 1, pp. 89–123). New York\London: Academic Press.
Noirot, C. (1985). The caste system in higher termites. In J. Watson, B. Okot-Kotber, & C. Noirot (Eds.), *Caste differentiation in social insects* (pp. 75–86). Oxford: Pergamon Press.
Noirot, C., & Darlington, J. P. E. C. (2000). Termite nests: Architecture, regulation and defence. In T. Abe, D. E. Bignell, & M. Higashi (Eds.), *Termites: Evolution, sociality, symbioses, ecology* (pp. 121–139). Dordrecht: Springer.
Okot-Kotber, B. M. (1981). Instars and polymorphism of castes in *Macrotermes michaelseni* (Isoptera, Macrotermitinae). *Insectes Sociaux, 28*, 233–246.
Osiemo, Z. B., Marten, A., Kaib, M., Gitonga, L. M., Boga, H. I., & Brandl, R. (2010). Open relationships in the castles of clay: High diversity and low host specificity of *Termitomyces* fungi associated with fungus-growing termites in Africa. *Insectes Sociaux, 57*, 351–363.
Otani, S., Mikaelyan, A., Nobre, T., Hansen, L. H., Koné, K. A., Sørensen, S. J., Aanen, D. K., Boomsma, J. J., Brune, A., & Poulsen, M. (2014). Identifying the core microbial community in the gut of fungus-growing termites. *Molecular Ecology, 23*, 4631–4644.
Otani, S., Hansen, L. H., Sorensen, S. J., & Poulsen, M. (2016). Bacterial communities in termite fungus combs are comprised of consistent gut deposits and contributions from the environment. *Microbial Ecology, 71*, 207–220.
Page, M., Nelson, L. J., Forschler, B. T., & Haverty, M. I. (2002). Cuticular hydrocarbons suggest three lineages in *Reticulitermes* (Isoptera: Rhinotermitidae) from North America. *Comparative Biochemistry and Physiology. B, 131*, 305–324.
Parihar, D. R. (1981). Termites affecting Eucalyptus plantations and their control in the arid regions of India. *Zeitschrift für Angewandte Entomologie, 92*, 106–111.
Parihar, D. R. (1985). Crop termite damage in Indian desert and its control in castor. *Zeitschrift fuer Angewandte Zoologie, 72*, 309–315.
Peppuy, A., Robert, A., Sémon, E., Bonnard, O., Truong Son, N., & Bordereau, C. (2001). Species specificity of trail pheromones of fungus-growing termites from northern Vietnam. *Insectes Sociaux, 48*, 245–250.
Poulsen, M. (2015). Towards an integrated understanding of the consequences of fungus domestication on the fungus-growing termite gut microbiota. *Environmental Microbiology, 17*, 2562–2572.
Poulsen, M., & Boomsma, J. J. (2005). Mutualistic fungi control crop diversity in fungus-growing ants. *Science, 307*, 741–744.
Poulsen, M., Hu, H., Li, C., Chen, Z., Xu, L., Otani, S., Nygaard, S., Nobre, T., Klaubauf, S., Schindler, P. M., Hauser, F., Pan, H., Yang, Z., Sonnenberg, A. S. M., de Beer, Z. W., Zhang, Y., Wingfield, M. J., Grimmelikhuijzen, C. J. P., de Vries, R. P., Korb, J., Aanen, D. K., Wang, J., Boomsmaa, J. J., & Zhang, G. (2014). Complementary symbiont contributions to plant decomposition in a fungus-farming termite. *PNAS, 111*, 14500–14505.

Qasim, M., Lin, Y., & Fang, D. (2015). Termites and microbial biological control strategies. *South Asia Journal of Multidisciplinary Studies, 1*, 1–27.

Quennedey, A., Sillam-Dussès, D., Robert, A., & Bordereau, C. (2008). The fine structural organization of sternal glands of pseudergates and workers in termites (Isoptera): A comparative survey. *Arthropod Structure and Development, 37*, 168–185.

Rajagopal, D. (1982). Relative incidence of termites on exotic species of Eucalyptus in India and their control. *Zeitschrift für Angewandte Entomologie, 98*, 225–230.

Rao, M. R., Singh, M. P., & Day, R. (2000). Insect pest problems in tropical agroforestry systems: Contributory factors and strategies for management. *Agroforestry Systems, 50*, 243–277.

Reddy, M. V. (1983). First record of *Odontotermes bellahunisensis* Holmg and Holmg feeding on *Cocos lucifera* Linn. *Journal of the Bombay Natural History Society, 80*, 653–656.

Reinhard, J., & Kaib, M. (2001). Trail communication during foraging and recruitment in the subterranean termite *Reticulitermes santonensis* De Feytaud (Isoptera, Rhinotermitidae). *Journal of Insect Behavior, 14*, 157–171.

Renoux, J., Rouland, C., Mora, P., & Dibangou, V. (1991). Les constructions du termite *P. spiniger* dans les champs de canne a sucre: Essai de lutte specifique. *AFCAS R Int Can Suc, 1*, 124–130.

Roberts, E. M., Todd, C. N., Aanen, D. K., Nobre, T., Hilbert-Wolf, H. L., O'Connor, P. M., Tapanila, L., Mtelela, C., & Stevens, N. J. (2016). Oligocene termite nests with in situ fungus gardens from the Rukwa Rift Basin, Tanzania, support a Paleogene African origin for insect agriculture. *PloS One, 11*, e0156847.

Rohrmann, G. F. (1978). The origin, structure, and nutritional importance of the comb in two species of *Macrotermitinae*. *Pedobiologia, 18*, 89–98.

Rouland, C., Ikhouane, A., & Nayalta, N. (1993). Etude biologique des populations d'*Ancistrotermes guineensis* présentes dans les plantations de la SONASUT. *Act Coll IUSSI, 8*, 79–87.

Rouland, C., Ben Moussa, D., Reversat, G., & Laumond, C. (1996). Etude de la sensibilite de sexués de termites Macrotermitinae a une infestation par des nématodes entomopathogènes des genres *Heterorhabditis* et *Steinernema*. *Comptes rendus de l'Académie des sciences. Série 3, Sciences de la vie, 319*, 997–1001.

Rouland-Lefevre, C. (2000). Symbiosis with fungi. In T. Abe, D. E. Bignell, & M. Higashi (Eds.), *Termites: Evolution, sociality, symbioses, ecology* (pp. 289–306). Dordrecht: Springer.

Rouland-Lefèvre, C., & Bignell, D. E. (2002). Cultivation of symbiotic fungi by termites of the subfamily Macrotermitinae. In J. Seckbach (Ed.), *Symbiosis: Mechanisms and model systems* (pp. 731–756). Dordrecht: Springer.

Rouland-Lefèvre, C., & Mora, P. (2002). Control of *Ancistrotermes guineensis* Silvestri (Termitidae: Macrotermitinae), a pest of sugarcane in Chad. *International Journal of Pest Management, 48*, 81–86.

Rouland-Lefevre, C., Diouf, M. N., Brauman, A., & Neyra, M. (2002). Phylogenetic relationships in *Termitomyces* (family Agaricaceae) based on the nucleotide sequence of ITS: A first approach to elucidate the evolutionary history of the symbiosis between fungus-growing termites and their fungi. *Molecular Phylogenetics and Evolution, 22*, 423–429.

Ruelle, J. (1985). Order Isoptera. In C. Scholtz & E. Holm (Eds.), *Insects of southern Africa* (p. 502). Durban: Butterworth.

Rust, M., & Saran, R. (2006). Toxicity, repellency, and transfer of chlorfenapyr against western subterranean termites (Isoptera: Rhinotermitidae). *Journal of Economic Entomology, 99*, 864–872.

Sands, W. A. (1960). *Termite* control in west African afforestation. Report. 7th Commonwealth Entomological Conference, pp. 91–95.

Sands, W. (1969). The association of termites and fungi. In K. Krishna & F. Weesner (Eds.), *Biology of termites* (Vol. 1, pp. 495–524). New York: Academic Press.

Sane, C. (2016). *Relations entre le fonctionnement des agrosystèmes a base de manguiers et les dégats des termites (Termitidae: Isoptera) dans les regions de Thiès et de Dakar, au Sénégal.* Thesis, Université Cheikh Anta Diop Dakar.

Sane, C., Rouland-Lefevre, C., Grechi, I., Rey, J., Vayssieres, J., Diame, L., & Diarra, K. (2016). Diversity, damages and management of termites (Isoptera) in Senegalese agrosystems. *International Journal of Biological and Chemical Sciences, 10*, 134–154.

Schuurman, G. W. (2012). Ecosystem influences of fungus-growing termites in the dry paleotropics. In D. H. Wall, R. D. Bardgett, V. Behan-Pelletier, J. E. Herrick, H. Jones, K. Ritz, J. Six, D. R. Strong, & W. H. van der Putten (Eds.), *Soil ecology and ecosystem services* (pp. 173–185). Oxford: Oxford University Press.

Sekamatte, B. M., & Okwakol, M. J. N. (2007). The present knowledge on soil pests and pathogens in Uganda. *African Journal of Ecology, 45*(Suppl. 2), 9–19.

Sekamatte, M., Latigo, M., & Smith, A. (2001). The effect of maize stover used as mulch on termite damage to maize and activity of predatory ants. *African Crop Science Journal, 9*, 411–419.

Sekamatte, B. M., Ogenga-Latigo, M., & Russell-Smith, A. (2003). Effects of maize-legume intercrops on termite damage to maize, activity of predatory ants and maize yields in Uganda. *Crop Protection, 22*, 87–93.

Sharma K, Kalpana KK, Sharma V, Gupta P, Jaya M, Kumar A, Singh B (2008) Persistence and vertical distribution of termiticide fipronil in modified ground board test. Environmental Monitoring and Assessment 137:179–184.

Sharma, R. K., Srinivasa Babu, K., Chhokar, R. S., & Sharma, A. K. (2004). Effect of tillage on termites, weed incidence and productivity of spring wheat in rice–wheat system of north western Indian plains. *Crop Protection, 23*, 1049–1054.

Shinzato, N., Muramatsu, M., Watanabe, Y., & Matsui, T. (2005). Termite-regulated fungal monoculture in fungus combs of a Macrotermitine termite *Odontotermes formosanus*. *Zoological Science, 22*, 917–922.

Sieber, R. (1983). Establishment of fungus comb in laboratory colonies of *Macrotermes michaelseni* and *Odontotermes montanus* (Isoptera, Macrotermitinae). *Insectes Sociaux, 30*, 204–209.

Sileshi, G., Mafongoya, P. L., Kwesiga, F., & Nkunika, P. (2005). Termite damage to maize grown in agroforestry systems, traditional fallows and monoculture on nitrogen-limited soils in eastern Zambia. *Agricultural and Forest Entomology, 7*, 61–69.

Sileshi, G. W., Kuntashula, E., Matakala, P., & Nkunika, P. O. (2008). Farmers' perceptions of tree mortality, pests and pest management practices in agroforestry in Malawi, Mozambique and Zambia. *Agroforestry Systems, 72*, 87–101.

Sillam-Dussès, D., Semon, E., Robert, A., Cancello, E., Lenz, M., ValterovA, I., & Bordereau, C. (2010). Identification of multi-component trail pheromones in the most evolutionarily derived termites, the Nasutitermitinae (Termitidae). *Biological Journal of the Linnean Society, 99*, 20–27.

Singh, B., Kular, J. S., Ram, H., & Mahal, M. S. (2014). Relative abundance and damage of some insect pests of wheat under different tillage practices in rice–wheat cropping in India. *Crop Protection, 61*, 16–22.

Starnes, R. L., Liu, C. L., & Marrone, P. G. (1993). History, use, and future of microbial insecticides. *American Entomologist, 39*, 83–91.

Sun, J., Fuxa, J., & Henderson, G. (2003). Virulence and in vitro characteristics of pathogenic fungi isolated from soil by baiting with *Coptotermes formosanus* (isopteran: Rhinotermitidae). *Journal of Entomological Science, 38*, 342–358.

Takahashi, S., & Gassa, A. (1995). Roles of cuticular hydrocarbons in intra-and interspecific recognition behavior of two rhinotermitidae species. *Journal of Chemical Ecology, 21*, 1837–1845.

Takematsu, Y., & Yamaoka, R. (1999). Cuticular hydrocarbons of *Reticulitermes* (Isoptera: Rhinotermitidae) in Japan and neighboring countries as chemo taxonomic characters. *Applied Entomology and Zoology, 34*, 179–188.

Tang, B., Tang, M., Chen, C., Qiu, P., Liu, Q., Wang, M., & Li, C. (2006). Characteristics of soil fauna community in the Dongjiao coconut plantation ecosystem in Hainan, China. *Acta Ecologica Sinica, 26*, 26–32.

Tiben, A., Pearce, M., & Wood, T. (1990). Damage to crops by *Microtermes najdensis* (Isoptera: Macrotermitinae) in irrigated semi-desert areas of the red sea coast. 2. Cotton in the Tokar Delta region of Sudan. *Trop Pest Manage, 36*, 296–304.

Traniello, J. F. A., & Leuthold, R. H. (2000). Behavior and ecology of foraging in termites. In T. Abe, D. E. Bignell, & M. Higashi (Eds.), *Termites: Evolution, sociality, symbioses, ecology* (pp. 141–168). Dordrecht: Springer.

Turner, J. (1994). Ventilation and thermal constancy of a colony of a southern African termite (*Odontotermes transvaalensis*, Macrotermitinae). *Journal of Arid Environments, 28*, 231–248.
Turner, J. S. (2001). On the mound of *Macrotermes michaelseni* as an organ of respiratory gas exchange. *Physiological and Biochemical Zoology, 74*, 798–822.
Um, S., Fraimout, A., Sapountzis, P., DC, O., & Poulsen, M. (2013). The fungus-growing termite *Macrotermes natalensis* harbors bacillaene-producing *Bacillus* sp. that inhibit potentially antagonistic fungi. *Scientific Reports, 3*, 3250.
UNEP-FAO. (2000). *Finding alternatives to persistent organic pollutants (POPs) for termite*. Global IPM facility expert group on termite biology and management (eds UNEP), pp. 47.
Verma, M., Sharma, S., & Prasad, R. (2009). Biological alternatives for termite control: A review. *International Biodeterioration and Biodegradation, 63*, 959–972.
Wardell, D. (1987). Control of termites in nurseries and young plantations in Africa: Established practices and alternative courses of action. *Commonwealth Forestry Review, 66*, 77–89.
Wood, T. (1996). The agricultural importance of termites in the tropics. *Agric Zool Rev, 7*, 117–155.
Wood, T., & Cowie, R. (1988). Assessment of on-farm losses in cereals in Africa due to soil insects. *International Journal of Tropical Insect Science, 9*, 709–716.
Wood, T., & Pearce, M. (1991). Termites in Africa: The environmental impact of control measures and damage to crops, trees, rangeland and rural buildings. *Sociobiology, 19*, 221–234.
Wood, T. G., & Sands, W. A. (1978). The role of termites in ecosystems. In M. Brian (Ed.), *Production ecology of ants and termites* (pp. 245–292). Cambridge: Cambridge University Press.
Wood, T., & Thomas, R. (1989). The mutualistic association between Macrotermitinae and *Termitomyces*. In N. Wilding, N. Collins, P. Hammond, & J. Webber (Eds.), *Insect-fungus interactions* (pp. 69–92). London: Academic Press.
Wood, T. G., Bednarzik, M., & Aden, H. (1987). Damage to crops by *Microtermes najdensis* (Isoptera, Macrotermitinae) in irrigated semi-desert areas of the red sea coast 1. The Tihama region of the Yemen Arab Republic. *Tropical Pest Management, 33*, 142–150.
Xie, Y., Wang, K., Huang, Q., & Lei, C. (2014). Evaluation toxicity of monoterpenes to subterranean termite, *Reticulitermes chinensis* Snyder. *Industrial Crops and Products, 53*, 163–166.
Ye, G., Li, K., Zhu, J., Zhu, G., & Hu, C. (2007). Cuticular hydrocarbon composition in pupal exuviae for taxonomic differentiation of six Necrophagous flies. *Journal of Medical Entomology, 44*, 450–456.
Zaremski, A., Fouquet, D., & Louppe, D. (2009). *Les termites dans le monde*. Editions Quae, pp. 94.
Zhang, M., Liu, N., Qian, C., Wang, Q., Wang, Q., Long, Y., Huang, Y., Zhou, Z., & Yan, X. (2014). Phylogenetic and functional analysis of gut microbiota of a fungus-growing higher termite: Bacteroidetes from higher termites are a rich source of β-Glucosidase genes. *Microbial Ecology, 68*, 416–425.
Zhong, J., & Liu, L. (2002). Termite fauna in China and their economic importance. *Sociobiology, 40*, 25–32.

Chapter 2
Termite Damage in Agriculture Areas and Implanted Forests: An Ecological Approach

Luciane Kern Junqueira and Daniela Faria Florencio

Contents

Abstract Termites are significant soil fauna components in tropical forests and play an essential role in organic matter decomposition, nutrient cycling, soil airing, and draining. They also contribute to the establishment of new soil in eroded areas. However, they can cause significant economic losses in commercial plantations and implanted forests. Therefore, the correct identification of the termite species and evaluation of the richness, abundance, and functional groups of the community is critical for control, of any species that acquired and/or could reach a pest status. This identification will contribute for the preservation and endurance of beneficial

L.K. Junqueira (✉)
Faculty of Biological Science, Pontifícia Universidade Católica de Campinas, Campinas, São Paulo, Brazil
e-mail: lkjunque@puc-campinas.edu.br

D.F. Florencio
Center of Social and Applied Human Sciences, Universidade Federal Rural do Semi-Árido, Mossoró, Rio Grande do Norte, Brazil

Post-Graduation in Ecology and Conservation, Universidade Federal Rural do Semi-Árido, Mossoró, Rio Grande do Norte, Brazil

M.A. Khan, W. Ahmad (eds.), *Termites and Sustainable Management*, Sustainability in Plant and Crop Protection,
https://doi.org/10.1007/978-3-319-68726-1_2

species, as some species appear more sensitive to soil usage and agricultural activities than others. In this chapter, the impact of termites in agricultural areas and implanted forests will be addressed. The importance of the taxonomic and ecological studies will be highlighted as a premise to the use of control agents and technical management.

Keywords Isoptera • Commercial plantations • Agroecosystem

2.1 Introduction

Termites (Blattodea: Isoptera), though less visible and attractive, are decisive for the functioning and yield of tropical terrestrial ecosystems, by increased soil water infiltration due to tunnels and enrichment in soil nitrogen. They are thus important for agricultural sustainability (Evans et al. 2011). These activities also sustain a greater biodiversity by an increase in richness and abundance of other organisms (Dangerfield et al. 1998; Florencio et al. 2013; Pringle et al. 2010).

Most species live in tropical environments and only a few ones survive in temperate regions. Krishna and Weesner (1970) identified some factors as important to the global distribution of termites: feeding exclusively on cellulosic materials; low sclerotin and consequently a soft body, a limiting factor for mobility; natural dispersion only during flutter period, although only for short distances; and high susceptibility to predators. Conversely, Wood and Johnson (1986) suggest the variety of dietary resources and different nesting areas favor the distribution of termites, over various regions of the world.

Due to these factors, termites are found in both tropical and temperate regions, between parallels 45°–48° N and 45° S (Wood 1975). In tropical ecosystems, termites cause impact due to their abundance, social behavior, and great diversity of species (Wood and Sands 1978). Roughly 10% of all described species have caused damage to human activity (McMahan 1986; Verma et al. 2009). In South America there are reports that 19% of species cause structural and/or agricultural damage (Constantino 2002).

Currently, the most appropriate integrated strategies to control pest termite species include preventive practices, such as less drastic changes in the environment and/or the introduction of termite-specific agents (see review by Rouland-Lefèvre 2011). These practices are complementary to reduce the use of toxic chemical products (insecticides).

In this review, the possible damages caused by termites in forest and agricultural areas are presented, focusing on commercial *Eucalyptus* forests and sugarcane plantations. Moreover, it includes a brief report on plantations of cocoa, corn, and rice. Finally, we highlight how the accurate identification of termite species and the evaluation of the community's ecological attributes (richness, abundance, functional groups, etc.) are a critical point for controlling those species that have acquired and/or may acquire pest status. At the same time, this approach sustains the preservation and maintenance of beneficial termite species, as some of them appear more sensitive than others to land use and agricultural or forestry activities.

2.2 Termites in Commercial *Eucalyptus* Forests

In forests, which make use of exotic species in tropical areas, termites have already caused such significant damage that, according to Harris (1971) and Cowie et al. (1989), they were considered as a limiting factor for the development of commercial *Eucalyptus* forests.

The termite species damaging *Eucalyptus* forests have been divided into two groups. In the first one, those that attack seedlings from plantation up to 1 year of age were included, also known as seedling, root, or soil termites (e.g., *Syntermes molestus*, *S. insidians*, and *Cornitermes cumulans*). They cause the destruction of the radicular system or the seedling girdling at the stem, damages that often cause the death of the plant. If the soil conditions are favorable, the seedlings may resist the attack and form a callus, which in turn originates a new radicular system over the destroyed tissue or even a sprout, forming a new aerial branch, in the case of girdling at the stem region. Consequently, the trees may develop a deficient radicular system and insufficient support, or will become dominated trees, due to their delayed initial development. In these cases mortality rates are high, accounting for the occurrence of damage up to 18% of *Eucalyptus grandis* seedlings in Brazil due to the attack of *Cornitermes* sp. For commercial plantations, the percentage of acceptable failure is from 2 to 5% since, and above this level, the replantation is costly (Wilcken 1992; Wilcken and Raetano 1995).

The other group is formed by the termites that attack *Eucalyptus* trees aged over 2 years and destroy their interiors, named heartwood termites (e.g., *Coptotermes testaceus*). They penetrate through the root, building their chambers within the trunk, destroying the heartwood, and making the trees hollow, thus causing loss of productivity. The internal damage only appears during harvest, and this hampers prevention (Wilcken and Raetano 1998). Nair and Varma (1985) already suggested that factors such as termite species present, population density, seasonal trends, buildup of litter fall and wood, soil conditions, physiological state of the plant, age, and establishment state were all related to the termite × *Eucalyptus* interaction. Wardell (1987) also pointed out that the lack of research programs to study the ecological relations between termites, host trees, and possible correlations with edaphic factors aggravated this situation.

According to Mill (1982), deforestation and isolation directly affected the ecology of termites from neotropical forests. Since most termites from primary forests are very sensitive to such effects and do not survive habitat changes, the few species which manage to adapt may become important pests in forestry. The available data, though limited, suggested that wood removal or certain agricultural practices reduce termite richness and lead to the selective loss of certain functional groups, especially the soil-feeding ones (Wood et al. 1982). However, the effects of forestry practices, which are considered less drastic, were still unknown (Mill 1982; Eggleton et al. 1995). These cited references described the termite problem in *Eucalyptus* plantations for roughly three decades. Conversely, more recent works have questioned the role of termites as a pest in this forest system.

Calderon and Constantino (2007), evaluating the termite diversity in a tillage of *Eucalyptus urophylla*, found only 0.2% of cut trees damaged at core, caused by *Coptotermes* sp., thus suggesting that termites do not cause significant problems in the studied region. Moreover, it was possible to highlight that the importance of termites as a pest to *Eucalyptus* has been overstated. One of the factors involved is due to the fact that the areas selected for study were therefore chosen due to the previous evidence of damage by termites.

Junqueira et al. (2009) studied the richness and abundance of termites in seven forest areas in Anhembi, São Paulo. The areas ranged from *Eucalyptus* plantations of different species/ages to forest fragments, in different stages of development and succession. The highest species richness (13) was found in the area in the stage of advanced succession and the lowest (8) in the implanted forest of *E. urophylla* of 3 years, with no sub-forest. The abundance of some species (Apicotermitinae sp.1, Apicotermitinae sp.2, Apicotermitinae sp.4, *Cornitermes cumulans*, *Diversitermes diversimiles*, and *Embiratermes* sp.) varied greatly between areas and indicated a lower diversity of termites in implanted *Eucalyptus* forests, when compared to forest fragments. The authors also pointed out that the presence of some termite species appeared to be associated to that of fallen wood on the forest floor. The permanency of this wood on the floor could favor the termite community in the area, without affecting seedlings or trees, since the termites identified feed and nest in the wood itself.

When evaluating the effects of *Eucalyptus* plantation on soil arthropod communities in a conservation area of the Atlantic forest, Camara et al. (2012) verified that 61% of the taxa found were common to *Eucalyptus* areas and to forest fragments, whereas the Isoptera group was only present in two plantations but did not have its species specified nor the possible damage caused by them. Nevertheless, it is important to highlight that the authors used pitfall traps, which is unsuitable for termite sampling, and this justifies the low presence of this group in the samples.

2.3 Termites in Sugarcane Commercial Plantations

In Brazil, planting of sugarcane is dedicated to the production of ethanol, which occupies an extensive land area of 4.9 Mha in 2011, with the estimate that by 2021 this area will reach 6.4 Mha (Goldemberg et al. 2014). Land use changes (LUC) and the agricultural methods undertaken for the introduction and upkeep of this monoculture contribute to the loss of biodiversity, especially of soil macrofauna, which includes termites (Franco et al. 2016). These losses result in soils of low fertility and sandy texture. These factors contribute to the occurrence of subterranean termites, a group which may cause damage to the various growth phases of the sugarcane culture (Campos et al. 1998).

Recently, a number of ecological aspects of the species found in sugarcane crops and other cultures, as well as their pest potential and the need of basic studies on the reproductive biology and population dynamics of various termite species, were discussed (Miranda et al. 2004; Junqueira et al. 2008; Menzel and Diehl 2008, 2010; Ackerman et al. 2007). According to Batista-Pereira et al. (2004), termites called the attention of the scientific community because they are important pests in sugarcane crops. At this regard, Mill (1992) already referred to a fall of up to 65% in production in Brazil, due to termite attack in sugarcane plantations.

In the Brazilian northeast, termites are considered as one of the main soil pests in sugarcane plantations, and a high occurrence of the *Heterotermes* and *Neocapritermes* genus is registered. Miranda et al. (2004) found four termite species in a single sugarcane plantation in the northeast, but only one type caused damage (*Cylindrotermes nordenskioeldi*), whereas a second one (*Amitermes nordestinus*) was possibly considered as a potential pest. According to the authors, the abundance and the vertical and horizontal spatial distribution of these insects were mainly influenced by the root biomass of the crops and by the amounts of organic matter in soil. In the southeast region, Novaretti and Fontes (1998) identified 14 termite species in sugarcane crops, while Almeida and Alves (1999), in studies carried out in northern São Paulo, considered *Heterotermes tenuis* as the main crop pest, due to its broad distribution and high number of individuals.

According to Junqueira et al. (2015), the lack of solid taxonomic and ecological data on termite communities in Brazilian sugarcane plantations led to the attribution of a pest category to species with low prevalence, with an indiscriminate use of insecticides for termite control, in spite of their central ecological role in soil fertility. The authors performed sampling of soil termites in commercial sugarcane crops in 53 counties in the state of São Paulo. The richness obtained accounted for 22 taxa, and the most frequent functional group was the humus feeders (37%), followed by wood (34%), litter (25%), and intermediaries feeders (4%). Finally, in contrast to what is pointed out in other studies, the results suggested that the greater part of the termite community registered in sugarcane plantations of São Paulo was potentially beneficial to the cultivation, considering the high frequency of soil-feeding species.

2.4 Termites in Other Cultures

Dibog et al. (1999) studied the impact of vegetation shelter on termite communities in commercial forests of *Terminalia ivorensis* (Combretaceae), aged from 6 to 18 years, in the southern regions of the Republic of Cameroon where the understory of these trees is used for growing banana and cocoa. In 18 years of *T. ivorensis* plantation, the highest abundance of termites occurred in locations with a denser canopy, regardless of the type of understory (banana or cocoa), the cultivation system (mixed or individual), or the soil preparation method (conserving or burning biomass, in this case specifically for growing bananas). For 6-year-old plantations, there was no significant difference in the abundance of termites when compared to lesser or greater canopy density, whereas in the latter, a higher frequency of termites did occur. Conversely, the crop yield was not directly related to the abundance of termite population, though the yield in cocoa production was positively related to the abundance of soil-feeding termites (the greater part of the community). This was probably due to improved soil conditions resulting from the presence of this group. Of the 82 total termite species found, 67 were soil feeders.

Some species of fungus-growing termites (Termitidae: Macrotermitinae) are one of the main pests of corn (*Zea mays* L.) in the African continent. Researchers have attempted to comprise suitable techniques in soil management, emphasizing the availability of nutrients (nitrogen) and termite activity (Sileshi and Mafongoya 2003; Sileshi et al. 2005; Nyagumbo et al. 2015). In a field experiment carried out at Msekera Research Station in Chipara (Zambia), Sileshi and Mafongoya (2003) studied the presence and activity of four termite types, namely, *Microtermes*, *Hodotermes*, *Odontotermes*, and *Pseudacanthotermes*. Root and stem base damage in the corn crops were identified following a short drought period. Productivity and total crop biomass were negatively associated to damage caused by termites. The losses resulting from termite action in conventional corn crops (natural fallow) were five to eleven times greater than corn crops associated to i) *Tephrosia* (*Tephrosia vogelii*) + pigeon pea (*Cajanus cajan*) and ii) *Sesbania* (*Sesbania sesban*) + pigeon pea (*C. cajan*), respectively. The authors suggested that the introduction of these legumes may alter the physical or chemical characteristics of soil, affecting the susceptibility of corn to pest attacks, therefore reducing losses.

In evaluating the occurrence of termites in corn through an experiment performed in the province of Manica, Mozambique, Nyagumbo et al. (2015) verified that the activity (number of individuals and proportion of tunnels) of *Macrotermes falciger*, *Pseudocanthotermes* sp., and *Odontotermes* sp. (Termitidae: Macrotermitinae) adjacent to the corn plantations was directly related to the availability of residues in furrows. Moreover, the increase in termite activity did not result in damage to the culture, and, provided that there were resources for the termites, these did not attack the crops. This research highlighted the importance of the development of management techniques that contribute to the quality and availability of soil nutrients. In rice (*Oryza sativa* L.), Mahapatro and Sreedevi (2014) registered *Odontotermes obesus* and *Microtermes obesi* (Termitidae: Macrotermitinae) as causing damage to crops, in dry land and irrigated farming, in India. The attack on live plants occurred mainly in the final phase of growth and in the absence of decomposing matter in the area.

2.5 Modification of Natural Habitat: Effects on Termite Community

The richness and abundance of soil invertebrates are influenced by the biogeographic history, latitude, altitude, and climate variables such as temperature, humidity, and rainfall and by the characteristics of the habitat involving soil type and depth (Hulugalle et al. 1997; Bignell and Eggleton 2000; Eggleton 2000; Ellis et al. 2001). Natural and anthropic disturbances can cause changes in environmental variables such as soil aeration, humidity, food resource availability, and height of the tree canopy. Consequently, there are changes in the composition, abundance, and/or richness of termites (DeSouza and Brown 1994; Jones et al. 2003). They may lead to a loss, a negative impact in the functioning of ecosystems, altering ecological processes as the carbon and nitrogen cycles, therefore reducing soil quality and its productivity (Black and Okwakol 1997; Okwakol 2000; Evans et al. 2011).

As the replacement of natural ecosystems with commercial forests occurred, reports emerged of damage caused by native and exotic termites. These modifications can favor certain limited species which, in the absence of predators, may reach a pest status (Mill 1982; Wood et al. 1982). According to Constantino (2002), 77 termite species have been reported as pests in South America, of which 53 were related to damage exclusively in agriculture and 15 were noxious both for structures and agriculture. Among them, *Heterotermes*, *Nasutitermes*, *Cornitermes*, *Procornitermes*, and *Syntermes* were highlighted as the main prompters of damage. Often, the damage caused by termites in commercial forests is indirect, and this hampers identification. In this sense, various authors report a damage by observing the termites feeding off parts of plants or just by their presence in that area. These reports cause part of the damage in commercial plantations to be incorrectly attributed to termites. Additionally, according to Eggleton (1999) and Constantino (2002), the taxonomic study is central to the understanding of the importance of these insects in ecosystems.

In Central Amazon area, Bandeira (1979) studied the effect of deforestation on termite populations by evaluating the distribution and diversity of these insects in primary forest areas, shrubland, and pastures. Most types presented equivalent distributions in all three areas. Members of *Nasutitermes* were the most common and diverse, with more frequency in pastures where a greater number of nests were observed. The soil termites were found in greater numbers in shrubland, followed by pastures and lastly in forests. The author suggested that the removal of the primary vegetation and the resulting microclimatic changes were responsible for the distribution of some groups.

Working in four locations in the surroundings of Manaus (Brazil), Mill (1982) found a greater density of termites in islands than in dry land, probably due to the competition for food. According to the author, the termites which are adapted to life in the bush, in islands, and in blackwater-flooded forests (Igapos) are the species that can become pests in commercial forests (e.g., *Coptotermes* and *Nasutitermes*). The termite fauna in dry land and whitewater-flooded forests (varzea) of the

Brazilian Amazon was distinct, with a low index of similarity. The composition of species and diversity varied greatly among locations, with no apparent correlation to climate or vegetation type. Part of these differences also occurred due to the collecting efforts and sampling methods. Constantino (1992) analyzed the termite fauna in primary forests of two locations in the Brazilian Amazon and observed that the subfamily *Nasutitermitinae* (especially the *Nasutitermes* type) was the predominant group, both in the number of species and abundance. The humus feeders were the second group as the number of species. The composition and diversity of species varied among the different locations and, apparently, did not present correlation to climate or vegetation type.

DeSouza and Brown (1994) studied the termite communities in the Amazon forest and in neighboring isolated, reservation fragments. The soil-feeding group was predominant in the forest, with greater species richness and lower proportion of rare species. In the fragments however, the species which use litter fall and those with intermediary feeding habits between soil and wood feeding were predominant. Moreover, the termites in the first area made a more equitable use of the forest, when compared to the fragments, suggesting a growing inadequateness of habitat due to fragmentation. The composition of termite communities in the fragments would hence be the result of an intrinsic pattern of forest and losses caused by fragmentation.

Eggleton et al. (1995) qualitatively evaluated the termite communities of five areas with different levels of forest disturbance in the Mbalmayo Forest Reserve, in the south of the Republic of Cameroon. When compared to primary forest, the areas with severe disturbances presented a significant reduction in richness of species, while the areas under process of regeneration showed a slight increase in richness. The soil-feeding termites predominated in the areas under process of regeneration and in primary forest, though the species richness decreased in areas that suffered severe disturbances. The wood-feeding type appeared to be more resistant to the disturbances than the soil-feeding type, although the richness of species was low in the more affected areas.

In the same forest reserve, Eggleton et al. (1996) evaluated the diversity, abundance, and biomass of termite communities in five locations with different levels of disturbance, over a 2-year period. The abundance and biomass were high in the locations where the forest was very similar to the primary area and in the forest in advanced process of succession. The disturbances had little effect on the abundance and biomass in the forest areas. There was, however, a clear reduction of these components in open areas. Differences were also found in the composition of taxonomic groups, in the abundance of different nesting areas, and in the composition of the functional groups among study areas, the latter affecting mainly soil-feeding termites. The area similar to the primary forest presented a more heterogeneous community of termites when compared to the areas with more disturbances, possibly resulting from the greater number of microhabitats available for the termites.

Davies et al. (1999) investigated the successional response of the termite community to the experimental disturbances in forests in the Republic of Cameroon, evaluating the implications to forest recovery. Even in treatments which involved severe disturbances to the soil and canopy cover, the richness and abundance of termites quickly recovered when dead wood was left on the forest floor. This wood availability also led to the occupation by a different group of termites of the compost sampled in other treatments, highlighting that this group included wood and soil-feeding termite species.

Bandeira et al. (2003), studying the termite fauna present in six environments with different disturbance levels, located at the Brejo dos Cavalos (PE, Brazil), observed a reduction in termite diversity as the disturbances increased, apart from not finding these insects in a monoculture area. The termites, which feed on humus, were more affected than those with an intermediary feeding. In parallel, the wood feeders presented more resilience, while some species, favored in secondary forest areas, tended to disappear in agricultural areas which presented little wood availability.

Sena et al. (2003) investigated the termite fauna in a fragment of savannah (cerrado) at the Guaribas Biological Reserve in Mamanguape (PB, Brazil). The richness was of 20 species, the majority of which was the wood feeders. The highest frequency in the sampling transects was for species which feed on humus. The richness was inferior to what reported in the savannah vegetation, probably due to the isolation time and distance of this specific fragment in the vast savannah areas in Brazil central region.

Jones et al. (2003) evaluated the impact of intensified soil use in termite communities in humid tropical forests in the province of Jambi, Sumatra (Indonesia). Therefore, the composition of the communities was identified using a gradient of seven areas with disturbances, which included primary forest, different commercial plantation systems, and areas with cassava plantations. In the seven environments, 54 species were collected: the primary forest presented the highest termite richness with 34 species and the cassava plantation the lowest richness, with only one species. The relative abundance of soil-feeding termites presented a greater fall over the gradient in comparison to the wood-feeding type. It was found that the basal area of trees was strongly correlated to the richness and relative abundance of termites, thus reflecting the adaptive response of these insects to the progressive simplification of the habitat physical structure. This was due to a reduced canopy cover, microclimatic changes, and reduction of feeding and nesting locations. The authors also analyzed work from other researchers where all, or a great part of communities, were evaluated over a local disturbance gradient. Data generally indicated a declining trend in richness and abundance of species, as land use increases. This trend was more apparent when more contrasting gradients were analyzed, ranging from primary forests to treeless areas.

Lavelle et al. (1997) refer to soil invertebrate as "engineers of soil ecosystems" since they ingest or manipulate organic and mineral material and form microstructures. The authors cite worms and termites as the most important engineers of earthly ecosystems because, as mediators of nutrient transformation, they influence the diversity and activity of the biota and subordinate trophic levels. They also consider the hypothesis that the vegetation affects not only the abundance but the richness of these organisms through the quality and quantity of litter fall, among other effects. Changes in vegetable communities would therefore affect soil engineers. According to Lavelle et al. (1997), land use and forest disturbances are responsible for the more immediate changes in the functional groups of these engineer communities. The disturbances affect the termites through the reduction of diversity, especially of those who feed off soil, and some species can become pests due to changes in the availability of organic material. The reduction in the abundance of these engineers leads to a decrease of carbon stock in soil and inequality among functional groups (compaction × decompaction) and may result in physical degradation of soil.

Jones et al. (2003) proposed actions toward the recovery of termite communities. The author suggests that by leaving dead wood in the forest floor following the disturbance, the recovery process of the termite communities is accelerated, due to the increased abundance and richness of wood- and soil-feeding termites. Moreover, the increase in fragment size and reduction of edge effects would have a positive effect in the survival of species that depend on the forest. Finally, the adoption of corridors connecting these fragments could assist termite recolonization and therefore increase their dispersion potential.

According to Calderon and Constantino (2007), the use of Brazilian savannah areas for planting *Eucalyptus* eliminated soil-feeding termites and favored the occurrence of wood and litter feeders. The authors suggest that the change in the structure of the termite community resulted from changes of microclimate, habitat structure, and quality and diversity of litter fall.

Carrijo et al. (2009) showed that the simplification of the savannah habitat led to the extinction of termite populations as well as other members of the fauna and flora, by the loss of specific resources due to the application of pesticides. Junqueira et al. (2008) published a wide bibliographic review that sought to discuss how intensified land use and deforestation for the implantation of commercial *Eucalyptus* forests caused changes in termite diversity, mostly affecting the soil-feeding group, and suggested that the decrease of this group could provoke likewise a decrease of organic matter in decomposition.

Finally, Dosso et al. (2013) investigated, in a transitional forest-savannah zone of Côte d'Ivoire, the termite assemblage structure across a sequence of differing land-use systems. The authors suggested that forest conversion to agricultural systems changes the termite assemblage structure, including loss of species and changes of feeding groups. The mean termite species richness declined from the semi-deciduous forest to the teak plantation, 4-year-old fallow, food crop, cocoa plantation, and *Jatropha* plantation. The soil and wood feeders showed response to disturbance with abundance of changes along the distinct land-use types: i) the soil feeders declined the semi-deciduous to all man-modified sites and ii) the wood-feeding species reduced their abundance to monospecific and modified sites without high trees. The fungus growers were considered the feeding group most adapted to disturbances, with dominant species *Ancistrotermes cavithorax*, *Microtermes toumodiensis*, and *Pseudacanthotermes militaris* found in all land-use systems. Thus, changes of land use can promote the loss of some termite species, reduce biological functions, and establish species that can become pests.

2.6 Conclusion

Currently, a number of works have highlighted that the damages caused by termites in planted forests and/or agriculture are directly related to the use and modification of soil for the implantation of crops. This promotes the simplification of the environment's physical structure, which alters the termite communities and leads to the maintenance and establishment of limited termite as well as pest species.

In plantations where soil is uncovered and the area is completely clean, the occurrence of termites can cause damages, since there will be no other resource apart the introduced culture. In contrast, plantations associated to the adoption of certain strategies which i) maintain the presence of sub-forest and adjacent forest fragments, ii) protect soil, and iii) present the availability of litter fall may reduce the loss of water and nutrients and promote the maintenance of soil-feeding species, as well as of wood and litter feeders. These species can consume other available resources and bring about improvements for the implanted culture with low impact on negative activities. In these environments, the use of pesticides and agents may be avoided, and the diversity of termites might contribute to productivity. In this sense, adequate handling practices, which promote the availability of cellulosic residue, may transform the termites into important components for the increase of productivity in agricultural areas, pastures, and commercial forests.

References

Ackerman, I. L., Teixeira, W. G., Rihab, S. J., Lehmann, J., & Fernandes, E. C. M. (2007). The impact of mound-building termites on surface soil properties in a secondary forest of Central Amazonia. *Applied Soil Ecology, 37*, 267–276.

Almeida, J. E. M., & Alves, S. B. (1999). Controle de *Heterotermes tenuis* (Hagen, 1858) (Isoptera: Rhinotermitidae) em cana-de-açúcar com iscas TERMITRAP® associadas ao fungo entomopatogênico *Beauveria bassiana* (Bals.) Vuill. e/ou a inseticidas em época de seca. *Arch Instituto Biol, 66*, 85–90.

Bandeira, A. G. (1979). Ecologia de cupins (Insecta: Isoptera) da Amazônia Central: efeitos do desmatamento sobre as populações. *Acta Amazonica, 9*, 481–499.

Bandeira, A. G., Vasconcellos, A., Silva, M. P., & Constantino, R. (2003). Effects of habitat disturbance on the termite fauna in a highland forest in the Caatinga domain, Brazil. *Sociobiology, 42*, 117–127.

Batista-Pereira, L. G., Santos, M. G., Correa, A. G., Fernandes, J. B., Arab, A., Costa-Leonardo, A. M., Dietrich, C. R. R. C., Pereira, D. A., & Bueno, O. C. (2004). Cuticular hydrocarbons of *Heterotermes tenuis* (Isoptera: Rhinotermitidae): Analyses and electrophysiological studies. *Zeitschrift für Naturforschung, 59*, 135–139.

Bignell, D. E., & Eggleton, P. (2000). Termites in ecosystems. In T. Abe, D. E. Bignell, & M. Higashi (Eds.), *Termites: Evolution, sociality, symbioses, ecology* (pp. 363–387). Dordrecht: Kluwer Academic Publishers.

Black, H. I. J., & Okwakol, M. J. N. (1997). Agricultural intensification, soil biodiversity and agroecosystem function in the tropics: The role of termites. *Applied Soil Ecology, 6*, 37–53.

Calderon, R. A., & Constantino, R. (2007). A survey of the termite fauna (Isoptera) of an eucalypt plantation in central Brazil. *Neotropical Entomology, 36*, 391–395.

Camara, R., Correia, M. E. F., & Villela, D. M. (2012). Effects of eucalyptus plantations on soil arthropod communities in a Brazilian Atlantic forest conservation unit. *Bioscience Journal, 28*, 445–455.
Campos, M. B. S., Alves, S. B., & Macedo, N. (1998). Seleção de iscas celulósicas para o cupim Heterotermestenuis (Isoptera: Rhinotermitidae) em cultura de cana-de-açúcar. *Scientia Agricola, 55*, 480–484.
Carrijo, T. F., Brandao, D., Oliveira, D. E., Costa, D. A., & Santos, T. (2009). Effects of pasture implantation on the termite (Isoptera) fauna in the Central Brazilian savanna (Cerrado). *Journal of Insect Conservation, 13*, 575–581.
Constantino, R. (1992). Abundance and diversity of termites (Insecta: Isoptera) in two sites of primary rain forest in Brazilian Amazonia. *Biotropica, 24*, 420–430.
Constantino, R. (2002). The pest termites of South America: Taxonomy, distribution and status. *Journal of Applied Entomology, 126*, 355–365.
Cowie, R. H., Logan, J. W. M., & Wood, T. G. (1989). Termite (Isoptera) damage and control in tropical forestry with special reference to Africa and indo-Malaysia: A review. *Bulletin of Entomological Research, 79*, 173–184.
Dangerfield, J. M., McCarthy, T. S., & Ellery, W. N. (1998). The mound building *Macrotermes michaelseni* as an ecosystem engineer. *Journal of Tropical Ecology, 14*, 507–520.
Davies, R. G., Eggleton, P., Dibog, L., Lawton, J. H., Bignell, D. E., Brauman, A., Hartmann, C., Nunes, L., Holt, J., & Rouland, C. (1999). Successional response of a tropical forest termite assemblage to experimental habitat perturbation. *Journal of Applied Ecology, 36*, 946–962.
DeSouza, O., & Brown, V. K. (1994). Effects of habitat fragmentation on Amazonian termite communities. *Journal of Tropical Ecology, 10*, 197–206.
Dibog, L., Eggleton, P., Norgrove, L., Bignell, D. E., & Hauser, S. (1999). Impacts of canopy cover on soil termite assemblages in an agrisilvicutural system in southern Cameroon. *Bulletin of Entomological Research, 89*, 125–132.
Dosso, K., Deligne, J., Yéo, K., Konaté, S., & Linsenmair, K. E. (2013). Changes in the termite assemblage across a sequence of land-use systems in the rural area around Lamto reserve in central Côte d'Ivoire. *Journal of Insect Conservation, 17*, 1047–1057.
Eggleton, P. (1999). Termite species description rates and the state of termite taxonomy. *Insectes Sociaux, 46*, 1–5.
Eggleton, P. (2000). Global patterns of termite diversity. In T. Abe, D. E. Bignell, & M. Higashi (Eds.), *Termites: Evolution, sociality, symbioses, ecology* (pp. 25–51). Dordrecht: Kluwer Academic Publishers.
Eggleton, P., Bignell, D. E., Sands, W. A., Waite, B., Wood, T. G., & Lawton, J. H. (1995). The species richness of termites (Isoptera) under differing levels of forest disturbance in the Mbalmayo Forest reserve, southern Cameroon. *Journal of Tropical Ecology, 11*, 85–98.
Eggleton, P., Bignell, D. E., Sands, W. A., Mawdsley, N. A., Lawton, J. H., Wood, T. G., & Bignell, N. C. (1996). The diversity, abundance and biomass of termites under differing levels of disturbance in the Mbalmayo Forest reserve, southern Cameroon. *Philosophical Transactions of the Royal Society of London. Series B, Biological Sciences, 351*, 51–68.
Ellis, L. M., CS Crawford, M. C., & Molles, J. R. (2001). Influence of annual flooding on terrestrial arthropod assemblages of a Rio Grande riparian forest. *Regulated Rivers: Research and Management, 17*, 1–20.
Evans, T. A., Dawes, T. Z., Ward, P. R., & Lo, N. (2011). Ants and termites increase crop yield in a dry climate. *Nature Communications, 2*, 262.
Florencio, D. F., Marins, A., Rosa, C. S., Cristaldo, P. F., Araujo, A. P. A., Silva, I. R., & DeSouza, O. (2013). Diet segregation between cohabiting builder and inquiline termite species. *PloS One, 8*(6), e66535.
Franco, A. L. C., Bartz, M. L. C., Cherubin, M. R., Baretta, D., Cerri, C. E. P., Feigl, B. J., Wall, D. H., Davies, C. A., & Cerri, C. C. (2016). Loss of soil (macro) fauna due to the expansion of Brazilian sugarcane acreage. *Science of the Total Environment, 563*, 160–168.

Goldemberg, J., Mello, F. F. C., Cerri, C. E. P., Davies, C. A., & Cerri, C. C. (2014). Meeting the global demand for biofuels in 2021 through sustainable land use change policy. *Energ Policy, 69*, 14–18.

Harris, W. V. (1971). *Termites: Their recognition and control* (2nd ed.p. 186). London: Longman.

Hulugalle, N. R., Lobry De Bruyn, L. A., & Entwistle, P. (1997). Residual effects of tillage and crop rotation on soil properties, soil invertebrates numbers and nutrient uptake in an irrigated vertisol sown to cotton. *Applied Soil Ecology, 7*, 11–30.

Jones, D. T., Susilo, F. X., Bignell, D. E., Hardiwinoto, S., Gillison, A. N., & Eggleton, P. (2003). Termite assemblage collapse along a land-use intensification gradient in lowland central Sumatra, Indonesia. *Journal of Applied Ecology, 40*, 380–391.

Junqueira, L. K., Diehl, E., & BertiFilho, E. (2008). Termites in eucalyptus forest plantations and forest remnants: An ecological approach. *Bioikos, 22*, 3–14.

Junqueira, L. K., Diehl, E. M. O., & BertiFilho, E. (2009). Termite (Isoptera) diversity in eucalyptus-growth areas and in forest fragments. *Sociobiology, 53*, 805–827.

Junqueira, L. K., Gonçalves, E. R., & Teixeira, L. M. C. (2015). Termite communities in sugarcane plantations in southeastern Brazil: An ecological approach. *Entomo Brasilis, 8*, 105–116.

Krishna, K., & Weesner, F. M. (1970). *Biology of termites (Vol 2)* (p. 643). New York: Academic Press.

Lavelle, P., Bignell, D., Lepage, M., Wolters, V., Roger, P., Ineson, P., Heal, O. W., & Dhillion, S. (1997). Soil function in a changing world: The role of invertebrate ecosystem engineers. *European Journal of Soil Biology, 33*, 159–193.

Mahapatro, G. K., & Sreedevi, K. (2014). Indigenous approaches for the management of termite and white grub in upland rice. *Current Biotica, 8*, 97–108.

McMahan, E. A. (1986). Beneficial aspects of termites. In S. B. Vinson (Ed.), *Economic impact and control of social insects* (pp. 144–164). New York: Praeger.

Menzel, L. V. A., & Diehl, E. (2008). Reproductive strategies of *Cortaritermes fulviceps* (Silvestri) (Isoptera, Termitidae, Nasutitermitinae) under laboratory conditions. *Sociobiology, 51*, 719–731.

Menzel LVA, Diehl E (2010) Soil choice for colony foundation by *Cortaritermes fulviceps* (Isoptera, Termitidae, Nasutitermitinae). Sociobiology 55:471–487.

Mill, A. E. (1982). Populações de térmitas (Insecta: Isoptera) em quatro habitats no baixo rio Negro. *Acta Amazonica, 12*, 53–60.

Mill, A. E. (1992). Termite as agricultural pests in Amazonia, Brazil. *Outlook and Agriculture, 21*, 41–46.

Miranda, C. S., Vasconcellos, A., & Bandeira, A. G. (2004). Termites in sugar cane in Northeast Brazil: Ecological aspects and pest status. *Neotropical Entomology, 33*, 237–241.

Nair, K. S. S., & Varma, R. V. (1985). Some ecological aspects of the termite problem in young eucalypt plantations in Kerala, India. *Forest Ecology and Management, 12*, 287–303.

Novaretti, W. R. T., & Fontes, L. R. (1998). Cupins: Uma grave ameaça à cana-de-açúcar no Nordeste do Brasil. In E. Berti Filho & L. R. Fontes (Eds.), *Cupins: O desafio do conhecimento. FEALQ, Piracicaba* (pp. 163–172).

Nyagumbo, I., Munamati, M., Mutsamba, E. F., Thierfelder, C., Cumbane, A., & Dias, D. (2015). The effects of tillage, mulching and termite control strategies on termite activity and maize yield under conservation agriculture in Mozambique. *Crop Protection, 78*, 54–62.

Okwakol, M. J. N. (2000). Changes in termite (Isoptera) communities due to the clearance and cultivation of tropical forest in Uganda. *African Journal of Ecology, 38*, 1–7.

Pringle, R. M., Doak, D. F., Brody, A. K., Jocqué, R., & Palmer, T. M. (2010). Spatial pattern enhances ecosystem functioning in an African savanna. *PLoS Biology, 8*(5), e1000377.

Rouland-Lefèvre, C. (2011). Termites as pests of agriculture. In D. E. Bignell, Y. Roisin, & N. Lo (Eds.), *Biology of termites: A modern synthesis* (pp. 499–517). London/New York: Springer.

Sena, J. M., Vasconcellos, A., Gusmao, M. A. B., & Bandeira, A. G. (2003). Assemblage of termites in a fragment of cerrado on the coast of Paraíba state, Northeast Brazil. *Sociobiology, 42*, 753–760.

Sileshi, G., & Mafongoya, P. L. (2003). Effect of rotational fallows on abundance of soil insects and weeds in maize crops in eastern Zambia. *Applied Soil Ecology, 23*, 211–222.

Sileshi, G., Mafongoya, P. L., Kwesiga, F., & Nkunika, P. (2005). Termite damage to maize grown in agroforestry systems, traditional fallows and monoculture on nitrogen-limited soils in eastern Zambia. *Agricultural and Forest Entomology, 7*, 61–69.

Verma, M., Sharma, S., & Prasad, R. (2009). Biological alternatives for termite control: A review. *International Biodeterioration & Biodegradation, 63*, 959–972.

Wardell, D. A. (1987). Control of termites in nurseries and young plantations in Africa: Established practices and alternative courses of action. *Commonwealth Forest Review, 66*, 77–89.

Wilcken, C. F. (1992). Danos de cupins subterrâneos *Cornitermes* sp. (Isoptera: Termitidae) em plantios de *Eucalyptus grandis* e controle com inseticidas no solo. *An Soc Entomol Bras, 21*, 329–338.

Wilcken, C. F., & Raetano, C. G. (1995). Controle de cupins em florestas. In E. Berti Filho & L. R. Fontes (Eds.), *Alguns aspectos atuais da biologia e controle de cupins* (pp. 141–154). Piracicaba: FEALQ.

Wilcken, C. F., & Raetano, C. G. (1998). Atualidades no controle de cupins em florestas de eucalipto. In L. R. Fontes & E. Berti Filho (Eds.), *Cupins: o desafio do conhecimento* (pp. 173–185). Piracicaba: FEALQ.

Wood, T. G. (1975). The effects of clearing and grazing on the termite fauna (Isoptera) of tropical savannas and woodlands. In J. Vanek (Ed.), *Progress in soil zoology* (pp. 409–418). Prague: Academia.

Wood, T. G., & Johnson, R. A. (1986). The biology, physiology and ecology of termites. In S. B. Vinson (Ed.), *Economic impact and control of social insects* (pp. 1–68). New York: Praeger.

Wood, T. G., & Sands, W. A. (1978). The role of termites in ecosystems. In M. V. Brian (Ed.), *Production ecology of ants and termites* (pp. 245–292). Cambridge: Cambridge University Press.

Wood, T. G., Johnson, R. A., Bacchus, S., Shittu, M. O., & Anderson, J. M. (1982). Abundance and distribution of termites (Isoptera) in a riparian forest in the southern Guinea savanna vegetation zone of Nigeria. *Biotropica, 14*, 25–39.

Chapter 3
Termites and Indian Agriculture

Bishwajeet Paul, Md. Aslam Khan, Sangeeta Paul, K. Shankarganesh, and Sarbasis Chakravorty

Contents

B. Paul (✉) • K. Shankarganesh
Division of Entomology, ICAR-Indian Agricultural Research Institute,
New Delhi 110012, India
e-mail: bishwajeet_paul2011@yahoo.com

M.A. Khan
Department of Biology, Faculty of Science, Jazan University, Jazan, Saudi Arabia

S. Paul
Division of Microbiology, ICAR-Indian Agricultural Research Institute,
New Delhi 110012, India

S. Chakravorty
Centre for Agricultural Technology Assessment & Transfer, ICAR-Indian Agricultural Research Institute, New Delhi 110012, India

M.A. Khan, W. Ahmad (eds.), *Termites and Sustainable Management*,
Sustainability in Plant and Crop Protection,
https://doi.org/10.1007/978-3-319-68726-1_3

Abstract Termites are the most dominant arthropod decomposers in the tropical forests and show high diversity and abundance. Within tropical ecosystems, they play a key role in modifying the biotic and abiotic environment. The areas of higher altitudes and extreme temperatures have restricted the distribution of termite fauna in India. The species richness is more in the north-eastern regions, compared to rest of India. Out of 337 species of termites known so far from India, about 35 have been reported damaging agricultural crops and buildings. *Odontotermes* is the major mound-builder, whereas *Coptotermes*, *Heterotermes*, *Microtermes*, *Microcerotermes* and *Trinervitermes* are the major subterranean genera occurring in India.

The losses caused amount to several hundred million of rupees per year. Termites damage crops from sowing till harvest, and it is difficult to detect damage in the field. Usually it is too late when the symptoms are noticed. In general, termite damage is seen more (20–25%) in rain-fed crops than irrigated ones (10%). Perennial crops are usually attacked during dry seasons and annual crops towards harvest time. Termite infestations have been reported in fruit crops, sugarcane, cotton, paddy, maize, pearl millet, pulses, citrus, vegetables, spices, groundnut and potato in arid zones of India.

Indian agriculture depends on unpredictable rains and is dominated by small and marginal farmers, with meagre resource amounts for insect pest management. The majority of farmers follow the age old practices for management of insect pests. The crop and species diversity often makes the issue more complicated. India is divided into 15 agroclimatic zones. Technologies need to be developed for each zone separately, as no single technology would be effective for all of them. Termite control is a herculean task and is not an advisable option, and management in cropped areas should be our goal. Complete elimination or prevention of termites is neither feasible nor advisable, as their complex biology in many regards poses complications in devising management strategies. Optimistically, prospects for the development of new or improved technologies as well as public acceptance of alternative management appear good. Least toxic and nonchemical methods have been and will continue to be developed. In this chapter we discuss issues related to Indian agriculture and the contemporary practices, being followed by the majority of Indian farmers.

Keywords Termite management • India • Biodiversity • Damage • Agroecosystems

3.1 Introduction

Termites represent the most important fraction of soil fauna in the semiarid tropics (Lobry de Bruyns and Conacher 1990). They can be found in a wide range of terrestrial environments and are distributed throughout the tropical, subtropical and temperate regions of the world (Freise 1949; Krishna and Weesner 1970; Pearce 1997a). They are white, tan or black orthopteroid social insects that can cause severe destruction to crops, constructions and wooden structures. Termite castes, viz. workers, soldiers and reproductives, live in small to large colonies, sometimes a single colony containing a million or more individuals. They belong to the insect order Isoptera, an ancient group that dates back more than 100 M yrs ago. The Latin name Isoptera means "equal wing" and refers to the fact that the front set of wings on a reproductive termite is similar in size and shape to the hind set. By historical convention, all but 2% of termite genera end in the suffix *termes*, the Latin word for termite. The limits of survival are between latitudes 45 and 50° north and south, respectively. The farthest north that termites are known to have reached is Hamburg (Germany), where they were found in a number of warehouses. The order Isoptera is divided into seven families. The most devastating species are distributed among four families, viz. Rhinotermitidae, Kalotermitidae, Hodotermitidae and Termitidae (UNEP Report 2000). Out of 300 species of termites known so far from India, about 35 have been reported damaging agricultural crops and buildings. The major mound-building species in India are *Odontotermes obesus* (Rambur), *Odontotermes redemanni* (Wasmann) and *Odontotermes wallonensis* (Wasmann), and the subterranean species are *Heterotermes indicola* (Wasmann), *Coptotermes ceylonicus* Holmgren, *Coptotermes heimi* (Wasmann), *Odontotermes horni* (Wasmann), *Microtermes obesi* Holmgren, *Trinervitermes biformis* (Wasmann) and *Microcerotermes beesoni* Snyder (Rajagopal 2002).

Termites are a serious threat to agriculture in tropical areas with high relative humidity. Depending on the habits and habitats, termites can be broadly classified into wood dwellers and ground dwellers. The wood dwellers are comprised of species inhabiting damp wood and dry wood. The ground dwellers are categorized into subterranean, carton-nest builders and mound-builders (Pearce 1997b). Subterranean termites, including mound-building and arboreal species, account for 80% (or 147 species) of the economically important species. The genus *Coptotermes* (Rhinotermitidae) contains the largest number of economically important subterranean termites (28 species). Unlike drywood termites that are easily transported from region to region, most subterranean species have a restricted distribution. Of the 147 economically important subterranean termite species, only two, *Coptotermes formosanus* Shiraki and *Coptotermes havilandi* Holmgren, have been introduced in more than five regions worldwide (Edwards and Mill 1986).

Termites are the most abundant soil invertebrate (Table 3.1). Their abundance in any ecosystem drastically changes the below ground biodiversity. They build nests in dead tree trunks/stumps and some build mounds. The subterranean nests are not detectable easily on the soil surface. They also construct mud galleries for movement and transport to the food sources, as workers may travel long distances in search of dead decaying wood.

As termites are primary consumers, they promote the mineralization of nutrients rich in cellulose (Cunha and Orlando 2011). They are most dominant arthropod decomposers in the tropical forests (Collins 1983) and show high diversity and abundance (Bignell and Eggleton 2000). Within tropical ecosystems, termites play a key role in modifying the biotic and abiotic environment. Their diversity and distribution are greatly influenced by factors such as vegetation type (Pardeshi et al. 2010), habitat disturbance (DeBlauwe et al. 2008) and habitat fragmentation (Davies et al. 2003).

The termites have evolved as eusocial organisms. The life-history traits of termites might have predisposed them for development of eusocial society. The factors leading to development of eusociality are slow development; overlapping

Table 3.1 Abundance of common soil invertebrates in different ecosystems (Rahman et al. 2012)

	Number of individuals per m^2							
Ecosystem	Earthworm	Ant	Termite	Beetles	Millipede	Centipede	Orthoptera	Spiders
Natural forests	97.6	115.2	1542.4	24.0	6.4	16.0	6.4	9.6
Agroforestry	83.2	75.2	40.0	3.2	9.6	4.8	3.2	1.6
Plantations	116.8	52.8	70.4	4.8	4.8	3.2	1.6	3.2
Annual crops	24.0	40.0	16.0	3.2	1.6	8.0	0.0	1.6
χ^2	13.8	18.4	81.9	7.8	9.3	29.3	16.8	24.1
Probability[a]	0.003	<0.001	<0.001	0.003	0.026	<0.001	<0.001	<0.001

[a]Probability of significant χ^2 assuming negative binomial error distribution of the counts

generations; monogamy; familial associations in cloistered, food-rich habitats; iteroparity; high-risk dispersal of individuals; opportunities for nest inheritance by offspring remaining in their natal nest; and advantage of group defence. These factors create a selection pressure on the organism and the evolutionary outcome is eusociality. Among the eusocial society, a limited number of individuals are fertile and fecund; others have reduced reproductive capacities or sometimes are completely sterile. The most important aspect of such group is rearing of the progenies of primary reproductives, by worker caste (Thorne 1997).

3.2 Taxonomic Diversity of Termites in India

All the known nine families of Isoptera exist since the late Mesozoic period. There are four monogeneric families of termites, viz. Mastotermitidae (holotype *Masotermes darwinensis* (Froggatt) in Australia), Serritermitidae (holotype *Serritermes serrifer* (Bates) in Brazil), Stylotermitidae (holotype *Stylotermes* containing eight species from Indian sub-continent) and Indotermitidae (holotype *Indotermes* containing seven species from Oriental region). The most diverse family of termites is Termitidae, comprising of 4 subfamilies and 145 genera (Roonwal and Chhotani 1989).

The diversity in termite fauna of India is restricted to only 337 species and subspecies under 54 genera belonging to seven families. The areas of higher altitudes and extreme temperatures have restricted the distribution of termite fauna in India (Fig. 3.1). The species richness is more in the north-eastern regions of India compared to rest of the country. Biodiversity of termites in north-western and central region is relatively poor. Table 3.2 represents the taxonomic diversity of termites in different forest ecosystems in India.

In humid plains of India, *Odontotermes* species is the most widely distributed mound-building termite, and *Odontotermes distans* Holmgren is reported to occur in higher altitudes, i.e. Kumaon hills. *Heterotermes* species is the common household termite in temperate areas, whereas *Reticulitermes* and *Archotermopsis* occur in the wild areas of these regions. In western arid areas, *Psammotermes* and *Anacanthotermes* are known to occur. *Calcaritermes* and *Rhynchotermes* are known to occur exclusively in Nicobar Islands and Manipur, respectively. *Prorhinotermes flavus* (Bugnion and Popoff) has been found to occur in the coastal areas of Mangalore, Andaman and Nicobar islands. Some of the termite species are endemic to different regions of India, viz. Himalayan 34, Gangetic 20, peninsular 73, North East borderland 19 and insular 19, with 16 species common to several divisions. Out of 253 termite species reported from India till date (1989), 73 are found in tropical rain forests of Western Ghats. From the Indian sub-continent, Roonwal and Chhotani (1989) listed and described 337 species of termites belonging to 59 genera, and Bose (1984) recorded 95 species comprised of 5 families from southern India. Chhotani (1997) listed 92 species of soil inhabiting or mound-building termites damaging agricultural crops, timbers and buildings.

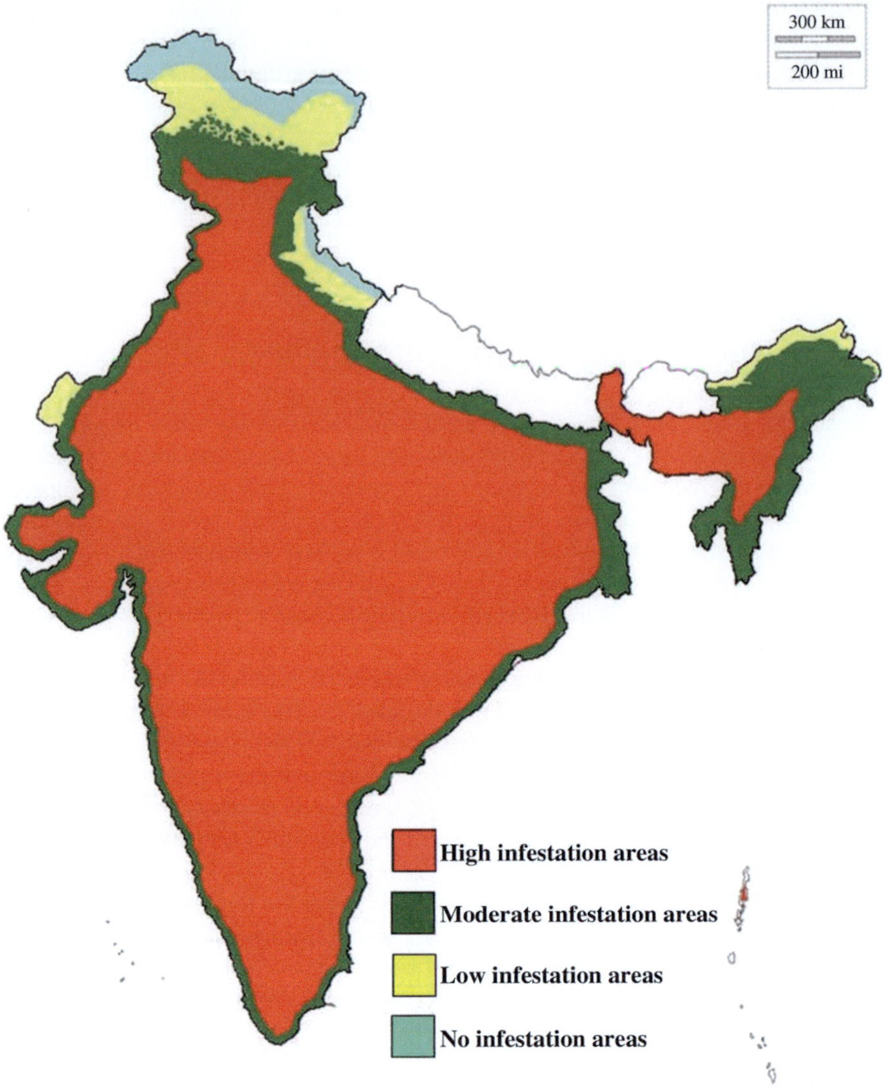

Fig. 3.1 Infestation of termites in India

3.3 Economic Importance

Termites cause extensive damage to agricultural and horticultural crops, agroforestry, stored timbers, books and records, woodworks in buildings and stored products containing cellulose (Rashmi and Sundararaj 2013). The losses caused in India alone run into several hundred million rupees per year, and the world loss must be more than $10,000 M. Howse (1970) observed that termites can damage man-made fabrics, plastics and some metal foils. A classical case of termite damage was reported from a Northern India town where termites damaged currency notes worth

Table 3.2 Taxonomic diversity of termites in different forest ecosystems in India (Maiti and Maiti 2011)

Tropical wet evergreen forest (Nilgiri hills)	23/14
Tropical moist deciduous forest (Nicobar)	22/10
Tropical dry deciduous forest (Chotanagpur)	17/9
Subtropical pine forests (Kumaon hills)	20/13
Tropical thorn forest (Coimbatore)	23/11
Desert vegetation (Thar desert)	22/11
Swamp forest (Sundarbans)	11/6
Insular Andaman evergreen forest	23/11

Rs 10 M (US $ 222,000) kept in a steel chest inside the State Bank of India branch, housed in an old building, infested with termites (Sacks 2011). Dwarika Prasad, a trader from Patna, Bihar, lost his life savings after termites infested his bank's safety deposit locker (Tewary 2008). In an estimate it was found that in Australia, 20% of homes are infested by termites. In China 90% of homes in south of Yangtze river are affected by termites (GEI 2005; MRP 2010). Roonwal (1955) reported that an entire township in India was gradually destroyed by the termite *Heterotermes indicola* (Wasmann) and eventually resembled a bombed-out ghost town. Annual losses caused by termites in the USA and Japan are 1000 and 800 M US$. In India the losses have been estimated around 35.12 M US$ (Joshi et al. 2005). Globally, the estimated loss due to termite damage is about 50 billion US$ annually (Subekti et al. 2015), although estimates vary considerably by the cropping systems followed in different geographical regions.

Roonwal and Chhotani (1967) reported that 58 species of termites cause major damage to wood. Sixty-four species of wood-destroying termites were reported by Sen-Sarma et al. (1975), 11 being major wood-destroying termites. Seventy-two wood-destroying termite species from Southeast Asia have been reported by Roonwal (1979). About 270 termite species were identified as injurious to economic plants, in South Asia (Srivastava 1996).

The economic impact due to termite damage worldwide is estimated to be increased to US$40 billion (Rust and Su 2012). Although many people think termites have only negative impacts, in nature they make many positive contributions to the world's ecosystems. Their greatest contribution is the role they play in recycling of wood and plant material. Their tunnelling efforts also help to ensure that soils are porous, contain nutrients and are healthy enough to support plant growth.

Significant yield losses are recorded on annual and perennial crops by termites in semiarid and sub-humid tropics. Damage is more severe during droughts and dry season, compared to irrigated crops. In rain-fed crops the plants experience moisture stress which predisposes them to termite infestations. Exotic crops are more susceptible to damage than indigenous ones.

3.3.1 Nature of Damage

There are several ways in which the termites attack plants. Nair and Varma (1981) discussed the different aspects of primary and secondary termite attack to *Eucalyptus* species. The termite made tunnels on the tree surface and built earthen runways on the surface indicating that the tree had been infested. However, under such case the tree appeared normal and healthy, as most of the living parts were not damaged, whereas *Odontotermes* species fed directly on the roots and killed the plant (Harris 1971). The plants attacked by the termite wilted before dying: this may be due to root damage, making the proper intake of water, minerals and nutrients difficult for the plant. In some cases, *Odontotermes* species infest maximum part of the plant particularly with shrubs and small plants. The attack of Termitidae (particularly of *Odontotermes* spp.) usually occurred in the form of earthen sheets and runways on the bark. The termite worker and soldiers continue their activity under the earthen layer, which covered the dead barks of almost any tree attacked (Roonwal 1979). The workers remained in a thin surface layer of the bark. However, the damage so caused was negligible although occasionally it can become serious. In some cases the termites formed nests among tree branches or trunk. The attack in most cases began from the root level and spread to the upper part. In older stem, the bark under lying tissue was eaten up gradually, reaching the pith hollowing out of the stem, resulting in the ultimate death of plant.

The nature of damage by termites varies in the different trees. The damages of *O. wallonensis* (Wassman) were recorded in the form of nibbling on dead, as well as live, bark of both stem and root underneath the cover of earthen sheet and runways. The attack of *Odontotermes* sp. occurred usually at the basal part of the trunk. The damage, although not severe, was localized, resulting in the formation of irregular cavities or grooves of various sizes on the surface of the trunk, which reached up to about 2–3 m height. The infestation generally originates and spread internally in the plant, as the termite move from soil into the roots. The damages were more or less similar in *Azadirachta indica* Juss., *Cocos nucifera* L., *Mangifera indica* L. and *Eucalyptus* species. It extended runways from the ground up to 2 m on the barks of the trunk. In some trees such as *C. nucifera*, *M. indica* runways were constructed in and around underneath bark. Mostly, these runways extended towards the dead portion of the living trees, and the damage was observed on the dried portion of the living plant. In *Eucalyptus* trees, it ate up dead wood as well as tissues adjacent to the stem, thus hollowing the stem. However, the trees remained in a live condition, as other parts were not damaged. The termite damages the stem of *C. nucifera* and *Eucalyptus* either by entering the roots or wounds present on the stem or feeding up to their middle portion.

The damages caused by *O. wallonensis* in *Acacia arabica* (Lam.) trees and *Tamarindus indica* L. are produced by extended runways of the stem above the bark, up to the branches at about 6 m in height, and were also recorded at the basal portion of the trunk. This species partially hollowed out the trunk of *Tectona grandis* L.f. from the base filling the excavated portion with mud and the cavity up to 2 m in the stem from the surface of the ground.

Nair and Varma (1981) did not find any correlation between the seasonal distribution of rainfall and incidence of termite attack nor any relationship between annual rainfall and annual loss due to termite. They even found no support to the general belief that termite attacks are more common during dry periods. They further reported that most of the damage occurred before the onset of the dry season. The damage intensities of *O. wallonensis* on *Eucalyptus* sp. and *A. indica* reached the maximum during winter season in southern India. With a decrease in temperature there was no significant correlation between its damage and seasonal temperature variations. These authors found seasonality in infestation of various trees, viz. *Casuarina equisetifolia* L., *C. nucifera* and *M. indica* with seasonal variations in temperature rainfall and relative humidity. In absence of relative humidity, temperature did not influence the damage intensity by *O. wallonensis*.

Predisposing Factors for Termite Attack in Crops
Man-made conditions that put crops under stress are the result of poor cultural practices. Some of the common practices inadvertently leading to termite attack are as follows:

- Unsuitable cropping site and climatic conditions; the crops would be stressed and weakened and are more liable to be attacked by termites.
- Accumulation of crop refuge, viz. stubbles, straw, uprooted dry weeds, etc., serves as additional food resources of termites.
- Unhealthy nursery raising practices in vegetables, horticultural and silvicultural plants, resulting in poor-quality crops.
- Non-removal of damaged bark would allow the termites to colonize the pruned dead ends.
- Root damage, due to intercultural operations in field/horticultural crops, attracts termites. The root exudates serve as attractants for termites.
- Root infection caused by soilborne diseases/nematodes weakens the plants and attracts termites.
- Any stress caused by drought, poorly drained soil, etc. favours termite attack

3.3.2 Symptoms of Damage

Wilting is the first sign of termite attacking roots on seedlings or older plants. Eventually some plants fall over or die. The presence of live termites is confirmed by pulling out the affected plants and examining the roots and lower stem. In some instances roots and stems may be completely hollowed out and soil-filled. Termites are often seen under the soil sheets prepared to escape biotic and abiotic agents. They move down deeper during the higher temperature by day and come back to the surface when the temperature becomes tolerable. In orchards, the termite damage often begins in an area of dead wood produced by pruning or any other damage. Termite attack on trees and bushes often begins with small cracks or tunnels made by other insects that may allow winged termites (reproductive stage) to enter. They may also travel up through the roots into the trunk and branches and eventually disrupt the movement of nutrients and water through the vascular system, resulting in the plant death.

3.4 Damages in Agroecosystem

Termite attacks on annual and perennial crops cause significant yield losses damaging the plants at all growth stages (Chhotani 1977). The attack usually begins on the roots and then spreads to upper parts. In older plants the bark and underlying tissue is eaten up, which are gradually exposed to attack by pathogenic microbes resulting ultimate death of the plant. Apart from crops, they also attack the crop leftover (root stubble), fallen leaves, twigs, bark, etc. So far there are no reports available on the losses caused, except a report of Rajagopal and Veeresh (1983). Termite damage has been recorded on the majority of crops grown in and around Bidar, Karnataka area. Plant may bc partially or severely attacked at the base such as maize, soybean, groundnut, sugarcane, finger millet and mango trees. Information on the economic losses caused by termites is difficult to obtain because the damages are often patchy in nature. The damage to crops in general is expressed as plant attacked or plant mortality. Termites mainly feed on woody materials, but some species are known to collect green grasses and seeds and store them inside their nest as food reserves.

Bark-eating termites cover the tree trunks and stems with a sheet of mud or make mud galleries on a wide range of crops, rendering them locally important pests. The galleries and tunnels are usually built with mud and saliva, but sometimes plant fragments are also used. They gnaw away the bark and wood underneath their tunnels and galleries on the roots and underground stems. Often the collected materials are transported to their nests. Due to such a feeding damage to trees is low, but the tree becomes weak and fruit bearing is seriously affected. The tree trunks often break under moderate wind speed.

In field crops, viz. cotton and groundnut, taproots of young seedlings are eaten up by termites just below the soil surface. The central root portion is damaged and filled with soil. The damaged plants wilt overnight and die within few days. Sorghum, maize and bajra plants often topple down due to termite attack at the collar region. When the plant lodges, the grains touch the ground, and soil fungi such as *Aspergillus* spp. may invade them. Termites threaten major crops, which form the basis of household nutrition in many countries such as wheat, maize, groundnut, sugarcane, yams and cassava. The most common termite species in India are *Microtermes* and *Odontotermes*.

In this chapter we would discuss only some most important crops damaged by termites. Most of the field crops grown under rain-fed agriculture are severely affected by termites, and considerable losses are observed (Tables 3.3, 3.4 and 3.5).

Table 3.3 Termite species found in various crops in India

Crops	Termite species	States
Paddy	*Anacanthotermes viarum* (Koenig)	Tamil Nadu
	Microtermes sp.	Delhi and Maharashtra
Wheat	*Odontotermes bangalorensis* (Holm. and Holm.)	Madhya Pradesh
	Odontotermes gurudaspurensis (Holm. and Holm.)	Rajasthan and N.W. India
	Odontotermes obesus (Rambur)	Bihar, Delhi, Punjab, U.P., Haryana, M.P., Rajasthan, Gujarat, A.P., Maharashtra, Karnataka, Kerala and Tamil Nadu
	Microtermes obesi Holm.	A.P., Bihar, Delhi, Punjab, U.P., M.P. and Rajasthan
	Microtermes mycophagus (Desneux)	Rajasthan
	Microtermes tenuignathus (Holm.)	
	Eremotermes sp.	
	Nasutitermes sp.	
Barley	*Odontotermes gurudaspurensis* (Holm. and Holm.)	Haryana
	Odontotermes latigula Snyder	
	Odontotermes latiguloides (Roonwal and Verma)	
	Microtermes mycophagus (Desneux)	
	Microtermes obesi Holm.	
	Odontotermes obesus (Rambur)	Haryana, U.P. and Rajasthan
Oats	*Microtermes obesi* Holm.	Bihar
Maize	*Odontotermes gurudaspurensis* (Holm. and Holm.)	Rajasthan and N.W. India
	Odontotermes wallonensis (Wasmann)	Karnataka
	Odontotermes obesus (Rambur)	Rajasthan
	Microtermes obesi Holm.	Bihar and Rajasthan
Bajra	*Microtermes mycophagus* (Desneux)	Delhi, Punjab and Rajasthan
	Odontotermes obesus (Rambur)	
	Microtermes obesi Holm.	
	Odontotermes guptai Roonwal and Bose	Rajasthan

(continued)

Table 3.3 (continued)

Crops	Termite species	States
Jowar	*Microtermes obesi* Holm.	Delhi, Punjab and Rajasthan
	Odontotermes obesus (Rambur)	Delhi, Haryana, Punjab and Rajasthan
Sugarcane	*Coptotermes heimi (Wasmann)*	U.P. and Bihar
	Eremotermes paradoxi Holm.	Bihar, Delhi, Punjab, U.P., M.P., Karnataka and Tamil Nadu
	Microtermes beesoni Snyder	U.P.
	Odontotermes taprobanes (Walker)	
	Microtermes mycophagus (Desneux)	Assam and T.N.
	Microtermes obesi Holm.	Assam, Delhi, Punjab, U.P., Rajasthan, Karnataka, Bihar and T.N.
	Odontotermes assmuthi Holm.	U.P., Maharashtra, A.P., Bihar and T.N.
	Odontotermes bangalorensis Holm.	M.P., Karnataka and W.B.
	Odontotermes obesus (Rambur)	Punjab, W.B. Assam, Bihar, U.P., Rajasthan, Delhi, Karnataka, M.P., Maharashtra, A.P. and T.N.
	Odontotermes obesus (Wasmann)	Karnataka, A.P., Bihar, M.P., Orissa and T.N.
	Trinervitermes biformis (Wasmann)	Bihar, U.P., M.P., Maharashtra, Orissa and T.N.
Groundnut	*Microtermes mycophagus* (Desneux)	Rajasthan
	Microtermes obesi Holm.	Rajasthan, U.P. and Delhi
	Odontotermes obesus (Rambur)	Punjab, U.P., Kerala, Rajasthan, Delhi, Gujarat, Haryana, Karnataka, M.P., Maharashtra, A.P. and T.N.
	Odontotermes wallonensis (Wasmann)	Karnataka
	Trinervitermes biformis (Wasmann)	U.P. and Maharashtra
Tea	*Coptotermes ceylonicus* Holm.	S. India and N.E. India
	Microtermes obesi Holm.	
	Odontotermes obesus (Rambur)	

Groundnut (*Arachis hypogaea* L.)

In semiarid tropical countries of Africa and India, groundnut crops are seriously damaged by *Microtermes*, *Odontotermes* and *Amitermes* spp., resulting in yield losses between 10 to 30%. Very little attention has been paid indeed, to the losses caused by termites in groundnut. Harris (1969) listed 17 termite species known to damage groundnuts in moderate to low rainfall areas of Africa and Asia. They are known to attack groundnut in all stages of growth and during storage too. Seedlings,

Table 3.4 Major crops attacked by termites causing extensive damage in different countries

Crop	Country
Cereals	
Maize	Argentina, Benin, Brazil, Democratic Republic of Congo, Ethiopia, India, Kenya, Malawi, Nigeria, Paraguay, South Africa, Saudi Arabian Peninsula, Swaziland, Tanzania, Uganda, Uruguay, Zambia, Zimbabwe, Yemen
Sorghum	Ethiopia, India, Malawi
Barley	Argentina, Brazil, India, Paraguay, Uruguay
Millets	China, Ethiopia, India, Yemen
Wheat	India, Yemen
Pulse crops	
Beans	India, Malawi
Cowpea	India, Malawi
Chickpea	India, Malawi
Oil crops	
Groundnut	Australia, Botswana, Brazil, Burkina Faso, China, Ethiopia, Gambia, Guyana, India, Malawi, Mali, Niger, Nigeria, Senegal, Sudan, Zambia, Zimbabwe
Sunflower	India, Yemen
Soybean	India, Brazil, Guyana
Sugarcane	Argentina, Australia, Bolivia, Brazil, Caribbean, Central African Republic, China, Colombia, Cuba, Dominican Republic, Guyana, India, Jamaica, Kenya, Mexico, Nicaragua, Nigeria, Pakistan, Panama, Paraguay, Philippines, Uruguay, Somalia, South Africa, Sudan, Venezuela
Root crops	
Sweet potatoes	India, Jamaica
Potatoes	Australia, India
Yam	Ghana, Nigeria
Cassava	Brazil, Guyana, West Africa, Malawi
Vegetables	
Tomato	Saudi Arabian Peninsula, Yemen
Okra	Saudi Arabian Peninsula, Yemen
Pepper	Saudi Arabian Peninsula, Yemen
Eggplant	Saudi Arabian Peninsula
Cabbage	India
Fruit trees	
Guava	India, Saudi Arabian Peninsula
Tea	India, Malawi, Pakistan, Peru
Coffee	Argentina, Brazil, Bolivia, Kenya
Citrus	Afghanistan, Algeria, Australia, Egypt, India, Iran, Iraq, Israel
Cocoa	Ghana
Passion fruit	Colombia, Trinidad, Venezuela
Banana	Malawi
Mango	Australia, India, Saudi Arabian Peninsula
Papaya	Saudi Arabian Peninsula

(continued)

Table 3.4 (continued)

Crop	Country
Grapes	Australia, India
Palm trees	
Oil palm	Ghana, Nigeria, South Asia, Pacific Island
Date palm	Afghanistan, Algeria, Egypt, Iran, Iraq, Israel, Jordan, Libya, Morocco, Sudan, Tunisia
Coconut	India, Malaysia, Nigeria, some South Pacific Islands
Field crops	
Pineapple	Argentina, Australia, Brazil, Kenya, Paraguay, Uruguay
Cotton	Central African Republic, India, Malawi, Sudan, Tanzania, Uganda, Yemen
Forestry plantations	
Rubber trees	Southeast Asia
Pine plantations	Australia, Southeast Asia
Hardwood plantations	Mahogany in South Pacific islands, eucalyptus in South America, street trees in France,

Table 3.5 Termite infestation (%) at different crop stages (Pardeshi et al. 2010)

		Crops			
Species	Stages	Castor	Cotton	Sugarcane	Wheat
Coptotermes heimi	Seedling	0.0	0.0	100.0	0.0
	Maturing	0.0	0.0	0.0	100.0
Odontotermes obesus	Seedling	58.6	44.2	82.4	40.5
	Maturing	4104	55.8	17.6	59.5
Odontotermes redemanni	Seedling	0.0	0.0	0.0	40.2
	Maturing	0.0	0.0	100.0	59.8
Microtermes obesi	Seedling	0.0	0.0	78.3	28.6
	Maturing	0.0	100.0	21.7	71.4
Microtermes mycophagus	Seedling	84.8	62.9	100.0	0.0
	Maturing	15.2	37.1	0.0	0.0

growing and mature plants are attacked by termites (Sands 1960; Kaushal and Deshpande 1967; Feakin 1973). Yield losses are direct and also indirect, reducing the quality of seed both for planting and human and animal consumption. The recommended management practices include use of resistant or tolerant varieties, cultural practices, botanical insecticides and minimum application of synthetic insecticides.

Chhotani (1980) observed that groundnut plants attacked by *O. wallenensis* show typical symptoms of stems covered with earthen sheet up to 5 cm high from the ground surface. Termites bored into the main stem just close to the ground level and then tunnelled down into the taproot or up to the stem. They damage pegs as well as mature pods, occasionally penetrating into their shells. The damage to pegs leads to their breaking during harvesting, thus leaving the pods in the ground. Termites remove the soft non-fibrous layer of the shell, leaving the veins exposed and causing scarification of the pods, which may become mouldy at the time of harvesting. More plants were attacked in the field area with low soil moisture content. However, there was no direct information on the relationship between field soil moisture content and the termite damage to groundnut plants. A significant relationship has been recorded between rainfall and *Microtermes* sp. infestation to the groundnut (Johnson et al. 1981). Termite damage may be serious in low rainfall area. The pods scarified are weaker and liable to crack, and scarification of pods is the most common type of damage caused by termites nearing maturity. *Aspergillus flavus* Link invades and colonizes the scarified pods and produces aflatoxin, a carcinogen that poses serious health hazards to farmers and consumers (McDonald 1970).

In India groundnut yields are low, rarely exceeding 700 kg/ha, and this is mainly due to suboptimal plant populations resulting from poor seedling emergence. Seed treatment with synthetic insecticides reduces the losses incurred in germination. In severely infested fields, the termites devour the seed sown, so substantial losses have been observed. As the crop matures the damage caused by termite increases. In general, seedling loss is relatively low, but sporadic cases with higher levels of damage have been recorded. Usually, under moisture-stressed conditions, the attacked plants die within a few days. However, upon irrigation they recover quickly, if the vascular tissues are not damaged.

Rawat et al. (1970) reported that in Madhya Pradesh (India), *O. obesus* attacks the crop more severely, leading to 35% plant mortality. Kaushal and Deshpande (1967) estimated direct pod losses to be more than 25%. Repeated mechanical cultivations reduce termite populations. Timely harvesting the crop as soon as they are mature and early removal of produce from the field also reduce termite damage.

Maize (*Zea mays* L.)

In the tropics, maize is often damaged by termites. Species of *Odontotermes* can defoliate maize seedlings or consume the entire plant. Field observations on *O. wallonensis* showed damage to seeds and seedlings. However, the termite attacked the stem of the maize plant at the ground level, covering with earthen sheet up to a 10 cm height from the base. When the earthen sheet was removed to examine the damage, a hole was found at the base which was completely eaten a few cm upward and downward and filled with soil. However, the outer covering of the stem remained intact. Some of plants were found covered with earthen sheet; these are severely damaged plants lodged on the ground even by a slight wind.

Odontotermes wallonensis caused severe damage to young maize crops. It doesn't attack the root until plants reach maturity. They may either remain standing or lodge, due to termite attacks resulting in the total destruction of cobs. In India it is observed that the *Microtermes* sp. attack on maize plants is maximum, as compared to other termites. Harvesting of lodged plants in commercial agriculture where the crop is mechanically harvested leads to high yield losses. In southern India, however, when harvesting is undertaken by hand, losses are considerably lower.

Preharvesting loss in maize due to termite is enormous. Agarwala (1955) noticed a gradual increase in the intensity of termite attacks from November, when rains were ceased. *Microtermes* sp. attacks maturing and mature maize plants, while *Macrotermes* sp. causes damage to seedlings. Maize plants attacked early in the season can compensate damage with new tiller growth. The management options include sowing at a higher seed rate and seed dressing with insecticides. Logan et al. (1990) listed measures to reduce termite damage to crop plants. They suggested attempts made to (1) initially prevent termite access to plants, (2) reduce termite numbers in the vicinity of plants and (3) reduce the susceptibility or increase resistance of the plants themselves.

For subsistence farmers, the use of intercropping to improve yields in low-input agriculture was proposed by Ofori and Stern (1987). In general intercropping is known to reduce damage caused by insect pests to the principal crop (Trenbath 1993).

Maize-legume intercropping system is the most widely recommended system in endemic areas of termite damage. Different legumes differ in their ability to influence the termite damage. In India soybean and groundnut intercropped with maize allow better yields of maize compared to intercropping with common beans. Subsistence and marginal farmers often include forage legumes as an intercrop. However, the socio-economic status of the farmer mainly decides whether he adopts intercropping or not.

Sugarcane (*Saccharum officinarum* L.)

Sugarcane is mainly damaged by termites belonging to five genera, viz. *Coptotermes, Macrotermes, Odontotermes, Microtermes* and *Eremotermes.* Their damage potential is very high in India, whereas in other countries such as Sudan and Central Africa, usually the losses are around 18% and 5–10%, respectively. The most common damage to sugarcane is the destruction of the planting material (setts).

Agarwala (1955) estimated 2.5% loss in sugarcane tonnage and 4.47% in sugar production in Bihar. Roonwal (1981) noticed that the most important termite species attacking wheat and cotton were *M. obesi* and *O. obesus*. He observed that intensity of damage to wheat by *M. obesi* was lower when the crop received two or three irrigations, vs nil or only one. Pardeshi et al. (2010) conducted investigations and recorded 15 species of termites belonging to two families and seven genera from Vadodra and Gujarat, during 2002–2005. They observed that sugarcane was attacked by maximum 5 species, followed by wheat (5), cotton (3) and castor (2). The incidence and attack of *C. heimi* was maximum (76%) and minimum (24%) in sugarcane and wheat, respectively. This particular species was found attacking the planting stalks of sugarcane but in wheat crop damage was mainly noticed in maturing stage.

In Gujarat, India, only a single species, *O. redemanni*, was recorded to damage the sugarcane at mature stage, while *M. obesi* and *O. obesus* were found to infest seedling and maturing stages. Sugarcane provides maximum shade and is more susceptible to termite attack. Shade, high sugar content and faster growth rate are some of the major reasons for the preference of this crop by a wide variety of termites. Food and habitat also greatly influence the termite activity. Termite-infested organic manure, when applied to field, also increased the intensity of the attack. Due to high evaporation rate and low water-holding capacity, the incidence of termite attack is lower in sandy loam soil. High evaporation rates of sandy loam soil pose in fact a desiccation threat to the soft-bodied termites, which probably restrict their distribution in those areas. Shady areas provide a good moisture level favouring termites, a factor that justifies the higher intensity of attacks in shady places than in open areas. In addition to shade and plant cover, objects like big boulders, manure heaps, wooden logs, tree stumps, etc. also provide shelter and moisture to the termites. Because of their affinity to shady and moist places, the termites make galleries in and around these objects. Thick vegetation provides the ground shade which in turn supplies more moisture and humidity to soil, one of the major factors promoting termite activity.

Attack by *O. wallonensis* after plantation prevents germination, resulting in a poor stand. Termite attacks the crop as it begins to mature. Secondary attacks also occur when termites gain access to soft pith through site damaged by rodent and stem borers. It also attacks the cane stalks in the year of scanty rainfall. The most common damage to crops occurs when setts are first planted in the field. The attack at this stage prevents germination, resulting in a poor stand. Termite also attacks crops as it begins to mature; further secondary attack also occurs when termite gains access to soft pith. They enter the cane laterally through one or more holes in the stalk (shoot) and bore downward as well as upward killing the growing points. Thus it cuts upward of the central leaf causing heavy yield losses as it affects the plant area which will be poor in juice with less cane weight. It damages the crops soon after internode formation, and its activity continues till harvest. The usual method of prevention is to dip the setts in various formulations of chlorinated hydrocarbons before planting or to spray them in the furrows before filling in.

Pardeshi et al. (2010) observed that *O. obesus* acted as pest to all crops, irrespective of the plant stages. However, attack was more prominent in sugarcane (43%) than in cotton (27%), wheat (19%) and castor (10%). This species caused more damage during sugarcane seedling stage (82%), as compared to maturing stage (18%). Such difference in occurrence between the seedling and the maturing of crop is only noticed in sugarcane. Wheat was mostly (73%) liked as food rather than sugarcane (27%) by *O. redemanni*. Attack severity was higher in maturing sugarcane. However, another subterranean species, *Microtermes mycophagus* (Desneux), showed preference for castor (53%) rather than cotton (29%) and sugarcane (18%). *Microtermes mycophagus* attacked the young plants of sugarcane, castor and cotton, and matured crops appeared less vulnerable. However, *M. mycophagus* was not recorded from wheat.

Microtermes obesi is a serious pest of sugarcane (58%), particularly at seedling stage, as well as of wheat (37%), whereas cotton (6%) was less preferred. However, matured wheat crops were more susceptible to the attacks of this species. Cotton was damaged only at its maturing stage. In terms of occurrence, *O. obesus* occupies the highest position (28%), followed by *M. obesi* (25%) and *M. mycophagus* (17%). *Odontotermes redemanni* and *C. heimi* are very rare, and a very few specimens were collected. Termite damage in sugarcane occurs both at seedling (setts) and maturing stage. In wheat, the infestation is seen more in the seedling than the maturing stage. Thakur (1996) recorded *O. obesus* and *M. obesi* as a major pest of sugarcane in India and Pakistan. *O. obesus*, *M. obesi* and *M. mycophagus* were found to be the most versatile species in this study, and besides their occurrence in the crop field, they also showed a marked presence in a number of other microhabitats.

Wheat (*Triticum aestivum* L.)

Among cereals, wheat is one of the most susceptible cultivated crops to termite attack at all stage of its growth, throughout the rain-fed and irrigated regions. *Odontotermes wallonensis* attacks all stages of wheat plants and has been reported as the most important termite species (Hussain 1935). The average annual losses of wheat in India were estimated to vary from 6 to 40%, at different places.

Other important species of termites attacking wheat crop are *O. obesus*, *M. obesi* and *Microcerotermes tenuignathus* Holmgren. Loss of wheat crop has been reported to be 7.15% by Parihar (1978). In western Rajasthan, on the whole, infestation is more severe in the rain-fed light soils than in irrigated, heavy soils. The infested plants wither and dry up, losing their anchorage and getting dislodged. Sometimes the attacks also occur in the earhead stage, resulting in chaffy earheads with little or no grain.

Soybean (*Glycine max* (L.) Merr.)

Under field conditions, *O. wallonensis* severely damages soybean plants below ground level and removes the internal tissue causing weakened plants. In addition, the termite also attacks roots, with maximum losses noticed on soybean root stubble.

Pearl Millet (*Pennisetum typhoides* (Burm) Stapf & Hubb.)

In Rajasthan, *bajra* or pearl millet (*Pennisetum typhoides* (Burm) Stapf & Hubb.) is subjected to attacks by termites such as *M. obesi.* This pest initially attacks roots, and later on the stem, resulting in wilting and ultimate drying of plants.

Cluster Beans (Guar) (*Cyamopsis tetragonoloba* (L.) Taub.)

The termite species responsible for damaging guar crops are *Microcerotermes baluchistanicus* (Ahmad), *Odontotermes guptai* (Roonwal and Bose) and *M. obesi.* They attack the crop at germination, flowering and fruiting stages in August and September (Parihar 1978). At the germination stage, they nibble the roots, while at the flowering and fruiting stages, they also enter the stem base by making a hole in it. They completely devour the inner portions, leaving only the rind, thus depriving the stem of its nutritional supply. The termite infestation ranges from 12.3 to 16.3% of plants.

Castor (*Ricinus communis* L.)

This crop is attacked by *M. mycophagus* both at the seedling and the growth stages. In the seedling stage, the attack is more prominent on roots, while the stem remains unaffected. In young plants the termites nibble the taproot. In grown-up plants the termites can be seen around the root zone and in certain cases up to 3 ft on the stem (Parihar 1977). The root bark, in each case, was found to be quite intact, while the hard cores were mostly tunnelled through. The fine roots (more than 2–4 mm in diameter) showed more damage. When severely damaged, the roots show galleries which were rather irregular and ran almost parallel to their length. The larger galleries are occasionally filled with earth and excreted wood. Owing to unique phyllotactic arrangement, castor leaves are mainly restricted to the apical region for which termites get very little shade around them. Since the termites are soft-bodied animals and are very much prone to desiccation, for obvious reasons they either stay away from the somewhat drier areas around castor plants or penetrate deep into the soil.

Primarily the termites attack young plants, immediately after planting or when they are very young, devouring the taproot. The injured plants become weak, and leaves turn yellowish. Young plants exhibit signs of drooping of tender leaves, followed by withering and death. Plants already weak due to drought, abnormally high and low moisture conditions, nutritional stress or pest attack became easily vulnerable to the termite attacks. Occasionally, soft plant parts, exposed because of mechanical injuries (strong wind, cattle grazing, several anthropogenic activities, etc.), become susceptible to termite attack (Wardle 1987; Thakur 1996).

Chillies (*Capsicum annuum* L.)

In Rajasthan, chillies in various growth stages are attacked by termites. The plants, both at and below ground level, are attacked, with a loss of 10–45% recorded at Mathania (Jodhpur). The damaging species were *O. obesus* and *M. obesi.* The attack is at the transplanting stage of the crop, when the termites nibble the growing root regions. At the flowering stage, they also enter the base of the stem and devour its inner portion by filling it with earth and excreted saliva.

Pulse Crops

Important crops like moong (*Phaseolus radiatus* L.), moth (*Phaseolus aconitifolius* Jacq.) and cowpea (*Vigna sinensis* L.) are sometimes attacked, at various stages of growth, by *O. obesus* and *Odontotermes parvidens* Holmgren and Holmgren. Losses between 25–30% (cowpea), 10–15% (moth) and 5–17% (moong) have been observed at Jodhpur.

Coconut Trees (*Cocos nucifera* L.)

Odontotermes wallonensis causes serious damage to coconut seedlings in nurseries and, particularly, on transplanted seedling in the earlier stages. More serious losses are caused when they nibble at the tender growing point. This species also damages coconut plants in nursery by constructing mud galleries. Trunk wilting of central shoot, stunted growth and presence of mud galleries are also seen on the trunk. The tender shoots of coconut seedlings have been observed to be cut off by termites, which also cause damages to roots and shoots of seedlings, as well as on young coconut trees. The attacked plants dry gradually and finally become wilted.

Yams and Cassava

Amitermes sp., a predominantly root-feeding termite, is known to attack the elephant foot yam (*Amorphophallus paeoniifolius* (Dennst.) Nicolson) and cassava (*Manihot esculenta* Crantz.), which are grown from tubers and stem cuttings in Africa and India. Mature crops are attacked at the stems by other termites belonging to *Macrotermes*, *Odontotermes* and *Microtermes* sp. At present, the best known management strategy involves treatment of tuber setts with synthetic insecticides, viz. chlorpyrifos.

Cotton (*Gossypium* sp.)
Cotton is grown in black clay soils in India, where a large area is grown under rain-fed agriculture. In India, the termite damage in cotton crop has been observed in states of Rajasthan, Gujarat, Haryana, Punjab and Madhya Pradesh. The crop is attacked mainly by species of three genera, viz. *Trinervitermes*, *Microtermes* and *Odontotermes*. The termites feed on the roots and make tunnels in them. The plants show symptoms of wilting because the root tissues are replaced with soil by the termites, eventually leading to the plant's death. Well-developed root systems of older plants help them to survive the attack. The most common management strategy consists of application of broad-scale insecticides or seed dressings and baiting with dry grass mulch, treated with insecticides.

Horticultural and Tree Crops
Eaten-bark symptom is the most common sight in orchards in tropical, arid and semiarid regions in India. In orchards, trees plastered with mud layer are very common. On removal of the mud plaster, live termites can be seen. Bark-eating termites are of local importance. The damage they cause is often negligible, but in severe infestations hollowed-out cavities are seen on the main trunk and branches, filled with soil, as well as on roots, filled with soil. Termites collect plant materials and transport them to their nests for their fungus gardens. Those feeding on dead bark gather no significance on established trees. It has been observed that sometimes the termites gain entry into the trees through pruned dead ends of branches and stumps. In the orchard, initiation stage termites have been reported to kill the saplings, damaging roots and stems.

Tea (*Camellia sinensis* (L.) Kuntze)
Tea, as a perennial crop, is attacked by 1031 species of arthropod pests across the world, of which only 300 species are recorded in India with about 190 fungi, reported from North East India (Das 1965; Rattan 1992; Sivapalan 1999). Among the tea pests, termites have a distinct niche by attacking plants from under soil or at collar region of a bush. They are important subterranean pests in tea, by limiting the establishment of newly planted young seedlings and by reducing the mature tea, by attacking their frame and killing bushes. These pests cause 11–55% losses in yield (Gurusubramanian et al. 2008).

3.5 Management of Termites

Occurrence of some species in a number of diverse microhabitats confers them an added advantage allowing survival in dry, arid or harsh environments, an important factor for a species to be considered as a pest. Most termite pests are subterranean, and their management primarily relies on soil treatment with termiticides at the site of attack. As a result of the adverse effects of the organochlorines on the environment, research trends shifted towards third- and fourth-generation insecticides such as carbamates, organophosphates and synthetic pyrethroids. Due to the well-understood and proven ill effects of synthetic pesticides on the environment, research trends have also shifted towards use of more environment-friendly techniques. Today farmers are looking for low impact measures of insect pest management.

While developing a strategy, the usefulness of termites must be kept in mind. They are the best decomposers and nutrient recyclers of dead plant material and an indispensable member of the food chain. However, a large number of termite species have an economic importance as pests. Effective foolproof management strategy against termites is not really available due to a fragmentary understanding of their biology. Moreover, the patchiness of infestations under field condition often makes the management strategies more complicated.

Various strategies have been developed over the last few decades to manage termites under field conditions in different field and horticultural crops. We would limit ourselves to the various practices that are being recommended and practised in India. There are several methods to manage termites. None of them is efficient and suitable enough to eradicate established colonies. There are typical shortcomings related to the development of chemical and biological control strategies that need to be solved. Termites are always hidden in galleries and nests. Any method to exterminate them needs to reach them, but usually these niches are inaccessible. Moreover, termites are known to cordon off or block the contaminated/treated area. They are known to bury the diseased individuals or carry them out of their nests or sometimes to eat away the dead ones. Termites are also known to produce certain antibiotics which allow only their beneficial fungi to grow in their colonies. The only effective remedy to the termite problem is hence to prevent their attacks. In orchards, termites are known to attack the weakened trees, so our aim should be to maintain the trees in healthy conditions and remove/destroy the dead or decaying ones.

3.5.1 Physical and Mechanical Methods

- Burning of crop residues on top of termite mounds is a common practice in Indian villages to suffocate these pests. However, neither heat nor smoke penetrates deep enough in the mounds so that the primary reproductives are killed.
- Destruction of mud galleries or tunnels by tillage or flooding gives only temporary solution to the problem. The termites eventually reinfest these areas.
- Destruction of mounds and killing the primary and secondary reproductives also give a temporary solution. This is due to the fact that the nests are often located deep inside the ground and are difficult to reach. Moreover, this method is highly labour intensive. Chances for some brood and workers to escape along with soldiers are very high, which would eventually develop into large colonies again, after a given time.
- Coconut nurseries are protected from foraging workers by covering a layer of sand over the nursery area rather than with soil (Kashyap et al. 1984). Digging deep trenches around the tree nurseries helps to protect the saplings from surface foraging workers of *Macrotermes* sp. which will not be able to construct galleries, (Beeson 1941).

Protection of Traditional Granaries

- Avoid construction of granaries in places infested with termites or with a close proximity to termite mounds. Before construction of granary, ensure removal of all organic matter and crop residues.
- Use termite-resistant timbers (e.g. teak) as poles for granaries. If termite-resistant wood is not available, apply a coating of engine oil on the poles and other wooden structures.
- Preferably use concrete or stone platforms resting on poles.
- Use pure mud walls instead of walls made up of mud and chopped straw.
- Underground pits and bunkers are easily invaded by termites. To avoid this situation, use thin galvanized metal sheets to line the pits and bunkers or line with clay or soil from termite mounds, which is then fired to harden.
- Always apply a layer of ash to the base of the granary or plant materials with insecticidal or repellent properties. This would not only prevent termite entry but also prevent common storage insect pests.

Cultural Practices

Preventive measures are a long-term exercise to avoid termite attacks, but cannot provide a suitable cure for any existing problem. Termites mostly attack diseased and stressed plants and rarely healthy plants (Sen-Sarma 1986). Removal of crop debris and residues will reduce termite food supplies, thereby leading to less foraging activity by workers and reduction in termite numbers and attacks (Brown 1962). Higher seed densities are recommended in areas where a termite attack is expected, so that even when attacked seedlings are thinned out some amount of economic losses are avoided (Wood and Cowie 1988). Deep summer ploughing is recommended before the onset of monsoon, in order to destroy the foraging tunnels and the workers present in the subterranean region (Kumar 1991). Only well-decomposed farmyard manure (FYM) has to be applied to the field. If partially decomposed FYM is applied, it will act as an attractant to foraging workers. Singh and Brar (1988) reported that optimum fertilizer application increases plant vigour, thereby reducing the crop susceptibility to attacks.

In plantations, debris and dead woods must be removed. Pruning has to be applied carefully, with clean cuts to minimize the area of exposure. The pruned areas and wounds should be painted with copper oxychloride to avoid termite attacks and dieback (Harris 1971; Sivapalan et al. 1977). Application of vetiver grass leaf mulch at the tree base has shown to prevent termite attack. Destruction of termite-infested trees and dead trees before the next rainy season helps to prevent release of swarmers from infested trees, also eliminating considerable amount of termite population in the infested area (Srivastava and Butani 1987). Sivapalan et al. (1977) reported that excessive use of nitrogenous fertilizer in tea encouraged growing soft tissues which are highly susceptible to attacks by *Glyptotermes dilatatus* Bugnion & Popoff. Care should be taken while establishing the orchard, avoid growing in sandy/red sandy loam soil. The pits must be also treated with soil insecticides before transplanting seedlings, removing the mud galleries in infested tree trunks and then swabbing the area with kerosene oil.

Clean Cultivation
The field or orchards must be maintained clean to avoid infestations, irrigating the cropped area regularly and removing all the dead and decaying plant/trees/weeds from the cropped areas and near vicinity. Removal of diseased and mechanically injured or damaged plants should also be done. Weeds surrounding the fields compete with crops for nutrients, light and water and may lead to stress and hence increased susceptibility to termite attacks. Inorganic fertilizers may be used to enhance plant vigour and eventually withstand pest damage. Timely proper application of nitrogen, phosphorus and potassium in wheat reduces termite incidence. Crop rotation results in better soil fertility and plant vigour and breaks continuous attack cycle of termites. Deep ploughing exposes the termites to desiccation and predators and thus helps in reducing their numbers. Harvesting the crop at the right time, without leaving the harvested plant material in the field, represents a useful practice.

Crop Rotation and Intercropping
Farmers should follow crop rotation especially including non-preferred crops, following a cropping system with a fallow period. This helps the soil to regain its fertility and also sustains the subsequent crop healthy growth, thereby developing some tolerance towards attacks. Intercropping maize with soybean or groundnut has reduced the termite activity and increased the predatory ant activity (Sekamatte et al. 2003).

Soil Management
Regular intercultural operations and pre-planting tillage destroy the tunnels and galleries built by termites. These operations restrict their foraging activities and also reduce damage to crops. In vertisols termite is not a problem due to frequent occurrence of small cracks and crevices that prevent maintenance of runways, galleries and mounds.

Water Stress
The healthy plant growth must be sustained to avoid termite damage, as these pests more often attack sickly or water-stressed plants than healthy ones. Frequent irrigation reduced attack by *M. obesi* in field crops, viz. maize, wheat, sugarcane and groundnut (Verma et al. 1980; Kumar and Veeresh 1990). On groundnut, Jayanthi et al. (1993) reported that the activity of termites recorded in drip-irrigated plots was higher than in surface-irrigated plots.

3.5.2 Biological Control Strategies

Biodiversity of natural enemies can be increased by applying less persistent insecticides and planting large trees around the agricultural fields. Efforts should be made to increase the presence of natural enemies preying on swarms of termites.

Predation

Wood and Sands (1978) reported that there are two different types of predation on termites, viz. on swarming alates and on foraging workers. In the former predation type, the antagonists are classified as arthropod predators (which include scorpions, spiders, centipedes, dragonflies, cockroaches, mantids, crickets, beetles, flies, ants and wasps) and vertebrate predators (including reptiles, amphibians, birds and mammals, sloth bear, echidna, ant-eaters, etc.). Sheppe (1970) reported that foraging termite workers are predated by ants such as *Myrmicaria cumenoides* Gerst and *Pheidole megacephala* (Fab.). Birds (drongo, bush lark, swallow, green bee eater, hoopoe, Indian roller) predate on termites during swarming. Reptiles such as lizards commonly feed on termites. Frogs are also an important predator of termites during swarming. Mammals such as the five-striped squirrel *Funambulus pennantii* (Wroughton) and mongoose *Herpestes edwardsi* (Geoffroy Saint-Hilaire) were recorded to feed upon a swarm of *Microtermes* sp. Predatory ants like *Pheidole* sp. and *Dorylus* sp. were observed feeding on *O. obesus*. Beeson (1941) noticed that termites were controlled when *Solenopsis* sp. was transplanted in nurseries of tree seedlings in India. The predatory ants attacking termites belong to family Ponerinae and Myrmicinae. Some of the predatory ant species recorded in India are *Leptogenys processionalis* (Jerdon), *Camponotus sericeus* (Fab.), *Anoplolepis longipes* (Jerdon) and *Oecophylla smaragdina* (Fab.) (Rajagopal 1979; Kumar 1991).

Microbial Control

For some past decades organochlorines have been used for termite management worldwide. Owing to their persistence, these pesticides were banned or withdrawn from the market for human health and environmental reasons in an increasing number of countries since the last two decades. The United Nations Environment Programme (UNEP) and the Food and Agriculture Organization (FAO) jointly made efforts to eliminate production and use of persistent organic pollutants including organochlorine pesticides. As a result, the focus on use of "greener" technologies increased tremendously.

In the recent decades, many researchers investigated the potential use of entomopathogens as microbial control agents for insects (Tanada and Kaya 1993). Much of this research focused on the use of *Beauveria bassiana* (Balsamo) Vuillemin and *Metarhizium anisopliae* (Metsch.) Sorokin. To date, the majority of work include evaluation of *M. anisopliae* for biological control of insects has focused on pests important for agriculture (Zimmerman 1993). Grace (1997) provided a review of biological control of termites and concluded that microbes, primarily entomopathogenic fungi, have some potential in the biological control of Isoptera. Termite pathogenic viruses have not been reported so far. However, some workers have published their findings in this respect, but the results do not indicate their possible use in biological control programmes. A potential candidate for development of microbial pesticides must have the capability to complete its life cycle and spread before the death of host (Chouvenc and Su 2010).

Before 1960, few reports noted the pathogenic effect of microorganisms on termites. Merrill and Ford (1916) and Pemberton (1928) first reported the presence of parasitic "head-inhabiting" nematodes in *Reticulitermes lucifugus* (Rossi) and *C. for-*

mosanus, respectively, but concluded that such nematodes could not kill termites in soil conditions. DeBach and McOmie (1939) later reported the existence of two bacterial species killing laboratory colonies of *Zootermopsis angusticollis* Hagen and identified them as *Bacterium* sp. and *Serratia marcescens* Bizio. A thorough perusal of the available literature indicated that plenty of research is being carried out using microbial agents for the management of termites. However, very few pathogens have shown promising results. Most of the experiments were conducted in Petri dishes only, and their results may not be applicable under field conditions. The most important reason for this uncertainty is the susceptibility of termites to light and humidity, as most of the workers die when exposed to fluorescent light for more than 30 min.

Myles (2002a) listed 2 viruses, 5 bacteria, 17 fungi, 5 nematodes and 4 mites as natural enemies of termites. Under natural conditions, diseased termite colonies are rarely encountered, as they maintain a very strict sanitary regimen, with absolute cleanliness, by removing the diseased and dead individuals from the colony. They may also produce selective antibiotics and ensure that only members of *Termitomyces* grow in their fungal gardens. When the colony is weakened by any other means, epizootics can be expected. Some of the present-day termiticides act synergistically with soil microbes and predispose termites to attacks by pathogenic microbes. High rate of fungal infection was observed in imidacloprid (sublethal dose)-treated insects. The stressed insects are also more susceptible to pathogen invasions (Neves and Alves 1999).

1. Nematodes

Entomopathogenic nematodes (EPNs) have been known to kill termites since decades. Under laboratory conditions, they are very effective against termites, but under field conditions their efficacy was not proved. Termites can recognize and wall off infected individuals, hence limiting the spread of nematodes throughout the colony. Furthermore, soil moisture and soil type appear to limit the nematode's ability to move in the soil and locate termites.With time and new frontiers in research, termite management using nematodes has increased. Reese (1971) studied the effectiveness of *Steinernema feltiae* (Filipjev) against large field colonies of *C. formosanus*. He opined that direct physical contact between termite and nematode species is required for colony control. Poinar (1979) noticed mortality among workers of *Coptotermes* and *Nasutitermes* when challenged with *Neoaplectana carpocapsae* (Breton strain) (=*Steinernema feltiae*) in laboratory experiments. Similarly, Georgis et al. (1982) recorded 96–98% mortality among *Zootermopsis* and *Reticulitermes* with 2000 infective stage of *N. carpocapsae* as well as *Heterorhabditis heliothidis* (Khan, Brooks & Hirschmann), respectively, 3 days after the treatment. The EPNs invade different body structures of termites, such as nervous and muscle tissue, fat body and salivary and sternal glands. Parasitism of termites was highly perceptible in Egyptian laboratories and field by *Heterorhabditis baujardi* Phan, Subbotin, Nguyen & Moens and *Heterorhabditis indica* Poinar, Karunakar & David (El-Bassiouny and El-Rahman 2011).

In Sri Lanka, Danthanarayana and Vitarana (1987) could successfully manage populations of the dampwood termite, *Glyptotermes dilatatus* (Bugnion & Popoff) (with colonies of several thousand members) applying *Heterorhabditis* sp. in tea plantations. Similarly, Lenz and Runko (1992) and Lenz et al. (2000) reported that in the South Pacific islands, nematodes have a potential to manage infestations of *Neotermes* sp. in the unbranched trunks of coconut palms. On the other hand, susceptibility of *Neotermes* sp. was reduced in branched trees of citrus, cocoa or American mahogany (*Swietenia macrophylla* King). Weeks and Baker (2004) recorded significant differences in survivability, detectability and ability to kill *Heterotermes aureus* (Snyder), when challenged with *Steinernema carpocapsae* (Weiser) and *Heterorhabditis bacteriophora* Poinar. Rich et al. (2006) indicated that efficacy of nematodes can be increased if they are applied in combination with some insecticide. In laboratory experiments, they observed that four EPNs were capable of killing termites. *Steinernema riobrave* (Cabanillas, Poinar and Raulston) caused more than 80% mortality of *H. aureus* and *Gnathamitermes perplexus* (Banks) on sand assays. However, *R. flavipes* was less susceptible to all nematodes (Yu et al. 2006). In Indian Agricultural Research Institute (IARI), the Division of Nematology has developed a nematode formulation against termite named as Nemagel. This formulation, tested in maize, gave an effective control of termites. The nematode used in this formulation is *Heterorhabditis thermophilum* Ganguly and Singh which harbours the symbiotic bacterium *Xenorhabdus indica.* The nematode releases the bacteria into the insect haemocoel causing a septicaemia that kills the termite within 24–48 h. Nemagel has to be dissolved in 20 l of water which has to be applied over an acre.

Till date there are 83 EPN species, which were able to parasitize insect pests during 2001 all over the world (Grewal et al. 2001). It was observed that the focus on the application of nematodes has increased progressively, and up to now 26 EPN species, along with more than 30 different isolates, have been recorded from all over the world. A list of nematode species parasitic on termites is presented below (Table 3.6).

2. Fungi

Fungi have been used all over the world in the management of insect pests (Glare and Milner 1991). More than 700 species of fungal pathogens of insects have been listed by Milner (2000). To cause infection, a fungus has to penetrate through the host cuticle, as infection would not occur only by ingestion of the spores/conidia. Milner et al. (1998) and Sun et al. (2003) isolated termite pathogenic fungi from attacked wood and mud galleries. The most common pathogenic fungi used in research are *B. bassiana*, *M. anisopliae* and *Paecilomyces fumosoroseus* (Wright 2005). *Metarhizium anisopliae* is the most widely used fungus for field evaluation against termites. A list of common fungal species pathogenic to termites is given below (Table 3.7).

Shortcomings in Utilization of Fungal Pathogens for Termite Management
Termites are highly sensitive to light, humidity and temperature. Their olfactory sense is also very well developed, to compensate for their blindness. Termites were capable to identify conidia of virulent strains of *M. anisopliae* and keep away by triggering alarm and aggregate around spore-dusted individuals (Staples and Milner 2000; Myles 2002b). The identified individual would be groomed by nest mates and then bitten and defecated upon. The dead termites are buried (Myles 2002b). Rosengaus et al. (1998) and Rosengaus and Traniello (2001) studied the

Table 3.6 Nematodes species parasitic to termites

Species	Reference
Heterorhabditis sonorensis, Stock, Rivera-Orduño & Flores-Lara, *H. indica* Poinar, Karunakar & David	Zadji et al. (2014a, c)
Steinernema sp.	Zadji et al. (2014b)
Steinernema carpocapsae (Weiser)	Divya and Sankar (2009)
Steinernema glaseri Konza	Murugan and Vasugi (2011)
Steinernema feltiae Filipjev	Mauldin and Beal (1989)
Steinernema longicadam	Zhu (2002)
Heterorhabditis bacteriophora (Poinar)	Yu et al. (2006)
Neosteinernema longicurvicauda Nguyen & Smart	Nguyen and Smart (1994)
Chroniodiplogaster aerivora (Cobb)	Merrill and Ford (1916); Poinar Jr (1990)
Diplogaster labiates (Pemberton)	Pemberton (1928)
Heterorhabditis baujardi Phan, Subbotin, Nguyen & Moens	El-Bassiouny and El-Rahman (2011)
Pseudaphelenchus yukiae Kanzaki & Giblin-Davis	Kanzaki et al. (2009b)
Pseudaphelenchus vindai Kanzaki, Giblin-Davis, Herre, Scheffrahn & Center	Kanzaki et al. (2010)
Pseudaphelenchus sui n. sp	Kanzaki et al. (2014)
Termirhabditis fastidiosus Massey	Massey (1971)
Rhabditis rainai n. sp.	Carta and Osbrink (2005)
Oigolaimella attenuata von Lieven & Sudhaus	von Lieven and Sudhaus (2008)
Poikilolaimus carsiops n. sp.	Kanzaki et al. (2011)
Poikilolaimus floridensis Kanzaki & Giblin-Davis	Kanzaki et al. (2009a)
Poikilolaimus ernstmayri Sudhaus & Koch	Sudhaus and Koch (2004)
Pelodera scrofulata sp. nov.	Tahseen et al. (2014)
Pelodera termitis sp. n.	Carta et al. (2010)
Pristionchus aerivorus (Cobb)	Christie (1941)
Hartertia gallinarum (Theiler)	Watson and Stenlake (1965)
Caenorhabditis sp.	Handoo et al. (2005)

behavioural defence mechanism and concluded that allogrooming among termites could make fungal spore/conidia treatments ineffective. They also observed that social grooming in bigger colonies would dislodge all the fungal spore/conidia from the spore-dusted individuals. The bigger the colony, the higher the number of individuals so that more grooming individuals would make fungal spore dusting ineffective. Milner (2003) observed that less virulent *M. anisopliae* strains are less repellant to termites. Rath and Tidbury (1996) noticed that when repellent conidia of *M. anisopliae* were formulated with attapulgite clay and surfactants, the challenged termites could not detect them. Milner (2003) suggested addition of attractants or reduction in spore dose to overcome the bottleneck of conidia detection.

Milner (2003) observed that foragers of *Nasutitermes exitiosus* (Hill), dusted with repellent spores at a feeding site, were denied entry in the nest, whereas individuals dusted with conidia of a less repellent strain mixed with masking agents were allowed to enter. Five fungal pathogens (*B. bassiana*, *M. anisopliae*,

Table 3.7 Fungal species pathogenic to termites

Species	References
Aspergillus sp.	Pandey et al. (2013)
Aspergillus flavus Link	Henderson (2007)
Aspergillus fumigatus Fresenius	Chai (1995)
Beauveria bassiana (Bals.-Criv.) Vuill.	Neves and Alves (1999a)
Conidiobolus sp.	Altson (1947)
Conidiobolus coronatus (Costantin) Batko	Sajap et al. (1997)
Cordycepioideus bisporus Stifler	Ochiel et al. (1996)
Entomophthora coronata Costantin , *E. Virulent* Hall and Dunn	Yendol and Paschke (1965)
Gliocladium virens (Miller, Giddens, and Foster)	Kramm and West (1982)
Gloeophyllum trabeum (Pers.) Murrill	Grace et al. (1992)
Hirsutella thompsonii F52 Fisher	James (2009)
Isaria fumosorosea Wize	Wright and Lax (2013)
Metarhizium anisopliae (Metchnikoff) Sorokin	Neves and Alves (1999)
Metarhizium anisopliae var. anisopliae (Metschn.) Sorokin	Khan et al. (1993)
Metarhizium anisopliae var. acridum (Driver & Milner) Bisch., Rehner & Humber	Jarrold et al. (2007)
Metarhizium anisopliae var. dcjhyium Dong, Jia M. Zhang, W.G. Chen & Y.Y. Hu	Dong et al. (2009)
Metarhizium flavoviride Gams & Rozsypal	Wells et al. (1995)
Metarhizium flavoviride var. Minus Rombach, Humber & Roberts	Khan et al. (1993)
Paecilomyces lilacinus (Thom) Samson	Khan et al. (1993)
Paecilomyces cicadae (Miq.) Samson	Chai (1995)

Metarhizium flavoviride, *Paecilomyces lilacinus* and *P. fumosoroseus*) were tested against *O. obesus*, showing that the termites were very susceptible to all fungi (Khan et al. 1993; Chouvenc et al. 2009a, b). *Aspergillus* sp. (Pandey et al. 2013) and *Isaria fumosorosea* (Wright and Lax 2013) caused prompt mortality by growing on the termite colony, and the worker caste became more susceptible due to an extensive exposure, as compared to other individuals. Coghlan (2004) developed a strategy to deliver pre-sporolytic phase mycelium in termite nests. His suggestion was to offer the termites a pre-sporolytic mycelium which is highly attractive to them. The termites need to carry the fungus and deposit it in their gardens, where it would sporulate and cause mycosis. This methodology would work and be effective only if the termites carry the mycelia and deposit them in their garden, rather consuming it.

Milner (2003) developed bait bioformulations containing *Metarhizium* conidia, for management of termites. In bait, the termite would consume the spores and pass it out encased in faecal matter. The encased spores are viable but lose the opportunity to germinate because termite faeces are known to have antifungal properties (Rosengaus et al. 1998). The ability of the spores to move out of the matrix and get attached to the termite body would only be able to cause infection. This process, however, took a fairly long time before the population in mounds of *N. exitiosus* were significantly reduced (Milner 2003). The infected workers and soldiers moved throughout the colony and get dispersed without any restriction, eventually planting small amount of inocula throughout the nest. The healthy workers would gather them and encase them with faecal matter and other building material, thus reducing the chances of disease spread in the colony.

Rosengaus et al. (1999) carried out studies and indicated development of resistance to various pathogens among *Zootermopsis angusticollis* (Hagen). However, *Reticulitermes flavipes* (Kollar) was effectively controlled (92%) by using commercial formulation of *M. anisopliae*, Bio-Blast™ (Quarles 1999). Maniania et al. (2002) reported that application of *M. anisopliae* at the seedling stage of maize effectively controlled termite attacks in Kenya.

3. Bacteria

The first candidates evaluated for use in termite biological control were some bacteria (Toumanoff and Toumanoff 1959; Smythe and Coppel 1965), although they never received serious consideration for field applications. They have been used for management of termites since the mid-1950s. Khan et al. (1977) isolated a strain of *Bacillus thuringiensis* Berliner from the termite species *Bifiditermes beesoni* (Gardener). Efficiency of bacterial pathogens may be accelerated by the warm, humid environment of the colony, trophallaxis as well as by their grooming contact with nest mates (Grace 1994). Fifteen bacterial species have been used to control *C. formosanus*, including *Serratia marcescens* Bizio which caused 100% mortality of hosts (Osbrink et al. 2001). Bacteria were shown to cause mortality of termite though inhibition of their respiration. *Pseudomonas fluorescens* (Flügge) Migula, when evaluated against termites, blocked their respiratory system by producing hydrogen cyanide (Devi and Kothamasi 2009). The pathogenicity of bacterial strains such as *B. thuringiensis* subsp. *israelensis* was assessed against *M. beesoni*, causing higher mortality at low concentrations, although under laboratory conditions (Singha et al. 2010).

Bacteria isolated from termites have also been recorded in previous research studies and include *Pseudomonas aeruginosa* (Schröter) Migula (Tsunoda et al. 1993), *S. marcescens* (Osbrink et al. 2001) and *Citrobacter* sp. (Harazono et al. 2003). A list of bacterial species pathogenic to termites is provided in Table 3.8.

3.5.3 Botanicals

Different plants contain some biologically active compounds that can be used in termite management (Table 3.9). Beeson (1941) reported the efficacy of two botanical mixtures (gambir mixture and gondal fluid) against termite attack. The gambir mixture is prepared by mixing the aqueous extract leaves of *Uncaria gambir*/*Acacia catechu* (L.) Willd., Oliv. along with *Canarium strictum* Roxb. oil. This mixture, when painted on wounds, provides effective protection against termites. The gondal fluid is prepared by mixing the castor oil cake with extracts of *Gardenia gummifera* L.f., *Ferula jaeschkeana* L. and *Aloe vera* (L.) Burm.f. This mixture, when painted around the base of a tree, offers protection against termites for 8 months. *Calotropis* latex is used to protect wooden pegs, offering protection up to 4 months (Giridhar et al. 1988). Singh et al. (2002a) suggested that sugarcane sett dipping in 15 or 20%

Table 3.8 Bacterial species pathogenic to termites

Species	Reference
Acinetobacter calcoaceticus Beijerinck	Osbrink et al. (2001)
Aeromonas caviae Popoff	Devi et al. (2007)
Bacillus cereus Frankland & Frankland	Khucharoenphaisan et al. (2012)
Bacillus licheniformis (Weigmann) Chester	Natsir and Dali (2014)
Bacillus subtilis (Ehrenberg) Cohn	Omoya and Kelly (2014)
Bacillus sphaericus Meyer and Neide	Toumanoff (1966)
Bacillus thuringiensis subsp. *israelensis*	Wang and Henderson (2013)
Burkholderia cepacia (Palleroni and Holmes) Yabuuchi et al.	Devi (2013)
Candida utilis (Henneberg)	Khucharoenphaisan et al. (2012)
Citrobacter sp.	Harazono et al. (2003)
Citrobacter freundii (Braak) Werkman and Gillen	Omoya and Kelly (2014)
Corynebacterium urealyticum Pitcher et al.	Osbrink et al. (2001a)
Enterobacter cloacae (Jordan) Hormaeche and Edwards	Husseneder and Grace (2005)
Enterobacter gergoviae Brenner et al.	Osbrink et al. (2001a)
Escherichia coli Throop	Khucharoenphaisan et al. (2012)
Photorhabdus luminescens (Thomas and Poinar) Boemare et al.	Shahina et al. (2011)
Pseudomonas aeruginosa (Schröter) Migula	Khucharoenphaisan et al. (2012)
Pseudomonas fluorescens (Flügge) Migula	Devi and Kothamasi (2009)
Rhizobium leguminosarum (Frank) Frank	Devi (2013)
Rhizobium radiobacter (Beijerinck and van Delden) Young et al.	Devi et al. (2007)
Serratia marcescens Bizio	Osbrink et al. (2001a)
Staphylococcus aureus Rosenbach	Khucharoenphaisan et al. (2012)
Xenorhabdus nematophila (Poinar and Thomas) Thomas and Poinar	Hiranwrongwera et al. (2007)

solution of *Calotropis procera* (Aiton) W.T.Aiton extract or soil treatment with 2% solution is effective in controlling termites in sugarcane.

Nakashima and Shimizu (1972) reported insecticidal activity of essential oils, known to have insect repellent activity along with contact and fumigant action against certain pests (Isman 2000). Vetiver oil was found to have long-lasting effects against *C. formosanus* (Zhu et al. 2001a). Zhu et al. (2001b) reported nootkatone (a sesquiterpene ketone), a component of vetiver grass oil, as a strong repellent and toxicant to *C. formosanus*. They act as arrestants, repellents and feeding deterrents. Nootkatone negatively affected termites for 12 months and appears more long-lasting then vetiver oil (Maistrello et al. 2003). It causes complete loss of *Pseudotrichonympha grassii koidzumi*, the most important flagellate required for cellulose digestion in *C. formasanus* (Maistrello et al. 2001). Mao et al. (2006) suggested use of vetiver oil and nootkatone into potting media for surface treatments to restrict the spread of *C. formasanus*. Nix et al. (2006) observed that vetiver grass-root mulch treatment decreased the tunnelling activity and wood consumption of *C. formosanus* and increased their mortality.

Table 3.9 Plants with antitermite properties

Plant species	Common name	Activity/active principle	Plant part
Acacia catechu (L.) Willd., Oliv.	Catechu, khair	Aqueous extract of tannins	Leaves and twigs
Acacia mearnsii De Wild.	Wattle tree	Ethyl acetate extract and water-soluble extract	Bark waste
Acacia crassicarpa Benth.	Red wattle	Ethyl acetate extract and water-soluble extract	Bark waste
Acacia modesta Wall.	Palosa	Methanol extract	Aerial parts
Acacia nilotica (L.) Hurter & Mabb.	Egyptian thorn	Anti-insect	Wood/pulp
Acacia polyacantha Willd.	Hook thorn	Insect repellent tanins	Roots
Afrormosia laxiflora (Benth. ex Baker) Harms	English satin wood		Wood/pulp
Afzelia cuanzensis Welw.	Pod mahogany		Seed oil
Agave americana L.	American aloe	Repellent, insecticidal	Whole plant
Albizia odoratissima (L.f.) Benth.	Tes shade tree		Wood/pulp
Albizia saman Muell.	Saman	Repellent	Wood/pulp
Albizia zygia (DC.) Macbr.	Nongo	Termite durable but not resistant	Wood
Allium sativum L.	Garlic	Anti-feedant, fungicidal, repellent	Bulbs
Anacardium occidentale L.	Cashew	Anti-insect, repellent	Seeds, oil
Argemone Mexicana L.	Mexican poppy	Insecticidal, repellent	Whole plant
Artemisia douglasiana Besser ex Besser	California mugwort	Insect repellent	Leaves
Azadirachta indica Juss	Neem	Termiticidal, anti-feedant	Leaves, seeds
Bidens pilosa L.	Blackjack	Anti-feedant, insecticidal, repellent	Whole plant, mature seeds
Borassus aethiopum Mart	African fan palm		The fibrous wood is highly resistant
Boswellia dalzielii Hutch.	English frankincense tree	Repellent	Gum/resin
Brachylaena hutchinsii Hutch. (Muhuhu)	Muhugu oil tree	Highly resistant, almost impenetrable to termites	Wood
Brugmansia candida Pers. (pro. sp.)	Angel's trumpet	Termiticidal activity in n-hexane, ethyl acetate and aquadest	leaves
Calotropis gigantea (L.) W. T. Aiton	Giant Milkweed	Anti-insect	Leaves, sap/latex/juice

(continued)

Table 3.9 (continued)

Plant species	Common name	Activity/active principle	Plant part
Calotropis procera (Aiton) W. T. Aiton	Rubber bush	Termiticidal	Latex
Camellia sinensis (L.) Kuntze	Tea	Anti-feedant, insectidical	Leaves and fruit
Capparis aphylla (Karien)	Bare Caper	Termite resistant shrub	Wood/pulp
Carica papaya L.	Pawpaw	Insecticidal	Fruit, fresh leaves and roots
Carya ovate (Mill.) K. Koch.	Shagbark hickory	Termiticidal	Bark
Cassia siamea Lam.	Yellow cassia, kassof tree	Repellent	Used as a leaf mulch
Catalpa bignonioides Walter	Common catawpa		Resistant to *Reticulitermes flavipes*
Cedrela odorata L.	West Indian cedar	Termiticidal	Wood
Chenopodium ambrosioides L.	Wormseed	Anti-feedant, insecticidal, repellant	Whole plant
Cleistanthus collinus (Roxb.) Benth. ex Hook.f.	Garari	Repellent	Bark
Commiphora africana (A.Rich.) Endl.	African myrrh	Termiticidal, repellent	Gum/resin
Consolida regalis Gray	Blue cloud	Termiticidal	Seeds
Corymbia citriodora (Hook.) K.D. Hill & L.A.S. Johnson	lemon-scented gum	Anti-termite property	crude volatile leaf oil and its methanol- and hexane fractions
Daniellia oliveri (Rolfe) Hutch. & Dalziel	African copaiba balsam tree	Anti-termite property	Gum/resin
Detarium senegalense J.F.Gmel.	Sweet detar	Oral poison	Wood/pulp
Diospyros ebenum J.Koenig ex Retz.	Ebony	Anti-insect	Roots
Dodonaea viscosea Jacq.	Purple hop bush	Termite resistant shrub	Wood/pulp
Erythropleum suaveolens (Guill. & Perr.) Brenan	Forest ordeal tree	Oral poison	Wood/pulp
Eucalyptus microcorys F. Muell.	Tallowwood	More resistant than other Eucalyptus	Wood
Grevillea robusta A.Cunn. ex R.Br.	Silky oak, silver oak	Termite tolerant in Tanzania	Wood
Hardwickia mannii (Harms) Oliv.	Indian black wood	Termiticidal	Stem/branches
Hyptis spicigera Lam. marubio	Black Sesame	Repellent	Aerial parts

(continued)

Table 3.9 (continued)

Plant species	Common name	Activity/active principle	Plant part
Intsia bijuga (Colebr.) Kuntze ifil. *or I. palembanica* Miq.	Merbau	Resistant	Wood
Juniperus procera Hochst. ex Endl.	E. African pencil cedar	Highly resistant	Wood
Juniperus virginiana L.	Eastern red cedar	Anti-insect	Wood
Leucaena leucocephala (Lam.) de Wit	Ipil ipil	Repellent	Used as a leaf mulch
Margaritaria discoidea (Baill.) G.L.Webster	*Common* pheasant-berry	Feeding deterrents	Bark and wood
Melia azedarach L.	Chinaberry, persian lilac	Oral poison, anti-feedant, contact poison, repellant	Bark, branches, leaves, fruit, oil
Mesua ferrea L.	Indian rose chestnut	Anti-insect	Wood
Ocimum basilicum L.	Sweet basil	Insecticidal, repellent	Whole plant
Ocimum canum L.	Wild basil	Insecticidal, repellent	Whole plant
Ocimum urticifolium Benth.	Basil		Water-based extracts
Pinus strobus (L.)	*Eastern white pine*	Termiticidal	Bark
Prosopis africana (Guill. & Perr.) Taub.	African *mesquite*	Anti-insect	Roots
Quassia indica (Gaertn.) Noot.	Bitter wood	Termiticidal	Leaves
Quercus prinus L.	Chestnut oak	Termiticidal	Bark
Samadera indica Gaertn.	Niepa Bark Tree	Termiticidal	Leaves
Santalum album L.	Sandalwood	Anti-insect	
Sassafras albidum (Nutt.) Nees.	Sassafras	Termiticidal	Bark
Semecarpus anacardium L.f.	Marking nut	Anti-insect	Seeds
Strychnos nux-vomica L.	Poison nut	Oral poison	Leaves
Swartzia madagascariensis Desv.	Snake bean	Repellent	Fruit
Tabebuia spp.	Ipe	Resistant	Wood
Tagetes minuta L.	Mexican marigold		Water-based extracts
Tectona grandis L. f.	Teak	Repellent	Wood/pulp
Terminalia chebula (Gaertn.) Retz.	Mirabolans	Antibacterial	Fruits
Uncaria gambier (Hunter) Roxb.	Gambier	Antifungal and antibacterial	Leaves
Zanthoxylum zanthoxyloides (Lam.)	Senegal prickly-ash		Wood/pulp

Essential oil extracted from leaves of *Tagetes erecta* L. rich in (Z)-β-ocimene caused mortality of *O. Obesus* after 24 h exposure (Singh et al. 2002b). *Calocedrus formosana* (Florin) Florin leaf essential oil and its main constituent, T-muurolol, caused 100% mortality of *C. formosanus* at the dosage of 5 mg g^{-1} (Cheng et al. 2004). Cheng et al. (2007) reported antitermitic activity of 11 essential oils from three species of coniferous tree against *C. formosanus*. Among all, heartwood of *Calocedrus macrolepis* var. *formosana* exhibited the strongest termiticidal property. Sakasegawa et al. (2003) indicated superiority of gelam oils over cajuput oils against termites. Park and Shin (2005) observed that garlic oil was more toxic than clove bud oil against termites.

Doolittle et al. (2007) reported that the number of microbes present in the hindgut of *C. formosanus* was reduced by treatments with neem extract, capsaicin and gleditschia. Neem extracts could significantly reduce the population of spirochaetes, leading to 100% mortality among termites. Rudman and Gay (1963) observed that anthracenes, anthrones, anthraquinones and xanthones act as deterrents against termites, whereas Cornelius et al. (1997) indicated the toxicity of monoterpenoids, alkaloids and hydrocarbons. Similarly, flavonoids and related compounds were found to be toxic and possess antifeedant properties (Boue and Raina 2003).

Grace and Yates (1992) found Margosan-O™, a neem-based formulation, containing 0.3% azadirachtin and 14% neem oil, to be toxic against the *C. formosanus*. Sharma et al. (1999) investigated and observed that *Acorus calamus* L., rhizomes, and aerial parts of *T. erecta*, were most toxic against *O. obesus*. Thambidurai (2002) observed that the fermented extract of *Musa paradisiaca* L. leaves at 100% concentration prevented termite attack for 50 days. Fokialakis et al. (2006) reported that four *Echinops* species had antitermite properties against *C. formosanus*. Verma and Verma (2006) studied termiticidal effects of 5% chloroform extract of *Lantana camara* var. *aculeata* L. leaves. Similarly Ganapaty et al. (2004) isolated plumbagin, isodiospyrin and microphyllone from the roots of *Diospyros sylvatica* L. and found them to be toxic against *O. obesus*.

The continued interest in search of greener pesticides led to the evaluation of different plant products all over the world. Workers around the globe are reporting activities of new compounds isolated from various plant parts. Under laboratory conditions, the plant products show promising results. However, often these results cannot be reproduced under field conditions. The reasons are plant products are an expensive option, the concentration of the active principle per unit varies from location to location and the avoidance behaviour of termites from the treated surface/substrate renders the treatments useless, under field conditions.

3.5.4 Chemical Methods

Because the termites remain hidden in the tree or under the soil surface/tree barks, chemical control is also not suitable in most of the cases. Practically, the insecticides fail to reach the target. Earlier, chlorinated hydrocarbons and cyclodienes were used for control, but with limited success. Today these insecticides are banned in most of the countries owing to their long persistence in soil. Management of subterranean termites primarily relies on soil treatment with termiticides at the site of active infestation. In India the most commonly used insecticide for termite management is chlorpyrifos. Farmers all over the country use chlorpyrifos with irrigation water irrespective of the crop grown. Scientists from different regions of India evaluated a plethora of synthetic insecticides with little or no effect. However, in some evaluations, a limited success was achieved for a short period of time. We presented a list of most commonly used insecticide (Table 3.10) for termite management in agricultural crops in India.

Table 3.10 Insecticides used in termite management in India

Crop	Insecticide	References
Sugarcane sett dipping	Imidacloprid 70 WS @ 0.1–0.2% solution	Singh and Singh (2002), Singh et al. (2003), and Jaipal and Singh (2003)
Spraying on sugarcane setts	Imidacloprid 200 SL @ 250 to 375 ml/ha	Santharam et al. (2002)
	Imidacloprid (17.8% SL) @ 375 ml/ha	Jaipal and Chaudhary (2010)
Soil application	Carbofuran @ 1 kg a.i./ha	Gangwar et al. (2003)
	Chlorpyrifos 20 EC @ 1.25 kg a.i./ha	Singh et al. (2001)
	Endosulfan 35 EC @ 1.25 kg a.i./ha	
	Phorate 10 G @ 2.5 kg a.i./ha	Singh and Singh (2002)
	Chlorpyrifos 20 EC @ 1 kg a.i./ha	Singh et al. (2003)
	Chlorpyrifos 15 G @ 2.5 kg a.i./ha	
	Chlorpyrifos 20 EC @ 1.25 kg a.i./ha	Madan et al. (1998)
		Sharma et al. (2002)
	Methyl parathion 2% dust @ 25 kg/ha	Kumawat (2001)
	Endosulfan 4% dust	
	Quinalphos 1.5% dust	
	Chlorpyrifos 10 G @ 1 kg a.i./ha	Mohapatra et al. (1995)
Wheat seed treatment	Chlorpyrifos 20 EC @ 4.5 ml/kg seed	Kumawat (2001)
	Endosulfan 35 EC @ 7.3 ml/kg seed	
	Lindane 20 EC at 0.5 ml/kg seed	
	Chlorpyrifos 20 EC @ 0.9 g a.i/kg seed	Rana et al. (2001)
	Endosulfan 35EC @ 2.4 g a.i/kg seed	
	Fipronil 5 SC @ 5 ml/kg seed	Gadhiya and Borad (2013)
	Imidacloprid 600 FS @ 3 ml/kg seed	
	Bifenthrin 10 EC @ 2 ml/kg seed	
Maize seed treatment	Endosulfan 35 EC @ 3 g per kg	Sharma et al. (2003)
Groundnut seed treatment	Chlorpyrifos 20 EC @ 12.5 ml/kg	
Side dressing of maize	Fipronil granules	Sharma et al. (2003)
	Chlorpyrifos granules	

3.6 Conclusion

Termite control is a herculean task, and their complete elimination or prevention in cropped areas is neither feasible nor advisable. Indian agriculture is a gamble with rains and is dominated by small and marginal farmers, with meagre amount of resources for insect pest management. The majority of the farmers follow the age old practices for management of insect pests. A good number of indigenous traditional practices are indeed available, but they result locally specific, with a limited general success. The biggest challenge for an Indian farmer is the availability of quality seeds and fertilizers, followed by availability of water, whenever needed. The pest management takes a back seat in their list of priorities; hence the allocation of resources for pest management is also minimal.

Scientists have generated a huge amount of data and recommended several practices for insect pest management, for all the crops grown in India. Several cutting-edge technologies have been developed for this purpose, but farmers are often unaware of these new technologies. There is a big extension gap between the laboratories and farms, and the crop and species diversity often makes this issue more complicated. India is a large country divided into 15 agroclimatic zones. Technologies need to be developed for each agroclimatic zone, separately. No single technology would be effective for all the agroclimatic zones. Optimistically, prospects for the development of new and improvements of existing technologies, as well as public acceptance of alternative termite management, appear positive and feasible. Least toxic and nonchemical methods have been, and will continue to be, developed. In India, termite baits are not available in the market, so there is a need and goal for development of termite bait technology, in particular for subterranean species, for which baits will play a major role in control. However, products increasing bait appeal and retention at application stations are still needed.

References

Agarwala, S. B. D. (1955). Control of sugarcane termites (1946–1953). *Journal of Economic Entomology, 48*, 533–537.

Altson, R. A. (1947). A fungus parasitic on *Coptotermes curvignathus*, Holmgr. *Nature, 160*, 120.

Beeson, C. F. C. (1941). A guide to the control of termites for forest officers. *Indian Forest Records (New Series) Entomology, 4*, 44–90.

Bignell, D. E., & Eggleton, P. (2000). Termites in ecosystems. In T. Abe, M. Higashi, & D. E. Bignell (Eds.), *Termites: Evolution, sociality, symbioses, ecology* (p. 466). Dordrecht: Springer.

Bose, G. (1984). Termite fauna of southern India. Rec. Zoological Survey of India, Occasional paper, 49, 1–270

Boue, S. M., & Raina, A. K. (2003). Effects of plant flavonoids on fecundity, survival, and feeding of the Formosan subterranean termite. *Journal of Chemical Ecology, 29*, 2575–2584.

Brown, K. W. (1962). *Termite control research in Uganda with particular reference to control of attacks in eucalyptus plantations* (p. 9). Eighth British commonwealth forestry conference, Entebbe, Government Publication, Uganda Protectorate.

Carta, L. K., & Osbrink, W. (2005). *Rhabditis rainai* n. sp. (Nematoda: Rhabditida) associated with the Formosan subterranean termite, *Coptotermes formosanus* (Isoptera: Rhinotermitidae). *Nematology, 7*, 863–879.

Carta, L. K., Handoo, Z. A., Lebedeva, N. I., Raina, A., Zhuginisov, T. I., & Khamraev, A. S. (2010). *Pelodera termitis* sp. n. and two other rhabditid nematode species associated with the Turkestan termite *Anacanthotermes turkestanicus* from Uzbekistan. *International Journal of Nematology, 20*, 125–134.

Chai, Y. Q. (1995). Preliminary studies on the pathogenicity of some entomopathogenous fungi to *Coptotermes formosanus*. *Chinese Journal of Biological Control, 11*, 68–69.

Cheng, S. S., Wu, C. L., Chang, H. T., Kao, Y. T., & Chang, S. T. (2004). Antitermitic and antifungal activities of essential oil of *Calocedrus formosana* leaf and its composition. *Journal of Chemical Ecology, 30*, 1957–1967.

Cheng, S. S., Chang, H. T., Wu, C. L., & Chang, S. T. (2007). Anti-termitic activities of essential oils from coniferous trees against *Coptotermes formosanus*. *Bioresource Technology, 98*, 456–459.

Chhotani, O. B. (1977). A review of taxonomy of Indian termites. *Records of the Zoological Survey of India, 9*, 1–36.

Chhotani, O. B. (1980). Termite pests of agriculture in Indian region and their control. *Technical Monograph, 4*, 1–84.

Chhotani, O. B. (1997). The fauna of India and the adjacent countries. Isoptera (Termites): (family Termitidae). *Zoological Survey of India, 2*, 750–800.

Chouvenc, T., & Su, N. Y. (2010). Apparent synergy among defense mechanisms in subterranean termites (Rhinotermitidae) against epizootic events – The limits and potential for biological control. *Journal of Economic Entomology, 103*, 1327–1337.

Chouvenc, T., Su, N. Y., & Robert, A. (2009a). Inhibition of *Metarhizium anisopliae* in the alimentary tract of the eastern subterranean termite *Reticulitermes flavipes*. *Journal of Invertebrate Pathology, 101*, 130–136.

Chouvenc, T., Su, N. Y., & Robert, A. (2009b). Cellular encapsulation in the eastern subterranean termite, *Reticulitermes flavipes* (Isoptera), against infection by the entomopathogenic fungus, *Metarhizium anisopliae*. *Journal of Invertebrate Pathology, 101*, 234–241.

Christie, J. R. (1941). Life history. General discussion. In *An introduction to nematology* (pp. 243–372). Section II. Chapters IV–XI.

Coghlan, A. (2004). Green pesticide is irresistible to ants. *New Scientist, 184*, 26.

Collins, N. M. (1983). Termite populations and their role in litter removal in Malaysian rain forests. In S. L. Sutton, T. C. Whitmore, & A. C. Chadwick (Eds.), *Tropical rain forest: Ecology and management* (pp. 311–325). Oxford: Blackwell Scientific Publications.

Cornelius, M. L., Grace, J. K., & Yates, J. R. (1997). Toxicity of monoterpenoids and other natural products to the formosan subterranean termite (Isoptera: Rhinotermitidae). *Journal of Economic Entomology, 87*, 705–708.

Cunha, H. F., & Orlando, T. Y. S. (2011). Functional composition of termite species in areas of abandoned pasture and in secondary succession of the parque estadual altamiro de moura pacheco, goiás, Brazil. *Bioscience Journal Uberlândia, 27*, 986–992.

Danthanarayana, W., & Vitarana, S. L. (1987). Control of the live wood tea termite. *Glyptotermes dilatatus* using *Heterorhabditis* sp. (Nemat.) *Agriculture Ecosystems and Environment, 19*, 333–342.

Das, G. M. (1965). *Pests of tea in North East India and their control* (p. 115). Memorandum No 27, Tocklai Experimental Station, Jorhat.

Davies, R. G., Hernandez, L. M., Didham, R. K., Fagan, L. L., & Winchester, N. N. (2003). Environmental and spacial influences upon species composition of a termite assemblage across neotropical forest islands. *Journal of Tropical Ecology, 19*, 509–524.

DeBach, P. H., & McOmie, W. A. (1939). New diseases of termites caused by bacteria. *Annals of the Entomological Society of America, 32*, 37–146.

DeBlauwe, I., Dibog, L., Missoup, A. D., Dupain, J., Van Elsacker, L., Dekoninck, W., Bonte, D., & Hendrickx, F. (2008). Spatial scales affecting termite diversity in tropical low land rainforest: A case study in southeast Cameroon. *African Journal of Ecology, 46*, 5–18.

Devi, K. K. (2013). Investigations on cyanide producing *Pseudomonas* bacterial species and their potential for application against termite *Odontotermes obesus*. University of Delhi.

Devi, K. K., & Kothamasi, D. (2009). *Pseudomonas fluorescens* CHA0 can kill subterranean termite *Odontotermes obesus* by inhibiting cytochrome c oxidase of the termite respiratory chain. *FEMS Microbiology Letters, 300*, 195–200.

Devi, K. K., Seth, N., Kothamasi, S., & Kothamasi, D. (2007). Hydrogen cyanide-producing rhizobacteria kill subterranean termite *Odontotermes obesus* (Rambur) by cyanide poisoning under *in Vitro* Conditions. *Current Microbiology, 54*, 74–78.

Divya, K., & Sankar, M. (2009). Entomopathogenic nematodes in pest management. *Indian Journal of Science and Technology, 2*, 53–60.

Dong, C., Zhang, J., Huang, H., Chen, W., & Hu, Y. (2009). Pathogenicity of a new China variety of *Metarhizium anisopliae* (M. anisopliae var.dcjhyium) to subterranean termite *Odontotermes formosanus*. *Microbiological Research, 164*, 27–35.

Doolittle, M., Raina, A., Lax, A., & Boopathy, R. (2007). Effect of natural products on gut microbes in Formosan subterranean termite, *Coptotermes formosanus*. *International Biodeterioration and Biodegradation, 59*, 69–71.

Edwards, W., & Mill, A. E. (1986). *Termites in buildings. Their biology and control*. East Grinstead: Rentokil Limited.

EL-Bassiouny, A. R., & El-Rahman, R. M. A. (2011). Susceptibility of egyptian subterranean termite to some entomopathogenic nematodes. *Egyptian Journal of Agricultural Research, 89*, 121–135.

Feakin, S. D. (1973). *Pest control in groundnuts*. PANS Manual No. 2 COPR, Foreign & Common. London: Office, Overseas Development Administration.

Fokialakis, N., Osbrink, W. L., Mamonov, L. K., Gemejieva, N. G., Mims, A. B., Skaltsounis, A. L., Lax, A. R., & Cantrell, C. L. (2006). Antifeedant and toxicity effects of thiophenes from four Echinops species against the Formosan subterranean termite, *Coptotermes formosanus*. *Pest Management Science, 62*, 832–838.

Freise, F. (1949). A significação de formigas e cupins nas matas tropicais e capoeiras. Observações feitas nas florestas litorâneas do Brasil. *Annl Brasil Econ Forestal, 2*, 145–154.

Gadhiya, V. C., & Borad, P. K. (2013). Effect of insecticidal seed treatment on reduction of termite damage and increase in wheat yield. *Pesticide Research Journal, 25*, 87–89.

Ganapaty, S., Thomas, P. S., Fotso, S., & Laatsch, H. (2004). Antitermitic quinones from *Diospyros sylvatica*. *Phytochemistry, 65*, 1265–1271.

Gangwar, S. K., Tewari, R. K., & Lakshman, L. (2003, August 22–24). Monitoring of insect pest incidence, yield losses and their management in late planted sugarcane. In *Proceedings of the 65th Annual Convention of the Sugar Technologists' Association of India* (pp. A186–A195). Bhubaneshwar.

GEI. (2005). *Demonstration project of alternatives to chlordane and mirex for termite control in China*. Bejing: Guangdong Entomological Institute. From The World Bank. http://web.worldbank.org/external/projects

Georgis, R., Poinar, G. O., Jr., & Wilson, A. P. (1982). Susceptibility of dampwood termites and soil and wood-dwelling termites to the entomogenous nematode *Neoplectana carpocapsae*. *International Research Communications Systems Medical Science: Microbiology Parasitology Infectious Diseases, 10*, 563.

Giridhar, G., Vesudevan, S., & Vesudevan, P. (1988). Antitermites properties of *Calotropis* latex. *Pesticides, 22*, 31–33.

Glare, T. R., & Milner, R. J. (1991). Ecology of entomopathogenic fungi. In D. K. Arora, K. G. Mukerji, & P. JGE (Eds.), *Handbook of applied mycology, Humans, animals, and insects* (Vol. 2, pp. 547–612). New York: Dekker.

Grace, J. K. (1994). Protocol for testing effects of microbial pest control agents on nontarget subterranean termites (Isoptera: Rhinotermitidae). *Journal of Economic Entomology, 87*, 269–274.

Grace, J. K. (1997). Biological control strategies for suppression of termites. *Journal of Agricultural Entomology, 14*, 281–289.

Grace, J. K., & Yates, J. R. (1992). Behavioural effects of a neem insecticide on *Coptotermes formosanus* (Isoptera: Rhinotermitidae). *Tropical Pest Management, 38*, 176–180.

Grace, J. K., Goodell, B. S., Jones, W. E., Chandhoke, V., & Jellison, J. (1992). *Inhibition of termite feeding by fungal siderophores* (pp. 1–4). The International Research Group on Wood Preservation. Document No: IRGAVP/1558-92. Biological Problems (Fauna).

Grewal, P. S., Nardo, E. D., & Aguillera, M. M. (2001). Entomopathogenic nematodes: Potential for exploration and use in South America. *Neotropical Entomology, 30*, 191–205.

Gurusubramanian, G., Sarmah, M., Rahman, A., Roy, S., & Bora, S. (2008). Pesticide usage pattern in tea ecosystem, their retrospects and alternative measures: A review. *Journal of Environmental Biology, 29*, 813–826.

Handoo, Z. A., Lebedeva, N. A., Carta, L. K., Khamraev, A. S., Zhuginisov, T. I., & Raina, A. K. (2005, October 16–22). A new species of *Caenorhabditis* (Nematoda: Rhabditida) found associated with termites (*Anacanthotermes turkestanicus*) in Uzbekistan. In *Proceedings of International workshop on termites of central Asia: Biology, Ecology and Control* (p. 38). Tashkent.

Harazono, K., Yamashita, N., Shinzato, N., Watanabe, Y., Fukatsu, T., & Kurane, R. (2003). Isolation and characterization of aromatics-degrading microorganisms from the gut of the lower termite *Coptotermes formosanus*. *Bioscience, Biotechnology, and Biochemistry, 67*, 889–892.

Harris, W. V. (1969). *Termites as pests of crops and trees*. Commonwealth Institute of Entomology HDRA – the organic organisation: Termite control without chemicals. www.gardenorganic.org.uk/pdfs/international_programme/Termite.pdf

Harris, W. V. (1971). *Termites, their recognition and control* (2nd ed.pp. 15–32). London: Longman Publishers.

Henderson, G. (2007). Effect of *Aspergillus flavus* and *Trichoderma harzianum* on survival of *Coptotermes formosanus* (Isoptera: Rhinotermitidae). *Sociobiology, 49*, 135–141.

Hiranwrongwera, C., Adisettakul, P., Tansirichaiya, S., Piyabun, O., & Somsuk, V. (2007, November 1–3). Efficiency of a nematode (*Steinernema carpocapsae*) and its symbiotic bacterium (*Xenorhabdus nematophila*) at eliminating the termite *Coptotermes curvignathus* that infests para rubber (*Hevea brasiliensis*). *The 5th international symposium on biocontrol and biotechnology* (p. 90). Nong Khai Campus, Nong Khai: Khon Kaen University.

Howse, P. E. (1970). *Termite: A study in social behaviour* (pp. 25–29). West Sussex: Rentokil Ltd. 163–176.

Hussain, M. A. (1935). Pests of wheat crop in India. In *Proceedings of the worlds grain exhibition and conference* (Vol. 2, pp. 562–564).

Husseneder, C., & Grace, J. K. (2005). Genetically engineered termite gut bacteria (*Enterobacter cloacae*) deliver and spread foreign genes in termite colonies. *Applied Microbiology and Biotechnology, 68*, 360–367.

Isman, M. B. (2000). Plant essential oils for pest and disease management. *Crop Protection, 19*, 10603–10608.

Jaipal, S., & Chaudhary, O. P. (2010). Imidacloprid as an effective insecticide against termites infesting sugarcane crop. *Indian Journal of Sugarcane Technology, 25*, 54–57.

Jaipal, S., & Singh, D. (2003). Bioefficacy of imidachlopirid and amrutgard against termites and shoot borer in sugarcane crop. *Indian Sugar, 59*, 709–716.

James, R. R. (2009). Microbial control for invasive arthropod pests of honey bees. In A. E. Hajek, T. R. Glare, & M. O'Callaghan (Eds.), *Use of microbes for control and eradication of invasive arthropods* (pp. 271–288). Dordrecht: Springer.

Jarrold, S. L., Moore, D., Potter, U., & Charnley, A. K. (2007). The contribution of surface waxes to pre-penetration growth of an entomopathogenic fungus on host cuticle. *Mycological Research, 111*, 240–249.

Jayanthi, M., Singh, K. M., & Singh, R. N. (1993). Succession of insect pest on high yielding groundnut variety under Delhi conditions. *Indian Journal of Entomology, 55*, 24–29.

Johnson, R. A., Lamb, R. W., & Wood, T. G. (1981). Termite damage and crop loss studies in Nigeria a survey to groundnuts. *Tropical Pest Management, 27*, 325–342.

Joshi, P. K., Singh, N. P., Singh, N. N., Gerpacio, R. V., & Pingali, P. L. (2005). *Maize in India: Production systems, constraints, and research priorities* (p. 42). Mexico: CIMMYT.

Kanzaki, N., Giblin-Davis, R. M., Scheffrahn, R. H., & Center, B. J. (2009a). *Poikilolaimus floridensis* n. sp. (Rhabditida: Rhabditidae) associated with termites (Kalotermitidae). *Nematology, 11*, 203–216.

Kanzaki, N., Giblin-Davis, R. M., Scheffrahn, R. H., Center, B. J., & Davies, K. A. (2009b). *Pseudaphelenchus yukiae* n. gen., n. sp. (Tylenchina: Aphelenchoididae) associated with *Cylindrotermes macrognathus* (Termitidae: Termitinae) in La Selva, Costa Rica. *Nematology, 11*, 869–881.

Kanzaki, N., Giblin-Davis, R. M., Herre, E. A., Scheffrahn, R. H., & Center, B. J. (2010). *Pseudaphelenchus vindai* n. sp. (Tylenchomorpha: Aphelenchoididae) associated with termites (Termitidae) in Barro Colorado Island, Panama. *Nematology, 12*, 905–914.

Kanzaki, N., Li, H. F., Lan, Y. C., Kosaka, G.-D. R. M., & Center, B. J. (2011). *Poikilolaimus carsiops* n. sp. (Rhabditida: Rhabditidae) associated with *Neotermes koshunensis* (Kalotermitidae) in Kenting National Park, Taiwan. *Nematology, 13*, 155–164.

Kanzaki, N., Li, H. F., Lan, Y. C., & Giblin-Davis, R. M. (2014). Description of two *Pseudaphelenchus* species (Tylenchomorpha: Aphelenchoididae) associated with Asian termites and proposal of Tylaphelenchinae n. subfam. *Nematology, 16*, 963–978.

Kashyap, R. K., Verma, A. N., & Bhanot, J. P. (1984). Termites of plantation crops, their damage and contol. *Journal of Plantation Crops, 12*, 1–10.

Kaushal, P. K., & Deshpande, R. R. (1967). Losses to groundnut by termites. *JNKVV Research Journal,* 92–93.

Khan, K. I., Fazal, Q., & Jafari, R. H. (1977). Pathogenicity of locally discovered B*acillus thuringiensis* strain to the termites *Heterorhabditis indica* (Wassman) and *Microtermes championi* (Snyder). *Pakistan Journal of Scientific Research, 29*, 12–13.

Khan, H. K., Jayaraj, S., & Gopalan, M. (1993). Muscardine fungi for the biological control of agroforestry termite *Odontotermes obesus* (Rambur). *Insect Science and Its Application, 14*, 529–535.

Khucharoenphaisan, K., Sripairoj, N., & Sinma, K. (2012). Isolation and identification of actinomycetes from termite's gut against human pathogen. *Asian Journal of Animal and Veterinary, 7*, 68–73.

Kramm, K. R., & West, D. F. (1982). Termite pathogens: Effects of ingested *Metarhizium*, *Beauveria*, and *Gliocladium* conidia on worker termites (*Reticulitermes* sp.) *Journal of Invertebrate Pathology, 40*, 7–11.

Krishna, K., & Weesner, F. M. (1970). Taxonomy, phylogeny and distribution. In K. Krishna & F. M. Weesner (Eds.), *Biology of termites* (p. 643). New York: Academic.

Kumar, N. G. (1991). Studies on the ecology of subterranean termite, *Odontotermes hornii* (Wasmann) (Isoptera: Termitidae) and its foraging effect on nutrient status. Ph.D. thesis, University of Agricultural Sciences, Bangalore. (p. 254).

Kumar, C. T. A., & Veeresh, G. K. (1990). Foraging activity of the subterranean termite *Microtermes obesi* Holmgren (Termitidae: Isoptera). In G. K. Veeresh, B. Mallik, & C. A. Viraktamath (Eds.), *Social insects and the environment: Proceedings of the 11th international congress of IUSSI*, 1990 (International Union for the Study of Social Insects): 575. Leiden: E.J. Brill, xxxi + 765 pp.

Kumawat, K. C. (2001). Evaluation of some insecticides against field termites, *Odontotermes obesus* and *Microtermes obesi* in wheat, *Triticum aestivum*. *Annals of Plant Protection Sciences, 9*, 51–53.

Lenz, M., & Runko, S. (1992). *Use of microorganisms to control colonies of the coconut termite Neotermes rainbowi (Hill) on Vaitupu, Tuvalu* (p. 47). Commonwealth Scientific and Industrial research Organisation, Division of Entomology, Termite Group Report No. 92/16.

Lenz, M., Kamath, M. K., Lal, S., & Senivasa, E. (2000). *Status of the tree-damaging Neotermes sp. in Fiji's American mahogany plantation and preliminary evaluation of the use of entomopathogens for their control*. ACIAR Small Project No. FST/96/205, Project Report (in part).

Lobry de Bruyns, L. A., & Conacher, A. J. (1990). The role of termites and ants in soil modification: A review. *Australian Journal of Soil Research, 28*, 55–93.

Logan, J. W. M., Cowie, R. H., & Wood, T. G. (1990). Termite (Isoptera) control in agriculture and forestry by non-chemical methods: A review. *Bulletin of Entomological Research, 80*, 309–330.
Madan, Y. P., Singh, M., & Singh, M. (1998). Evaluation of some soil insecticides for termites and shoot borer control in sugarcane. *Indian Sugar, 49*, 515–518.
Maistrello, L., Henderson, G., & Laine, R. A. (2001). Effects of nootkatone and a borate compound on formosan subterranean termite (Isoptera: Rhinotermitidae) and its symbiont protozoa. *Journal of Entomological Science, 36*, 229–236.
Maistrello, L., Henderson, G., & Laine, R. A. (2003). Comparative effects of vetiver oil, nootkatone and disodium octaborate tetrahydrate on *Coptotermes formosanus* and its symbiotic fauna. *Pest Management Science, 59*, 58–68.
Maiti, P. K., & Maiti, P. (2011). *Biodiversity: Peril and Preservation* (p. 537). New Delhi: PHI Learning.
Maniania, N. K., Ekesi, S., & Songa, J. M. (2002). Managing termites in maize with the entomopathogenic fungus *Metarhizium anisopliae*. *Insect Science and Its Application, 22*, 41–46.
Mao, L., Henderson, G., Bourgeois, W. J., Vaughn, J. A., & Laine, R. A. (2006). Vetiver oil and nootkatone effects on the growth of pea and citrus. *Industrial Crops and Products, 23*, 327–332.
Massey, C. L. (1971). Two new genera of nematodes parasitic in the eastern subterranean termite, *Reticulitermes flavipes*. *Journal of Invertebrate Pathology, 17*, 238–242.
Mauldin, J. K., & Beal, R. H. (1989). Entomogenous nematodes for control of subterranean termites, *Reticulitermes* sp. (Isoptera: Rhinotermitidae). *Journal of Economic Entomology, 82*, 1638–1642.
McDonald, D. (1970). Fungal infection of groundnut fruit after maturity and during drying. *Transactions of the British Mycological Society, 54*, 461–472.
Merrill, J. H., & Ford, A. L. (1916). Life history and habits of two new nematodes parasitic on insects. *Journal of Agricultural Research, 6*, 115–127.
Milner, R. J. (2000). Current status of *Metarhizium* as a mycoinsecticide in Australia. *Biocontrol News and Information, 21*, 47N–50N.
Milner, R. J. (2003). Application of biological control agents in mound building termites (Isoptera: Termitidae) – Experiences with *Metarhizium* in Australia. *Sociobiology, 41*, 419–428.
Milner, R. J., Staples, J. A., & Lutton, G. G. (1998). The selection of an isolate of the hyphomycete fungus, *Metarhizium anisopliae,* for control of termites in Australia. *Biological Control, 11*, 240–247.
Mohapatra, H. K., Padhi, J., Samalo, A. P., & Patra, G. J. (1995). Screening promising groundnut varieties against termite damage at Bhubaneshwar, Orissa, India. *IAN, 15*, 59–60.
MRP (2010) *Maxwell Robinson Phelps Termite Report*. Maxwell, Robinson and Phelps. http://www.pestcontrol-perth.com/wp-content/uploads/2010/06/Maxwell-Robinson-Phelps-MRP-Termite-Report.pdf
Murugan, K., & Vasugi, C. (2011). Combined effect of *Azadirachta indica* and the entomopathogenic nematode *Steinernema glaseri* against subterranean termite, *Reticulitermes flavipes*. *Journal of Entomological and Acarological Research, 43*, 253–259.
Myles, T. G. (2002a). Isolation of *Metarhizium anisopliae* (Deuteromycotina: Hyphomycetes) from *Reticulitermes flavipes* (Isoptera: Rhinotermitidae) with convenient methods for its culture and collection of conidia. *Sociobiology, 40*, 257–262.
Myles, T. G. (2002b). Alarm, aggregation, and defense by *Reticulitermes flavipes* in response to a naturally occurring isolate of *Metarhizium anisopliae*. *Sociobiology, 40*, 243–255.
Nair, K. S. S., & Varma, R. V. (1981). *Termite control in eucalyptus plantations, KFRI Res Report No 6* (p. 48). Peechi: Kerala Forest Research Institute.
Nakashima, Y., & Shimizu, K. (1972). Antitermitic activity of *Thujopsis dolabrata* var *hondai*. III. Components with a termiticidal activity. *Miyazaki Daigaku Nogakubu Kenkyu Hokoku, 19*, 251–259.
Natsir, H., & Dali, S. (2014). Production and application of chitin deacetylase from *Bacillus licheniformis* HSA3-1a as Biotermicide. *Marina Chimica Acta, 15*, 8–12.

Neves, P. J., & Alves, S. B. (1999). Associated control of *Cornitermes cumulans* (Kollar, 1832) (Isoptera: Termitidae) with *Metarhizium anisopliae*, *Beauveria bassiana* and imidacloprid. *Scientia Agricola, 56*, 305–311.

Nguyen, K. B., & Smart, G. C. (1994). *Neosteinernema longicurvicauda* n. gen., n. sp. (Rhabditida: Steinernematidae), a Parasite of the Termite *Reticuldermes flavipes* (Koller). *Journal of Nematology, 26*, 162–174.

Nix, K. E., Henderson, G., Zhu, B. C. R., & Laine, R. A. (2006). Evaluation of vetiver grass root growth, oil distribution, and repellency against formosan subterranean termites. *Horticultural Science, 41*, 167–171.

Ochiel, G. S., Eilenberg, J., Gitonga, W., Bresciani, J., & Toft, L. (1996). *Cordycepioideus bisporus*, a naturally occurring fungal pathogen on termite alates in Kenya. *IOBC-Wprs Bulletin, 19*, 172–178.

Ofori, F., & Stern, W. R. (1987). Cereal-legume intercropping systems. *Advances in Agronomy, 41*, 41–49.

Omoya, F. O., & Kelly, B. A. (2014). Variability of the potency of some selected entomopathogenic bacteria (*Bacillus* sp. and *Serratia* sp.) on termites, *Macrotermes bellicosus* (Isoptera: Termitidae) after exposure to magnetic fields. *International Journal of Tropical Insect Science, 34*, 98–105.

Osbrink, W. L. A., Williams, K. S., Connick, W. J., Wright, M. S., & Lax, A. R. (2001). Virulence of bacteria associated with the Formosan subterranean termite (Isoptera: Rhinotermitidae) in New Orleans, LA, USA. *Environmental Entomology, 30*, 443–448.

Pandey, P., Singha, L. P., & Singha, B. (2013). Colonization and antagonistic activity of entomopathogenic *Aspergillus* sp. against tea termite (*Microcerotermes beesoni* Snyder). *Current Science, 105*, 1216–1219.

Pardeshi, M. K., Kumar, D., & Bhattacharyya, A. K. (2010). Termite (Insecta: Isoptera) fauna of some agricultural crops of Vadodara, Gujarat (India). *Records of the Zoological Survey of India, 110*, 47–59.

Parihar, D. R. (1977). Note on some termites of Rajasthan desert. *Geobios, 4*, 173.

Parihar, D. R. (1978). Field observations on the nature and extent of damage by Indian desert termites and their control. *Annals of Arid Zone, 17*, 192–199.

Park, I. L. K., & Shin, S. C. (2005). Fumigant activity of plant essential oils and components from garlic (*Allium sativum*) and clove bud (*Eugenia caryophyllata*) oils against the Japanese termite (*Reticulitermes speratus* kolbe). *Journal of Agricultural and Food Chemistry, 53*, 4388–4392.

Pearce, M. J. (1997a). *Laboratory culture and experimental techniques using termites* (p. 52). Chatham: Natural Resources Institute.

Pearce, M. J. (1997b). *Termites: Biology and pest management* (1st ed.p. 192). Chatham: CAB International.

Pemberton, C. E. (1928). Nematodes associated with termites in Hawaii, Borneo and Celebes. *Proceedings of the Hawaiian Entomological Society, 7*, 148–150.

Poinar, G. O., Jr. (1979). *Nematodes for Biological Control of Insects*. Boca Raton: CRC Press.

Poinar, G. O., Jr. (1990). Redescription of *Chroniodiplogaster aerivora* (Cobb) gen. n., comb. n. (Rhabditida: Diplogasteridae) from Termites. *Journal of the Helminthological Society of Washington, 57*, 26–30.

Quarles, W. (1999). Non-toxic control of drywood termites. *IPM Practitioner, 21*, 1–10.

Rahman, P. M., Varma, R. V., & Sileshi, G. W. (2012). Abundance and diversity of soil invertebrates in annual crops, agroforestry and forest ecosystems in the Nilgiri biosphere reserve of Western Ghats, India. *Agroforestry Systems, 85*, 165–177.

Rajagopal, D. (1979). Ecological studies of the mound building termites, *Odontotermes wallowensis* (Wasmann) (Isoptera: Termitidae) (p. 205). Ph.D. thesis, submitted to the UAS, Bangalore.

Rajagopal, D. (2002). Economically important termite species in India. *Sociobiology, 41*, 33–46.

Rajagopal, D., & Veeresh, G. K. (1983). Swarming behaviour and colony establishment in *Odontotermes walloensis* (Wasmann) (Isoptera: Termitidae) in South India. *Journal of Soil Biology and Ecology, 3*, 29–34.

Rana, J. S., Ombir, & Dahiya, K. K. (2001). Management of termite, *Microtermes obesi* (Holm.) in wheat, *Triticum aestivum* through seed treatment. *Annales Biologiques, 17*, 207–209.

Rashmi, R. S., & Sundararaj, R. (2013). Host range, pest status and distribution of wood destroying termites of india. *Journal of Tropical Asian Entomology, 2*, 12–27.

Rath, A. C., & Tidbury, C. A. (1996). Susceptibility of *Coptotermes acinaciformis* (Isoptera: Rhinotermitidae) and *Nasutitermes exitiosus* (Isoptera: Termitidae) to two commercial isolates of *Metarhizium anisopliae*. *Sociobiology, 28*, 67–72.

Rattan, P. S. (1992). Pest and disease control in Africa. In K. C. Wilson & M. N. Clifford (Eds.), *Tea: Cultivation to Consumption* (pp. 331–352). London: Chapman and Hall.

Rawat, R. R., Deshpande, R. R., & Kaushal, P. K. (1970). Comparative efficacy of different modern insecticides and their methods of application for the control of termites *Odontotennes obesus* Rambur in groundnut. *The Madras Agricultural Journal, 57*, 83–87.

Reese, K. M. (1971). Navy fights Formosan termites in Hawaii. *Chemical and Engineering News, 49*, 52.

Rich, W. N., Stuart, J. R., & Rosanna, R. G. (2006). Susceptibility and behavioral responses of the dampwood termite *Zootermopsis angusticollis* to the entomopathogenic nematodes *Steinernema carpocapsae*. *Journal of Invertebrate Pathology, 95*, 17–25.

Roonwal, M. L. (1955). Termites ruining a township. *Zeitschrift für Angewandte Entomologie, 38*, 103–104.

Roonwal, M. L. (1979). *Termite life and termite control in tropical South Asia* (p. 177). Jodhpur: Scientific Publishers.

Roonwal, M. L. (1981). Termites of agricultural importance in India and their importance. In G. K. Veeresh (Ed.), *Progress in soil biology and ecology in India* (pp. 253–265). Bangalore: UAS, Tech. Ser. No. 37.

Roonwal, M. L., & Chhotani, O. B. (1967). Indian wood destroying termites. *Journal of Bombay Natural History Society, 632*, 354–364.

Roonwal, M. L., & Chhotani, O. B. (1989). *The fauna of India and the adjacent countries, Isoptera Termites* (Vol. 1, p. 671). Calcutta: Zoological Survey of India.

Rosengaus, R. B., & Traniello, J. F. A. (2001). Disease susceptibility and the adaptive nature of colony demography in the dampwood termite *Zootermopsis angusticollis*. *Behavioral Ecology and Sociobiology, 50*, 546–556.

Rosengaus, R. B., Maxmen, A. B., Coates, L. E., & Traniello, J. F. A. (1998). Disease resistance: A benefit of sociality in the dampwood termite *Zootermopsis angusticollis* (Isoptera: Termopsidae). *Behavioral Ecology and Sociobiology, 44*, 125–134.

Rosengaus, R. B., Traniello, J. F. A., Chen, T., & Brown, J. J. (1999). Immunity in a social insect. *Naturwissenschaften, 86*, 588–591.

Rudman, P., & Gay, F. J. (1963). Causes of natural durability in timber. X. Deterrent properties of some three-ringed carboxylic and heterocyclic substances to the subterranean termite, *Nasutitermes exitiosus*. C.S.I.R.O. Div. Forest Prod., Melbourne. *Holzforschung, 17*, 21–25.

Rust, M. K., & Su, N. Y. (2012). Managing Social Insects of Urban Importance. *Annual Review of Entomology, 57*, 355.

Sacks, E. (2011). *Termites eat through $222,000 worth of rupee notes in Indian bank*. Daily News. Available from: http://www.mydailynews.com/news/world/termites-eat-222-000-worth-rupee-notes-indian-bank-article-1.111054

Sajap, A. S., Atim, A. B., Husim, H., & Wahab, Y. A. (1997). Isolation of *Conidiobolus coronatus* (Zygomycetes: Entomophthorales) from soil and its effect on *Coptotermes curvignathus* (Isoptera: Rhinotermitidae). *Sociobiology, 30*, 257–262.

Sakasegawa, M., Hori, K., & Yatagai, M. (2003). Composition and antitermite activities of essential oils from *Melaleuca* species. *Journal of Wood Science, 49*, 81–187.

Sands, W. A. (1960). *Observations on termites destructive to trees and crops in W. Africa* (pp. 14–66). Commonwealth Institute of Entomology, Colonial Termite Research Report, London.

Santharam, G., Kumar, K., Kuttalam, S., & Chandrasekaran, S. (2002). Bioefficacy of imidacloprid against termites in sugarcane. *Sugar Tech, 4*, 161–163.

Sekamatte, M. B., Ogenga, L. M., & Russell, S. A. (2003). Effects of maize–legume intercrops on termite damage to maize, activity of predatory ants and maize yields in Uganda. *Insect Science and Its Application, 22*, 87–93.
Sen-Sarma, P. K. (1986). Economically important termites and their management in the oriental region. In S. B. Vinson (Ed.), *Economic impact and control of social insects* (pp. 69–102). New York: Prager.
Sen-Sarma, P. K., Thakur, M. L., Misra, S.C., & Gupta, B. K. (1975). *Wood destroying termites of india* (p. 190). FRI Publication.
Shahina, F., Tabassum, K. A., Salma, J., & Mahreen, G. (2011). Biopesticidal affect of *Photorhabdus luminescens* against *Galleria mellonella* larvae and subterranean termite (Termitidae: Macrotermis). *Pakistan Journal of Nematology, 29*, 35–43.
Sharma, R. N., Tare, V., & Pawan, P. (1999). Toxic action of some plant extracts against selected insect pest and vectors. *Pestology, 23*, 30–37.
Sharma, D. C., Katoch, K. K., & Kashyap, N. P. (2002). Relative efficacy of different insecticides to *Odontotermes* sp. and *Agrotis* sp. in wheat. *Insect Environment, 8*, 10–11.
Sharma, R. K., Sharma, K., & Sekhar, J. C. (2003). Evaluation of plant protectants on damage and yield of rainfed maize by termites. *Pesticide Research Journal, 15*, 36–39.
Sheppe, W. (1970). Invertebrate predation on termites of the African savanna. *Insectes Sociaux, 17*, 205–218.
Singh, D., & Brar, D. S. (1988). Growth and yield of rainfed wheat as affected by seed treatment with aldrin and fertilizer use. *Journal of Research Punjab Agricultural University, 25*, 188–192.
Singh, S. K., & Singh, G. (2002). Comparative evaluation of chemical and botanical insecticides against termites. *Entomon, 27*, 153–160.
Singh, M., Singh, D., & Madan, Y. P. (2001). Evaluation of different soil insecticides for the control of termites in sugarcane. *Indian Sugar, 51*, 365–368.
Singh, G., Singh, O. P., Lampasona, M. P., & Cesar, A. N. (2002a). Studies on essential oils. Part 35: Chemical and biocidal investigations on *Tagetes erecta* leaf volatile oil. *Flavour and Fragrance, 18*, 62–65.
Singh, M., Singh, N. B., & Singh, M. (2002b). Effect of certain insecticides on termite infestations in planted setts of sugar cane. *Cooperative Sugar, 34*, 311–315.
Singh, M., Singh, N. B., & Singh, M. (2003). Bud damage due to termites in sugarcane. *Cooperative Sugar, 33*, 655–658.
Singha, D., Singha, B., & Dutta, B. K. (2010). Ultrastructural details of the morphological changes in termite (*Microtermes obesi* Holmgren) pest of tea exposed to entomopathogenic fungi *in vitro*. *Assam University Journal of Science and Technology, 5*, 100–104.
Sivapalan, P. (1999). Pest management in tea. In N. K. Jain (Ed.), *Global advances in tea science* (pp. 625–646). New Delhi: Aravali Books.
Sivapalan, P., Senaratne, K. A. D. W., & Karunaratne, A. A. C. (1977). Observations on the occurrence and behaviour of live wood termites (*Glyptotermes dilatatus*) in low country tea fields. *Pest Articles News Summaries, 23*, 5–8.
Smythe, R. V., & Coppel, H. C. (1965). The susceptibility of *Reticulitermes flavipes* (Kollar) and other termite species to an experimental preparation of *Bacillus thuringiensis* Berliner. *Journal of Invertebrate Pathology, 7*, 423–426.
Srivastava, K. P. (1996). *A text book of applied entomology* (Vol. I & II). Ludhiana: Kalyani Publishers.
Srivastava, K. P., & Butani, D. K. (1987). Insect pests of tea in India and their contol. *Pesticides, 21*, 16–21.
Staples, J. A., & Milner, R. J. (2000). A laboratory evaluation of the repellency of *Metarhizium anisopliae* conidia to *Coptotermes lacteus* (Isoptera: Rhinotermitidae). *Sociobiology, 36*, 133–146.
Subektia, N., Yoshimurab, T., Rokhmanc, F., & Masturd, Z. (2015). Potential for subterranean termite attack against five bamboo species in correlation with chemical components. *Procedia Environmental Sciences, 28*, 783–788.

Sudhaus, W., & Koch, C. (2004). The new nematode species *Poikilolaimus ernstmayri* sp n associated with termites, with a discussion on the phylogeny of *Poikilolaimus* (Rhabditida). *Russian Journal of Nematology, 12*, 143–156.

Sun, J., Fuxa, J. R., & Henderson, G. (2003). Virulence and in vitro characteristics of pathogenic fungi isolated from soil by baiting with *Coptotermes formosanus* (Isoptera: Rhinotermitidae). *Journal of Entomological Science, 38*, 342–358.

Tahseen, Q., Akram, M., Mustaqim, M., & Ahlawat, S. (2014). Descriptions of *Pelodera scrofulata* sp. nov. and *Pelodera aligarhensis* sp. nov.(Nematoda: Rhabditidae) with supplementary information on *Pelodera teres* (Schneider, 1866). *Journal of Natural History, 48*, 1027–1053.

Tanada, Y., & Kaya, H. K. (1993). *Insect pathology*. San Diego: Academic Press.

Tewary, A. (2008). *Termites feast on trader's money, life savings*. Daily news. Available from: http://news.bbc.co.uk/2/hi/south_asia/7334033.stm

Thakur, R. K. (1996). Termite problems in arid zones and their management. *Indian Forester, 122*, 161–169.

Thambidurai, S. (2002). Termite control using natural plant products. *Indigenous Agriculture News, 1*, 9.

Thorne, B. L. (1997). Evolution of eusociality in termites. *Annual Review of Ecology and Systematics, 28*, 27–54.

Toumanoff, C. (1966). Observations sur les affections bactériennes des termites en Saintonge (*Reticulitermes santonensis* de Feytaud). *Insectes Sociaux, 13*, 155–163.

Toumanoff, C., & Toumanoff, C. H. (1959). Les épizooties dues à Serratia marcescens Bizio chez un termite (Reticulitermes santonensis de Feytaud). *Comptes Rendus Hébdomadaires de l'Academie Agricole Française, 45*, 216–218.

Trenbath, B. R. (1993). Intercropping for the management of pests and diseases. *Field Crops Research, 34*, 381–405.

Tsunoda, K., Ohmura, W., & Yoshimura, T. (1993). Methane emissions by the termite, *Coptotermes formosanus* (Isoptera: Rhinotermitidae). (II) Presence of methanogenic bacteria and effect of food on methane emission rates. *Japanese Journal of Environmental Entomology and Zoology, 27*, 45–49.

UNEP/FAO/Global IPM Facility Expert Group on Termite Biology and Management. (2000). *Finding alternatives to persistent organic pollutants (POPs) for termite management*. Online at: www.chem.unep.ch/pops/termites/termite_ch4.htm

Verma, R. K., & Verma, S. K. (2006). Phytochemical and termiticidal study of *Lantana camara* var. *aculeata* leaves. *Fitoterapia, 77*, 466–468.

Verma, A. N., Bhanot, J. P., & Khurana, A. D. (1980). Effect of different dates of sowing of aldrin treated and untreated wheat seed on germination, termite damage and yield of wheat crop. *Haryana Agricultural University Journal of Research, 10*, 41–44.

VonLieven, A. F., & Sudhaus, W. (2008). Description of *Oigolaimella attenuata* n. sp. (Diplogastridae) associated with termites (Reticulitermes) and remarks on life cycle, giant spermatozoa, gut-inhabiting flagellates and other associates. *Nematology, 10*, 501–523.

Wang, C., & Henderson, G. (2013). Evidence of Formosan subterranean termite group size and associated bacteria in the suppression of entomopathogenic bacteria, *Bacillus thuringiensis* subspecies *israelensis* and *thuringiensis*. *Annals of the Entomological Society of America, 106*, 454–462.

Wardle, D. A. (1987). Control of termites in nurseries and young plantations in Africa: Established practices and alternative courses of action. *Commonwealth Forestry Review, 66*, 77–89.

Watson, J. M., & Stenlake, J. B. (1965). *An introduction to parasitology* (p. 22). London: William Heinemann Medical Books Ltd.

Weeks, B., & Baker, P. (2004). *Subterranean Termite (Isoptera: Rhinotermitidae) Mortality Due to Entomopathogenic Nematodes (Nematoda: Steinernematidae, Heterorhabditidae)*. University of Arizona College of Agriculture. 2004 Turfgrass and Ornamental Research Report, Online at: http://cals.arizona.edu/pubs/crops/az1359/

Wells, J. D., Fuxa, J. R., & Henderson, G. (1995). Virulence of four fungal pathogens to *Coptotermes formosanus* (Isoptera: Rhinotermitidae). *Journal of Entomological Science, 30*, 208–215.
Wood, T. G., & Cowie, R. H. (1988). Assessment of on-farm losses in cereals in Africa due to soil insects. *Insect Science and Its Application, 9*, 709–716.
Wood, T. G., & Sands, W. A. (1978). The role of termites in ecosystems. In M. V. Brian (Ed.), *Production ecology of ants and termites* (pp. 245–292). Cambridge: Cambridge University Press.
Wright, M. S. (2005). A strain of the fungus *Metarhizium anisopliae* for controlling subterranean termites. *Journal of Economic Entomology, 98*, 1451–1458.
Wright, M. S., & Lax, A. R. (2013). Combined effect of microbial and chemical control agents on subterranean termites. *Journal of Microbiology, 51*, 578–583.
Yendol, W. G., & Paschke, J. D. (1965). Pathology of an *Entomophthora* infection in the eastern subterranean termite *Reticulitermes flavipes* (Kollar). *Journal of Invertebrate Pathology, 7*, 414–422.
Yu, H., Gouge, D., & Baker, P. (2006). Parasitism of subterranean termites (Isoptera: Rhinotermitidae: Termitidae) by entomopathogenic nematodes (Rhabditida: Steinernematidae; Heterorhabditidae). *Journal of Economic Entomology, 99*, 1112–1119.
Zadji, L., Baimey, H., Afouda, L., Moens, M., & Decraemer, W. (2014a). Characterization of biocontrol traits of heterorhabditid entomopathogenic nematode isolates from South Benin targeting the termite pest *Macrotermes bellicosus*. *BioControl, 59*, 333–344.
Zadji, L., Baimey, H., Afouda, L., Moens, M., & Decraemer, W. (2014b). Comparative susceptibility of *Macrotermes bellicosus* and *Trinervitermes occidentalis* (Isoptera: Termitidae) to entomopathogenic nematodes from Benin. *Nematology, 16*, 719–727.
Zadji, L., Baimey, H., Afouda, L., Moens, M., & Decraemer, W. (2014c). Effectiveness of different *Heterorhabditis* isolates from Southern Benin for biocontrol of the subterranean termite, *Macrotermes bellicosus* (Isoptera: Macrotermitinae), in laboratory trials. *Nematology, 16*, 109–120.
Zhu, J. H. (2002). Study on application of entomopathogenic nematodes to control *Odontotermes formosanus* Shiraki on eucalyptus. *Journal of Fujian College of Forestry, 22*, 366–370.
Zhu, B. C. R., Henderson, G., Chen, F., Fei, H., & Laine, R. A. (2001a). Evaluation of vetiver oil and seven insect-active essential oils against the formosan subterranean termite. *Journal of Chemical Ecology, 27*, 1617–1625.
Zhu, B. C. R., Henderson, G., Chen, F., Maistrello, L., & Laine, R. A. (2001b). Nootkatone is a repellent for formosan subterranean termite (*Coptotermes formosanus*). *Journal of Chemical Ecology, 27*, 523–531.
Zimmerman, G. (1993). The entomopathogenic fungus *Metarhizium anisopliae* and its potential as a biocontrol agent. *Pesticide Science, 37*, 375–379.

Chapter 4
Termites Infesting Malaysian Forests: Case Study from Bornean Forest, Sabah, Malaysia

Homathevi Rahman, Kevin Fernandez, and Nivaarani Arumugam

Contents

H. Rahman (✉) • K. Fernandez
Institute for Tropical Biology and Conservation, Universiti Malaysia Sabah,
Jalan UMS, 88400 Kota Kinabalu, Sabah, Malaysia
e-mail: homa.ums@gmail.com

N. Arumugam
Faculty of Earth Science, Universiti Malaysia Kelantan,
Jeli Campus, Locked Bag No.100, 17600 Jeli, Kelantan, Malaysia

M.A. Khan, W. Ahmad (eds.), *Termites and Sustainable Management*,
Sustainability in Plant and Crop Protection,
https://doi.org/10.1007/978-3-319-68726-1_4

Abstract The tropical rainforest of Borneo is known to be one of the hotspots for its biodiversity. This ancient lowland tropical forest has been continuously investigated and studied for its various ecosystem functions. One of the many researches that have been conducted here includes the study pertaining to termite's importance in the ecosystem, several of which have been documented, especially in Sabah. Termites play important ecological roles in the tropical forest and their function or species composition changes with the alteration of the habitats. Species composition together with feeding and nesting behaviors can be important indicators to determine the pest species, in view of control or prevention methods. In this chapter, we consider the termite assemblages of few primary forest stands in Sabah, the effect of forest disturbance which may promote pest species, the patterns of termite infestation in natural and disturbed habitats, and some control practices in replanted forests and plantations.

Keywords Termite assemblages • Forest disturbance • Termite pest • Borneo • Oil palm plantation

4.1 Introduction

Sabah is Malaysia's second largest state, lying approximately between 4.5° and 7°N and 116° and 118°E in the northeastern region of the Borneo Island. It covers an area of about 73,371 km sq., occupying about one tenth of the island of Borneo (Marsh and Greer 1992). This rugged, hilly country is one of the few places in the world where it is still possible to find pristine tropical rainforest with very diverse flora and fauna and many endemic and rare species. The rainforest of Borneo as a whole is known to support between 10,000 and 12,000 species of flowering plants of which 40–50% are endemic to the island (Soepadmo 1995). Of these, about 80% are endemic to Sabah and Sarawak. In Sabah alone, about 3000 species of trees are known as endemic species (Sabah Forestry Department 1989). The feature that distinguishes the forests of Sabah or Southeast Asia from the other main tropical rainforest areas of the world is the dominance of one family of trees, the Dipterocarpaceae, commonly known as dipterocarps. Out of their 386 known species, 291 (75%) are recorded in Borneo and about 257 (66%) occur in Sabah (Soepadmo 1995). These normally huge trees give the Malaysian lowland forests some of the highest canopies in the world. As for the faunal composition, about 580 species of birds (Davison and Fook 1996), 145 of snakes (Stuebing 1991), 100 of frogs (Inger and Stuebing 1989), and 220 of terrestrial mammals (Payne et al. 1985) have been recorded. Arthropods are known to dominate the faunal composition, but their exact number of species remains unknown. The rainforest of Southeast Asia is the oldest and most diverse place on earth (Whitmore 1998) and is an ideal place for ecological studies.

4.1.1 Forest of Sabah

In Sabah, about 63% (4.7 million ha) of the total land surface is covered with forest (Sabah Forestry Department 1989). This includes all the forest reserves, state land, and park forests. Forests in Sabah can be described as a continuation from the coastal beach forest, mangrove, and swamp forest (coastal line, floodplain, and riverbanks) to lowland dipterocarp forest (0–900 m a.s.l.) and hill dipterocarp forest (750–1200 m a.s.l) and progressing to montane forest (1200 m a.s.l. and above) (Ashton 1995).

Beach forest is common along the coastline as well as on some offshore islands (Sabah Forestry Department 1989). This forest is dominated by *Casuarina equisetifolia*, known to produce excellent firewood. Mangrove, peat swamp, and freshwater swamp forests occur in the wet areas between 0 and 500 m a.s.l. Structurally these forests are not as complex as the dryland ones and are reduced in overall species richness. Mangrove forests are dominated by *Rhizopora* and *Bruguiera* tree species and mostly occur on the east and southeast of Sabah, where waters are brackish and wave and tidal conditions are conducive for mud accumulation. There are approximately 364,800 ha of mangrove forest in Sabah that provide physical protection to the coastline. The seaward edges are usually important as food sources for marine groups and local fisheries. In some of the mangrove forests, especially where more fresh water is present, nipa palms are found.

Swamp forest includes peat swamp and fresh water swamp forests, occurring in river deltas or near riversides further upstream (Sabah Forestry Department 1989). Peat swamp forest grows on peaty substrate and is normally flooded for short periods in the year, mainly in the northwestern coast, eastwards to Papar (Ashton 1995). The swamp forests of the west coast are mainly a mixture of *Dryobalanops rappa* and *Shorea platycarpa*, *S. scabrida* and *S. teysmannia*, and *Hopea pentanervia* (Wood and Meijer 1964). The east coast swamp forests are quite similar to those of dipterocarp forests on high-level alluvium with common species such as *Dryobalanops beccarii*, *Shorea parvifolia*, *S. leprosula*, and *Hopea beccarii* (Wood and Meijer 1964). In drier areas between mangroves, nipa and swamp forest is the transitional forest, including plants such as nibong palms, rattan, and *Intsia palembanica* (Merbau) (Sabah Forestry Department 1989).

The lowland forest of Sabah is characterized mainly by dipterocarp forest with canopy reaching 25–45 m in height, plus some emergent reaching 80 m. Most of the trees have large buttresses and many woody climbers. In the drier parts, the forest is rich with dipterocarps, structurally complex, with many woody climbers, palms, parasitic plants, and epiphytes. Dipterocarps usually grow from sea level up to 600 m, with small number of species reaching altitudes of 1300–1600 m or even higher (Wood and Meijer 1964). The vegetation changes with altitudes, and each type shows its own characteristic association of dominant trees, dictated by location and terrain. Dipterocarp forests, also known as mixed dipterocarp forest, can be divided into lowland, hill, and upper dipterocarp forest (Ashton 1995).

The most important commercial forest is the dryland one of the lowlands and hills, occurring between 0 and 600 m a.s.l. Variation in species composition depends mainly on soils and altitude. Trees such as selangan batu and yellow seraya are frequent on loamy sandstones on the hills and ridges compared to flatter areas. Various species of keruing (*Dipterocarpus acutangulus*, *D. grandiflorus*, and *D. pachyphyllus*) prefer sand soils. Other examples are *Parashorea malaanonan* (urat mata licin), *Parashorea tomentella* (urat mata beludu), and *Shorea leptoclados* (majau), common in richer soils especially in alluvial parts in sandstone areas.

Vegetation type's change on soils is derived from ultrabasic rocks, totally different in structure from other forests. It is composed of trees with smaller crowns and leaves and a characteristically even tree canopy cover. Species that are common to this type of soils are *Dipterocarpus lowii*, *D. geniculatus*, *Shorea kunstleri*, *S. venulosa*, *S. laxa*, and *S. andulensis* (Wood and Meijer 1964). In Sabah, there are about 180 species of dipterocarps, contributing to almost 90% of the timber production (Sabah Forestry Department 1989).

Around 300 m a.s.l. and above, there is a transition from "lowland" to "hill" dipterocarp forest and the species composition changes. Upper dipterocarp forests comprising trees such as *Shorea atrinervosa*, *Shorea monticola*, *S. nebulosa*, *Hopea dyeri*, *H. montana*, and *Dipterocarpus ochraceus* are found between 600 and 1300 m a.s.l. (Wood and Meijer 1964).

Heath forests or "Kerangas" are characterized by small and thin trees (less than 8 cm dbh) (Jones et al. 1998) with small, leathery leaves and few climbing plants associated with ants and without emergents (Ashton 1995). These forests are confined to organic white sandy, nutrient-poor soils. Examples may be seen on the forested hills in the Sandakan Peninsula and Maliau Basin. They are relatively poor in timber trees and wildlife. The forest is characterized by specialized plants such as *Nepenthes* sp., *Dischidia* sp., *Myrmecodia* sp., and *Hydnophytum* sp. (Jones et al. 1998).

Around 1200 m a.s.l or more, the dipterocarps give way to montane forests, found mostly in the Crocker Range, Mount Kinabalu, and also Maliau Basin. The lower montane forest (LMF) (1200–1500 m a.s.l.) also known as oak-laurel forest is characterized by its smoother canopy, between 15 and 33 m tall, and occasional emergent reaching 35 m. Trees of this forest are often with small buttresses and lack the extensive woody climbers of the lowland forests. It has a well-developed organic-rich topsoil and far less moss at ground level than the heath forest. Some of the common trees are *Shorea platyclados*, *Agathis borneensis*, and *Casuarina sumatrana*, while at the ground level, rattan (*Calamus* sp.), *Pandanus* sp., and ferns are common (Jones et al. 1998). Upper montane forest (above 1500 m a.s.l.) includes smaller, shrubby trees (canopy 15–18 m tall), including conifers, occasional emergent trees up to 26 m tall (Sabah Forestry Department 1989; Ashton 1995). The trees lack buttresses and are without climbers. Also, there is less undergrowth and an overall smaller number of species.

Other types of lowland forest that can be found in Sabah are riverine and limestone forests. Riverine dipterocarp forests are common along streams and flood lines. This forest varies in species composition but it is relatively poor in timber tree species. Some of the dipterocarp trees found here are *Dipterocarpus warburgii*,

D. exaltatus, *D. oblongifolius*, and *Shorea gysbertsiana* (Wood and Meijer 1964). Riverine forest is important in absorbing the effects of flooding, river bank protection, and reducing soil erosion into rivers from the bare hill slopes (Sabah Forestry Department 1989). Limestone flora found in east Sabah is rich in endemics, especially among herbaceous species (Ashton 1995).

Forests in Sabah have been a main resource for its wood and production and also food, shelter, and recreation. Under the new forest rules and regulations, the Forest Enactment 1968 and Forest Rules 1969 have been introduced for a better management (Sabah Forestry Department 1989). In accordance with the National Forestry Policy, Sabah now firmly pursues reforestation and rehabilitation of degraded or logged-over forests.

4.1.2 Termites of Sabah

Termites are well known for their high diversities in the tropical rainforest especially in Africa, South America, and Southeast Asia (Eggleton et al. 1994). However, their assemblage and species composition varies within similar ecosystems, as well as among different ecosystems. The distribution and dominance of these eusocial and polymorphic insects are strongly associated with their evolution and origin. The current differences in termite species composition could have resulted from isolation and the process of evolution, migration, and extinction against a backdrop of rising and falling sea levels. Termites' faunal compositions within the Indo-Malayan region tend to be similar. About 323 species from 52 genera have been recorded in the Indo-Malayan (Oriental) region (Tho 1992), and about 103 species (33 genera) have been recorded from Sabah (Thapa 1981). Tho (1992) mentioned that the termite faunas of Borneo and Sumatra are similar to that of West Malaysia (also known as Peninsular Malaysia). The termites' community of Sabah forests comprises three families of lower termite, viz., Kalotermitidae, Stylotermitidae, Rhinotermitidae, and a higher termite family, Termitidae. These families are further divided into subfamilies of which Termitidae is represented by four subfamilies, Amitermitinae, Termitinae, Macrotermitinae, and Nasutitermitinae.

Family Kalotermitidae

This group of termites, commonly known as dry-wood termites, nests inside dead branches of the forest trees. There are also some species from this family which live inside structural timber and even in furniture, identified as urban pests (Genet et al. 2001). In tropical countries some species set colonies in the trunks of dead standing trees. These dry-wood termites rely on symbiotic flagellate protozoa in their intestine, which are responsible for the wood cellulose digestion (Hill and Abang 2010). In Sabah, 14 species of 3 genera were recorded (Thapa 1981).

Family Stylotermitidae

This group of termites is among the smallest family recorded from Borneo. It is represented by only one species in Sabah, namely, *Stylotermes roonwali* (Thapa

1981). Krishna et al. (2013) have recorded two fossil genera which are *Parastylotermes* and *Prostylotermes*, with six fossil species. The genus of *Stylotermes* has a total of 45 species distributed in the Nearctic, Oriental, and Palaearctic regions.

Family Rhinotermitidae
Termites of this family, known as wet-wood termites, live underground and also have symbiotic intestinal microorganisms that allow the digestion of cellulose. The name "wet-wood termites" is referred to the eating habit of these termites, which feed on moist wood (Genet et al. 2001). This family has been identified to damage various tree crops, forest trees, rice, and vegetable crops. *Coptotermes curvignathus* is known as rubber tree termite, as it is a common species which attacks the rubber tree (*Hevea brasiliensis*). This family is available in the tropical rainforest of Borneo as it is renowned for its ability to attack healthy trees, of a wide range of species (Hill and Abang 2010).

Family Termitidae
Termitidae is the largest family of termites and are found throughout the tropics. Some species build mounds that can reach two to three meters in height (Genet et al. 2001). If a termite mound is found, it can be seen that the boarder bed is raised about 10–15 cm above ground level depending on location of the mounds. This is to avoid the worst effects due to periodical water logging. In some locations, the mound would be further raised to 20–30 cm above the bed (Hill and Abang 2010). The majority of termite species in this family are subterranean, which do not build conspicuous nests.

4.1.2.1 Termite Trophic Group

Termite feeding groups are derived from a recent Bornean study (Eggleton et al. 1999). Actual feeding group classification splits termites into four groups (I–IV) depending on their position along a humification gradient (Donovan et al. 2001). Groups III and IV are classified as "soil feeders," while groups I and II were combined and subsequently reassigned as "saprophages" (if predominantly litter feeders) or "wood feeders." However, to make the termite data directly comparable to the other taxa, an earlier substrate-based feeding group classification is referred. As for Southeast Asian tropical rainforest, the common functional groups are wood, soil, wood-soil interface, litter, microepiphyte, and fungus feeders (Collins 1989; Jones 1990; Eggleton et al. 1998).

Wood Feeders
Termites in this group feed on wood and woody litter, including dead branches of stems or tree trunks (Donovan et al. 2001). For some species, the feeding galleries within wood may act as a colony center or habitat for them (Wardle 1987; Eggleton et al. 1996, 1997). Many species in families Kalotermitidae, Rhinotermitidae, and Serritermitidae are wood feeders (Bignell and Eggleton 2000). Some of them, such

as *Coptotermes* sp. and *Microcerotermes* sp., could be a plantation pest that colonizes the rubble and oil palm trees, as well as wooden structures (Doi et al. 1997).

Soil Feeders
This is the most abundant group of termites in tropical rainforests. The main sources of foods for these termites are the minerals found in soil (Donovan et al. 2001). Soil and wood feeders are morphologically and behaviorally different, due to their different feeding activities (Brauman et al. 2000). Most soil feeders are represented by species in the subfamilies of Apicotermitinae, Termitinae, and Nasutitermitinae (Thapa 1981; Bignell and Eggleton 2000).

Soil-Wood Interface Feeders
Termites in this group feed in decaying wood which has become friable and soil-like or predominantly within soil under logs. They also feed on the soil plastered on the surface or inside of rotting logs or even mixed with leaf litter in stilt-root complexes. This group of termites can be found in the subfamilies of Termitinae, Apicotermitinae, and Nasutitermitinae. Most of their habitats are within dead logs, epigeal nests, or even colonized soil (Eggleton et al. 1997).

Litter Feeders
Termites in this group feed on leaf litters and small woody items of the forest floor. The litters fed by them mostly are in various stages of decay. The foods are often taken back to be temporarily stored in the nest. Litter feeders are found within the subfamilies of Macrotermitinae and Nasutitermitinae. They feed on the ground surface or in litter layers (Bignell et al. 1997; Eggleton et al. 1997).

Microepiphyte Feeders
Termites in this group consume moss, algae, fungi on tree barks, and lichens. *Hospitalitermes hospitalis* was found in Southeast Asia where it feeds on lichens (Jones and Gathorne-Hardy 1995; Eggleton et al. 1997).

Fungus Feeders
Also known as fungus-growing termites or fungus grower, they are known to feed on the fungi cultivated by themselves. Fungus-growing termites are only found in subfamily of Macrotermitinae. The tasks inside the mound, such as cultivating the fungus combs (also known as fungus garden) and tending the larvae and the reproductive queen, are conducted by the smaller workers which genetically are females. The worker termites usually collect grass, hay, straw, and vegetable litters. Food is processed collectively in the mound, using the symbiotic fungus *Termitomyces* sp. The termites feed either on the plant material or eat the fungi as a source of protein.

4.1.2.2 Termite Nesting Group

Basically nesting behavior can be organized into four different groups (Eggleton et al. 1997). These are wood nesters, hypogeal or subterranean nesters, epigeal mound builders, and arboreal nesters. Termite nesting behavior is closely related to their feeding habit. For example, wood feeders only nest on woods, because this is for them the main food supply used.

Wood Nesters
Termites in this group live in or around the standing trees or dead logs. Members of this group are also adapted to live in the wood carton or a woody substance, with low nutrient concentrations and high levels of lignin or other undigested components (Collins 1989). Termites in this group include Kalotermids, Rhinotermids, and also some members of Termitidae.

Hypogeal or Subterranean Nesters
The hypogeal or subterranean nesters are referred to termites whose colony centers are below ground without any indication of their presence (Wood and Johnson 1986). The hypogeal or subterranean termites use their excrement or a mixture of excrement and mineral soil to build their nest. In some Macrotermitinae, *Apicotermes* and *Homallotermes*, surface holes or a little internal structure is present together with their complex underground nests. The function of the hole is to enable the foragers to forage on aboveground vegetation.

Epigeal Mound Builders
This group of termites, commonly called as mound builders, refers to the termites whose colony centers are associated with living or dead woody plants above ground (Jones 1990; Eggleton et al. 1996). Epigeal mound structure can vary for different genera and also between regions within widely distributed species, showing different sizes and structures. Nest material includes subsoil with relatively low organic content, added with salivary secretion (*Macrotermes* and *Cornitermes*), wood carton for most wood feeders, and a mixture of feces and organic-rich topsoil, for most soil feeder. Some species of Macrotermitinae and *Odontotermes* build huge mounds using selected clay-rich subsoil (Wood and Johnson 1986).

Arboreal Nesters
Arboreal nests are referred to nests attached outside the trees at different heights (usually two meters above ground). These nests usually are made with wood carton (digested woody fecal and saliva) and usually connected to the ground with covered runways (Homathevi and Noel 2003). The nests vary in size and structure for the different genera or species.

4.2 Case Study from Sabah Forest

Termite studies in Sabah were designed to illuminate their assemblages and contribution to ecosystem processes by applying sampling techniques in relatively well-documented forest ecosystems. It was mainly aimed at characterizing termite assemblages in the primary forest and the effects of forest disturbance. In this context the latter is defined as changes in vegetation, soil, and microclimate due to logging activities at selected study sites and also due to forest conversion to plantations.

Rapid assessment biodiversity sampling techniques, employed in a large number of locations throughout the humid tropics (Eggleton et al. 1997; Jones and Eggleton 2000; Davies 2001), were applied to evaluate termite species richness and functional group composition, in a range of forest stands, ranging from lowland dipterocarp to hill montane forests, with altitudes of 50–1000 m a.s.l. The termites sampled were allocated into trophic and nesting group as suggested by Eggleton et al. (1997). Meanwhile, the taxonomic composition of the termite assemblages is treated as a possible pointer to ecological functions, reflected by forest disturbance.

4.2.1 Study Sites

Termite assemblages were studied from seven primary sites, located in lowland, hill dipterocarp, and lower montane forests, at altitudes ranging from 50 to 1000 m a.s.l. These included Danum Valley Conservation Area, Tabin Wildlife Reserve, Tawau Hills Park, Kabili-Sepilok Forest Reserve, Gunung Rara Forest Reserve, Sayap Substation, and Serinsim (Homathevi et al. 2002). All sampling sites within each forest type were located about 1–1.5 km away from any kind of human disturbance (i.e., trails, logging activities, or conversion to plantation). Sampling sites were chosen randomly, with the assumption that the selected sites represented an undisturbed primary forest. Two types of disturbed habitats, recently logged forest and regenerating forest within the Danum Valley Field Centre, were chosen (Eggleton et al. 1996; Fernandez 2015) to study the effect of disturbance on termite assemblages and distribution of wood-feeding termites.

4.2.2 Sampling

At each selected site, termites were sampled using protocol described by Eggleton et al. (1995, 1997). One or two transects, each of 100 × 2 m, were marked and further divided into 20 contiguous 5 × 2 m sections. Each section was sampled sequentially by two trained persons for 30 min per section (alternatively one person per hour). During the sampling, all known microhabitats such as surface soil, litter to a depth of about 5 cm, root mats, dead wood in all stages of decay, tree root systems and buttresses, runways on tree trunks and aboveground vegetation, and mounds/nests (subterranean, epigeal, and arboreal) were investigated for termites.

Termites were preserved in 80% alcohol and subsequently identified or assigned to morphospecies by referring to Ahmad (1976), Ahmad and Akhtar (1981), Thapa (1981), Collins (1984), and Tho (1992) or the accessed collections of the Natural History Museum (London, UK). Identification of workers was aided by examining the mandibles (Ahmad 1950) and gut structure (Johnson 1979). Identification of the worker caste is important, especially in soldier-less termites and genera in which soldiers are very rare, in order to get accurate estimates of species richness.

4.3 Termite Assemblages in Sabah Forest

The diversity of termites in Sabah was investigated based on qualitative data gathered from various primary forest sites. The highest species richness was observed in the lowland primary forest of Danum Valley and decreased with increasing elevation and with small (fragment) size (Homathevi et al. 2002). Differences of diversity at lower elevations were mainly due to forest or soil conditions. Variation in elevation (altitude, climate, and habitat area), soil types, competition, and food resources are some of the factors that may contribute to changes in termite assemblages. Other termite assemblage studies in Borneo include work by Collins (1980, 1983, 1984) in Gunung Mulu National Park (Sarawak) and by Eggleton et al. (1997, 1999) and Homathevi (1999) in Danum Valley Conservation Area (Sabah). Published work has mainly covered lowland forest, including lower and upper montane forests (Jones et al. 1998; Jones 2000).

4.3.1 Composition of Taxonomic Group

Termite distribution and dominance of any taxonomic groups are strongly associated with evolution and origin. In the case of Sabah forests, Termitidae was the most important family, with Termitinae (29 species) dominating the assemblage followed by Nasutitermitinae (27 species) and Macrotermitinae (15 species) (Thapa 1981; Homathevi et al. 2002). The highest numbers of species found in the primary forest were from subfamily Termitinae, with majority of the species being soil and soil-wood interface feeders. Numbers of species in this subfamily were almost similar in all forest types. The next most abundant group was the subfamily Nasutitermitinae, where most of the species were wood feeders, followed by fungus-growing Macrotermitinae.

The subfamilies Termitinae and Nasutitermitinae are common in most tropical forests. However, Termitinae is more dominant in Southeast Asian forests meanwhile Nasutitermitinae has more species in the Amazonian rainforest. In the Cameroon forests, the subfamily Apicotermitinae is dominant followed by Termitinae. As for the Apicotermitinae, it is believed to have originated from the Afrotropical region. The Macrotermitinae are less important in regions with low latitude and high rainfall and are absent from the Neotropical sites (Collins 1989).

As for lower termites, the family Rhinotermitidae is the most important family and very common in the tropical rainforest of southern and southeastern Asia. This family has greater number of species in the Oriental sites than in the Neotropical or Afrotropical. Some important genera recorded in large numbers in Malaysian forests as well as in Sabah are *Prorhinotermes*, *Parrhinotermes*, and *Schedorhinotermes*. Rhinotermitidae are also abundant in areas where there are plenty of damp logs.

4.3.2 Composition of Trophic Groups

Termite assemblages in the forest ecosystem comprised of five trophic groups, varying in abundance and species composition. Common at all study sites, wood and soil feeders dominated the assemblages compared to the other trophic groups and were equivalently better represented in the primary forest, compared to disturbed area, while the wood feeders dominate the disturbed habitat (Eggleton et al. 1997; Homathevi 1999; Jones 2000; Homathevi et al. 2002).

Soil feeders are common and dominant in most of the tropical forests, but they were poorly represented in the Southeast Asian rainforest. The proportion of species representing this group differs between biogeographical regions. In the Oriental region, e.g., the Southeast Asian tropical rainforest, soil feeders are dominated by the *Termes-Capritermes complex* (Homathevi 1999). Termites recorded in this group include *Mirocapritermes connectens*, *Pericapritermes* sp., *Procapritermes* sp., *Malaysiocapritermes* sp., *Oriencapritermes* sp., *Kemneritermes* sp., and *Dicuspiditermes* spp. Soil-wood interface feeders also had the highest abundance and are well represented in the primary forests. In many cases, this feeding group is represented by *Protohamitermes* sp., *Homallotermes* sp., *Prohamitermes mirabilis*, *Termes borneensis*, and *T. propinquus*.

Apart from the soil- and wood-feeding termites, other types of termite functional groups were also recorded. This includes litter-feeding termites such as *Havilanditermes atripennis*. This species does not harm or infest the tree; rather, it forages for food sources around the dead tree barks. In addition, termites are also known to be good wood decomposers. Other feeding group is wood-litter feeders, also known as fungus grower, commonly represented by *Macrotermes* sp. and *Odontotermes* sp. Microepiphyte feeders, usually represented by *Hospitalitermes* sp., are also common in primary forest at all study sites.

Termite assemblages and species composition vary consistently between ecosystems. Physical factors such as latitude (Eggleton et al. 1994), rainfall, altitude (termites are restricted at higher elevations (Bignell and Eggleton 1998)), temperature and climate (Collins 1989; Emerson 1955), soil organic matter, and disturbance (Wood 1996; Eggleton et al. 1997, 1998) have great influence in their distribution and assemblage structure. Regions with wetter climates and less variability in diurnal and seasonal changes (e.g., tropical and subtropical region forests) usually accommodate vast numbers of termite populations compared to arid savannahs and temperate regions (Abe and Matsumoto 1979; Collins 1989; Bignell and Eggleton 2000). In addition, some regional patterns of termite generic richness are closely related to historical factors such as the origin, evolution, dispersal ability, and geological history (Emerson 1955; Eggleton et al. 1994; Eggleton and Bignell 1995). Termites are poor fliers not able to disperse far from their nests, and this may result in high endemism in their own biogeographical region (Emerson 1955; Bignell and Eggleton 2000).

Knowledge on termite species dispersal patterns in different habitat types would be an aid to determine the size and degree of fragmented habitats in which populations can persist, as well as to know whether the fragmentation process is harmful, beneficial, or neutral for biodiversity and, by extension, to ecosystem function. Such informations contribute to our understanding and description of habitat degradation and land-use changes.

4.3.3 Composition of Nesting Groups

Nesting groups found at all sites were of four types: arboreal, wood, hypogeal, and epigeal. The hypogeal nesting group was the most abundant in the primary forest and also the most species-rich, compared to other nesting groups, and common in all forest types. Hypogeal nesters were mostly of subfamilies Termitinae, some Macrotermitinae, and soil-feeding Nasutitermitinae. Though there were more species of hypogeal nesters in regenerating logged forest, in comparison to primary forest, their abundance was relatively low. Arboreal nesters, represented mostly by *Glyptotermes* sp., *Nasutitermes* sp., and *Bulbitermes* sp., were abundant in the regenerating logged forest but not significantly different compared to primary forest. Wood nesters were also abundant in regenerating logged forest. A number of species of wood nesters were almost similar in all forest sites. Epigeal nesting termites, represented by *Dicuspiditermes* sp., were relatively more frequent in the primary forest. *Discuspiditermes* sp. is a type of soil-feeding group which builds towerlike nests around trees. It does not affect the growth of the tree as it foragers on soil around the standing tree. The arboreal, epigeal, and wood nests have conspicuous nest structures (Fig. 4.1) made of fecal, saliva, and soil or wood material, depending very much on the feeding behavior of the termites.

Fig. 4.1 Different types of nest structures: (**a**) arboreal nest belonging to *Bulbitermes* sp., (**b**) epigeal nest belonging to *Dicuspiditermes* sp., (**c**) wood nest belonging to *Microcerotermes* sp.

4.4 Effect of Forest Disturbance on Termite Assemblages

Numerous studies showed that termite assemblages are very sensitive to environmental variables (Eggleton et al. 1996; Jones and Prasetyo 2002; Jones et al. 2003). Gradual and sudden collapse is recorded in the termite assemblages after increased land-use intensification such as logging and conversion of natural forest ecosystems to plantations or subsistence crop fields (Chey 1996; Bignell et al. 1997; Eggleton et al. 1997, 1999; Luke et al. 2014). Forest disturbance, especially logging and plowing, affects the habitat and composition of termite species that occur in an area (Collins 1980; Wood et al. 1982; Lavelle et al. 1997). Such disturbance reduces many species and may promote others to pest status (Eggleton et al. 1996, 1997; Bignell and Eggleton 2000). Forest clearance is known to reduce termite abundance, biomass, and species richness, for example, as seen in dipterocarp forests in Sarawak, riparian forest in Nigeria, and M'balmayo forest in Cameroon (Collins 1980; Wood et al. 1982; Eggleton et al. 1995, 1996).

Assessment of the impact of forest disturbance on termite assemblages in Danum Valley showed a considerable overlap in species representation across the three (primary, young secondary, and old secondary) forest types. Although forest disturbance without clear-felling statistically did not affect termite abundance and biomass, there was tendency for changes in species composition of trophic and nesting groups, following the degree of disturbance. Soil feeders and soil-wood interface feeders were equivalently better represented in the primary forest, while wood feeders were more successful in disturbed forest sites. Species that are able to utilize patchy resources were preadapted to survive logging and may dominate the logged-over forest. Unless their resources are reduced too far, some of these species or guilds may disappear or increase in abundance.

Forest disturbance was also noted to promote the presence of certain taxonomic groups, which contribute to differences in nutrient contents and decomposition rates. Such example is the subfamily Macrotermitinae, being the most important decomposer in Malaysian forests. In certain circumstances, the species which are able to support the disturbance adapt themselves to the man-modified environment and, in absence of latent competitors, increase their abundance and become significant pests (Wood and Johnson 1986).

Forest disturbance may have direct impact on termite nesting habit by causing damage to the nests/mounds, soil compaction, and depletion to nesting sites (i.e., removal of potential trees or vegetation used as nesting sites) or indirectly by leaving the nest exposed to predators or natural disaster. The abundance of soil-inhabiting termites in this system could be highly influenced by soil disturbance, as lower abundance was observed in disturbed habitats. As most of the hypogeal nesters are soil feeders, the decrease in abundance of soil feeders reduces abundance in this nesting group. Soil compaction associated with logging reduces the activity of soil-nesting and soil-feeding types. In such areas, termite activities were completely absent in compacted soil, waterlogged area, and soil that were exposed to direct sunlight.

Selective logging may have a long-term continuous effect on the soil-dwelling colonies, without immediately causing them all to die. As the forest regenerates, the termite assemblage structure will change, depending on the severity of the disturbance. Though the colonies might recover as the forest regenerates, over time, some termite colonies might die and the mounds will fall into disrepair or occupied by inquilines.

Forest fragments in selectively logged forest or plantations may serve as a temporary reservoir for species undergoing habitat changes. However, not all populations isolated in high forest fragments can remain for many generations. Some evidence suggests that in forest fragmentation, abiotic factors (especially edge effects) lead to changes in insect population levels (Didham et al. 1996). Abiotic factors are known to have differential effects on different species, leading to important impacts on ecosystem functioning.

Under certain degrees of disturbance or other man-made conditions, some termite species may be able to withstand the disturbance and become pests, causing enormous amount of damage to plantations (e.g., trees such as rubber, pine, and teak), agriculture (crops such as tea and sugarcane), and timber structures, according to species, habitat, and feeding habits (Dhanarajan 1969; Collins 1988).

Also the absence of potential competitors may cause some species to increase their abundance and thus become significant pests (Wood and Johnson 1986). Though most of the damages are confined to tropical and subtropical regions, some pest species attain worldwide distribution mainly through human transportation, for example, species living in timbers used for buildings, furniture, transmission poles, and fencing (Bignell and Eggleton 2000).

4.5 Termite Infestation in Forest and Type of Damage Caused with Habitat Alteration

In general, the word "termite" refers to destruction and devastation, but surprisingly not all termite species are pest. Indeed, only a small number of species are responsible for this situation. Out of the 3000 living species of termites, only 371 (12.4%) have been identified as a pest species or considered as a serious threat (Krishna et al. 2013). Although termite pest species comprised of either a wood dwelling or a subterranean termite, not all wood dwelling or subterranean species are categorized as a pest. This was shown by the 23 species of termites which were sampled from the forest of Danum Valley, of which 17 were wood feeders (Fernandez 2015). In comparison to the termite pest species checklist by Krishna et al. (2013), only eight species (34.8%) are classified as termite pests. In comparison to the termite pest species present around the world, Danum Valley showed only 2.2% of the world's termite pests (Fernandez 2015). The nine remaining species do not cause mortality toward the tree although sometimes they can be found on living trees. Instead, these termites convert dead trees, dung, woody material, and soil material into organic matter (Yohannes 2006).

Survey on wood termites by Fernandez (2015) showed that pest species adapt well toward changes. In comparison to primary forest, the regenerated forest has more decaying material for wood-feeding termites. These conditions provide a suitable surrounding for the pest termite species to nest (Fernandez 2012). The termite pest species sampled from the forest of Danum Valley and logged forest belonged to five subfamilies: Coptotermitinae, Rhinotermitinae, Amitermitinae, Macrotermitinae, and Nasutitermitinae. The pests sampled included *Coptotermes curvignathus*, *Schedorhinotermes medioobscurus*, *Schedorhinotermes sarawakensis*, *Microcerotermes dubius*, *Macrotermes gilvus*, *Odontotermes javanicus*, and *Nasutitermes matangensis*. Sampled termite species were found on both living and dead trees (Fig. 4.2).

Well-established termite species in the forest of Southeast Asia are *Coptotermes curvignathus* and *Microcerotermes dubius* (Chey 1996; Speight and Wylie 2001). *Coptotermes curvignathus* is also known to have a competitive edge over other termite species especially in rubber plantation regions, as it is able to infest rubber trees (Yohannes 2006). According to Tho (1982), *M. dubius* is the cause for gap formation within the virgin lowland rainforest of Sungei Menyala and Pasoh. Another major damaging termite species belongs to the fungus-growing members of genera *Macrotermes* and *Odontotermes*. These termites depend on the fungus cultivated within the nest for digestion of their food (Abdurahman 2000). Some *Odontotermes* spp. occasionally feed upon the outer bark of living trees in Malaysia (Speight and Wylie 2001). Termites kill trees by damaging the tree bole and attacking the wood by consuming the heartwood. This would hollow the trunk, thus reducing the value of the tree as a timber source.

Tree species which were infested the most were in Dipterocarpaceae, *Shorea johorensis* (primary forest), and in Euphorbiaceae, *Macaranga gigantea* (regenerating forest) (Fernandez 2015). Both families of Dipterocarpaceae and Euphorbiaceae are dominant in the forest of Sabah. However, *Macaranga* is labeled

Fig. 4.2 Termite infestation on forest trees (Fernandez 2015)

as a pioneer genus and is mostly found regenerating in the secondary forest of Sabah (Kamaruzaman 1998). The termite species mostly infesting these two tree genera is *Schedorhinotermes sarawakensis* within the regenerated forest, a species that could adapt well toward changing environments, especially after logging. This species is also known to be a pest in cocoa plantations (Krishna et al. 2013).

The termite and tree distribution showed that the attacks of pest species are not dependent on the type of trees. In forest ecosystem and plantations, living trees are vulnerable to termite attacks. The termite species recorded on living trees includes *Coptotermes curvignathus*, *C. kalshoveni*, *Macrotermes gilvus*, *Microcerotermes dubius*, *Nasutitermes longinasus*, *N. matangensis*, and *S. sarawakensis*. In comparison to *Microcerotermes dubius*, *Macrotermes gilvus* and *C. curvignathus* are among the serious pests in Malaysia (Krishna et al. 2013). *Microcerotermes dubius* can be easily found in the forest as they build spiky mounds on the tree bark (Tho 1982; Fernandez 2015). *Coptotermes curvignathus* is known as a major pest species toward oil palm plantation as it attacks immature palms (Zulkefli 2007). This species has also been reported attacking *Acacia mangium*, rubber tree, and other fruit trees such as coconut and mango. This species damages fresh tissue rather than scavenging and feeding on woody material (Chan et al. 2011).

4.6 Pest Control Management in Oil Palm Plantation

The oil palm, *Elaeis guineensis* Jacq originating from Africa, is one of the crops that has been fully utilized and has become an important contributor to the economy of Malaysia. The oil palm development in Malaysia has been increasing over the years, and until 2006 it spread throughout the country with 2.3 million ha planted in Peninsular Malaysia, 1.2 million ha in Sabah, and 0.5 ha in Sarawak (Malaysian Palm Oil Board 2006). As any other crop, oil palm plantation also faces many pest problems and among the important pests there are the termites. The termite problem in oil palm plantation has been reported on mineral soils and later becomes a major problem on oil palms planted on peat (Wood 1968; Cheng et al. 2008). Since then several studies were carried out in treating and controlling this pest.

The common termite species found in oil palm plantations includes *Coptotermes curvignathus* and *Macrotermes gilvus* (Wong and Homathevi 2014). *Coptotermes curvignathus* is the major pest of oil palm known to destroy the palm leading to its death; meanwhile, *Macrotermes gilvus* is known to build mound attached to oil palm trees (Fig. 4.3), with a tendency to disturb the root system only (Sudharto et al. 1991).

Termite infestation on palm trees can be seen as early as 12 months after planted. The damage caused may lead to death of more than 5.3% of standing palms, over a 1-year period (Basri et al. 2003; Zulkefli 2007). At the early stages of planting, the infestation may reach palm petiole causing fatal damage to the newly planted palm (Zulkefli 2007). At this stage, termite detection is basically from their mud-work trail formed on the palm trunk, used to tunnel through the palm for the apical meristematic tissue.

Fig. 4.3 (**a**) Mound of *Macrotermes gilvus* attached to oil palm tree, (**b**) *M. gilvus* forming trails on bark of oil palm tree, (**c**) nest constructed on a living oil palm tree, and (**d**) newly constructed *M. gilvus* mound

Termite control in oil palm plantations has undergone several changes, and most of the methods are adopted from pest control approaches. The commonly used techniques are treating with chemicals such as organophosphate, carbamates, and synthetic pyrethroids (Langewald et al. 2003). The early methods used for termite control are either using repellent or fast-acting termiticides, with application modes based on the stages of infestation and palm age (Zulkefli 2007). The mud trails are scraped off prior to spraying, for better penetration of the termiticides. The latter are usually sprayed at the palm shoot, frond base, and trunk. Drenching and spraying with chlorpyrifos and fipronil are the most effective methods recognized by planters as the most effective chemicals to control termites (Zulkefli 2007). Meanwhile, to eliminate the existing termite colonies, a nonrepellent, slow-acting termiticide with baiting methods is used. For termite control in peat soil, it is suggested that high water table and good water management in the plantation sites can reduce the damage by *C. curvignathus* (Melling 1999).

4.7 Conclusion

The termite assemblage of primary forest stands in Sabah is relatively lower compared to any other biogeographical regions. The forest is mainly dominated by soil- and wood-feeding termites. Studies on their assemblages from different habitats showed a considerable overlap in species representation, across different forest types. Forest disturbance without clear-felling contributes to changes in species composition of trophic and nesting groups. Wood and litter feeders are less affected by disturbance and benefit from the major input of dead wood created by logging activities. Habitat degradation and land-use changes may promote termite pest species. Classified wood-feeding termites, namely, include *Microcerotermes dubius*, *Coptotermes curvignathus*, *Macrotermes gilvus*, and *Nasutitermes matangensis*, commonly known as pests in Malaysian forest. *Coptotermes curvignathus* and *M. gilvus* are also known to be notorious pests in oil palm plantations. Termite control practices in oil palm plantations include combination of chemical and water management (spraying and drenching) strategies.

References

Abdurahman, A. (2000). Termites as structural pests in Ethiopia. In *Termite biology and management*. Rpt. UNEP/FAO/Global IPM Facility Workshop 2000, Geneva. UNEP Chemicals, 16 pp.

Abe, T., & Matsumoto, T. (1979). Studies on the distribution and ecological role of termites in a lowland rain forest of West Malaysia. (3) Distribution and abundance of termites in Pasoh Forest reserve. *Japanese Journal Ecology, 29*, 337–351.

Ahmad, M. (1950). The phylogeny of termite genera based on imago-worker mandibles. *Bulletin of the American Museum of Natural History, 95*, 37–86.

Ahmad, M. (1976). The soldierless termite genera of the oriental region, with a note on their phylogeny (Isoptera: Termitidae). *Pakistan Journal of Zoology, 8*, 105–123.

Ahmad, M., & Akhtar, M. S. (1981). New termite genera of the *Capritermes complex* from Malaysia, with a note on the status of *Pseudocapritermes* (Isoptera: Termitidae). *Pakistan Journal of Zoology, 13*, 1–21.

Ashton, P. S. (1995). Biogeography and ecology. In E. Soepadmo & K. M. Wong (Eds.), *Tree flora of Sabah and Sarawak* (Vol. 1, pp. XLIII–XLILI). Kuala Lumpur: Forest Research Institute Malaysia Kepong.

Basri, M. W., Norman, K., Idris, A. S., Ariffin, D., Shamala, S., Ramle, M., & Ramlah, S. A. A. (2003). *Hand book of pests and diseases of oil palm* (pp. 5–7). Malaysia Palm Oil Board.

Bignell, D. E., & Eggleton, P. (1998). Termites. In P. P. Calow (Ed.), *Encyclopedia of ecology and environmental management* (pp. 744–746). New York: Wiley-Blackwell.

Bignell, D. E., & Eggleton, P. (2000). Termites in ecosystems. In T. Abe, D. E. Bignell, & M. Higashi (Eds.), *Termites: Evolution, sociality, symbiosis, ecology* (pp. 363–387). Dordrecht: Kluwer Academic Publishers.

Bignell, D. E., Eggleton, P., Nunes, L., & Thomas, K. L. (1997). Termites as mediators of carbon fluxes in tropical forest: Budgets for carbon dioxide and methane emissions. In A. D. Watt, N. E. Stork, & M. D. Hunter (Eds.), *Forest and insects* (pp. 109–134). London: Chapman and Hall.

Brauman, A., Fall, S., & Chotte, J. L. (2000). Caracteristiques organiques, physique et microbiologique du sol soumis a l'influence des termites. Etude comparative de deux especes dominantes

des sols en jachere (Haute Casamance, Senegal). In F. Ch & R. Pontanier (Eds.), *La Jachere en Afrique tropical* (pp. 308–316). Paris: John Libbey Eurotext.
Chan, S. P., Bong, C. F. J., & Lau, W. H. (2011). Damage patterns and nesting characteristics of *Coptotermes curvignathus* (Isoptera: Rhinotermitidae) in oil palm on peat. *American Journal of Applied Sciences, 8*, 420–427.
Cheng, S., Kirton, L. G., & Gurmit, S. (2008). Termite attack on oil palm grown on peat soil: identification of Pest species and factors contributing to the problem. *The Planter, Kuala Lumpur, 84*(991), 200–210.
Chey, V. K. (1996). *Forest pest in Sabah*. Sandakan Sabah: Sabah Forestry Department, Forest Research Centre. 110 pp.
Collins, N. M. (1980). The effect of logging on termite (Isoptera) diversity and decomposition processes in lowland dipterocarp forests. *Tropical Development Ecology, 198*, 113–121.
Collins, N. M. (1983). Termite populations and their role in litter removal in Malaysian rain forests. In S. L. Sutton, T. C. Whitmore, & A. C. Chadwick (Eds.), *Tropical rain forest: Ecology and management* (pp. 311–325). Oxford: Blackwell Scientific Publications.
Collins, N. M. (1984). The termites (Isoptera) of the Gunung Mulu National Park, with a key to the genera known from Sarawak. *Sarawak Museum Journal, 30*, 65–87.
Collins, N. M. (1988). Termites. In E. Cranbrook (Ed.), *Key environments Malaysia* (pp. 196–211). Oxford: Pergamon Press.
Collins, N. M. (1989). Termites. In H. Lieth & M. J. A. Werger (Eds.), *Tropical rain forest ecosystems* (pp. 455–471). Amsterdam: Elsevier Science Publishers.
Davies, R. G. (2001). *Patterns of termite functional diversity: From local ecology to continental history*. Ph.D thesis, University of London, 238 pp.
Davison, G. W. H., & Fook, C. Y. (1996). *A photographic guide to birds of Borneo*. London: New Holland Publishers Ltd. 144 pp.
Dhanarajan, G. (1969). The termite fauna of Malaya and its economic significance. *Malay Forester, 32*, 274–278.
Didham, R. K., Ghazoul, J., Stork, N. E., & Davis, A. J. (1996). Insects in fragmented forests: A functional approach. *Trends in Ecology & Evolution, 11*, 255–260.
Doi, S., Kurimoto, Y., Takahashi, M., & Yoshimura, T. (1997). *Effects of steaming heat treatment on the stimulation of termite feeding*. International Research Group on Wood Preservation, Doc No IRG/WP 97–10212.
Donovan, S. E., Eggleton, P., & Bignell, D. E. (2001). Gut content analysis and a new feeding group classification of termites (Isoptera). *Ecological Entomology, 26*, 356–366.
Eggleton, P., & Bignell, D. E. (1995). Monitoring the responses of tropical insects to changes in the environment: Troubles with termites. In R. Harrington & N. E. Stork (Eds.), *Insects in a changing environment* (pp. 434–497). London: Academic.
Eggleton, P., Williams, P. H., & Gaston, K. J. (1994). Explaining global termite diversity: productivity or history? *Biodiversity and Conservation, 3*, 318–330.
Eggleton, P. E., Bignell, D. E., Sands, W. A., Waite, B., Wood, T. G., & Lawton, J. H. (1995). The species richness of termites (Isoptera) under differing levels of forest disturbance in the Mbalmayo Forest Reserve, Southern Cameroon. *Journal of Tropical Ecology, 11*, 85–98.
Eggleton, P., Bignell, D. E., Sands, W. A., Mawdsley, N. A., Lawton, J. H., Wood, T. G., & Bignell, N. C. (1996). The diversity, abundance and biomass of termites under differing levels of disturbance in the Mbalmayo Forest Reserve, southern Cameroon. *Philosophical Transactions of the Royal Society of London. Series B, Biological Sciences, 351*, 51–68.
Eggleton, P., Homathevi, R., Jeeva, D., Jones, D. T., Davies, R. G., & Maryati, M. (1997). The species richness and composition of termites (Isoptera) in primary and regenerating lowland dipterocarp forest in Sabah, East Malaysia. *Ecotropica, 3*, 119–128.
Eggleton, P., Davies, R. G. D., & Bignell, D. E. (1998). Body size and energy use in termites (Isoptera): The responses of soil feeders and wood feeders differ in a tropical forest assemblage. *Oikos, 81*, 525–530.

Eggleton, P., Homathevi, R., Jones, D. T., MacDonald, J. A., Jeeva, D., Davies, R. G., Bignell, D. E., & Maryati, M. (1999). Termite assemblages, forest disturbance, and greenhouse gas fluxes in Sabah, East Malaysia. *Philosophical Transactions of the Royal Society of London. Series B, Biological Sciences, 354*, 1791–1802.

Emerson, A. E. (1955). Geographical origins and dispersions of termite genera. *Fieldiana Zool, 37*, 465–521.

Fernandez, K. (2012). *Comparison of termite fauna between primary forest and regenerated forest in Deramakot Forest Reserve. Kota Kinabalu Malaysia*. B.Sc. dissertation, School of International Tropical Forestry, Universiti Malaysia, Sabah, 97 pp.

Fernandez, K. (2015). *Distribution of wood feeding termites in the primary and regenerating forests of Danum Valley, Lahad Datu, Sabah*. M.Sc. dissertation, Institute for Tropical Biology and Conservation, Universiti Malaysia Sabah, 120 pp.

Genet, J. A., Genet, K. S., Burton, T. M., Murphy, P. G., & Lugo, A. E. (2001). Response of termite community and wood decomposition rates to habitat fragmentation in a subtropical dry forest. *Tropical Ecology, 42*, 35–49.

Hill, D. S., & Abang, F. (2010). *The insects of Borneo* (pp. 88–95). Kota Samarahan: University Malaysia Sarawak.

Homathevi, R. (1999). Diversity and ecology of forest termite (Isoptera) populations in Sabah, East Malaysia, with special reference to the *Termes-Capritermes-clade* (pp. 1–335). PhD. thesis, School of Science and Technology, Universiti Malaysia Sabah, Kota Kinabalu.

Homathevi, R., & Noel, T. (2003). *Isoptera (Termite): Inventory and collection, introduction to entomology* (pp. 121–133). Kota Kinabalu: Universiti Malaysia Sabah.

Homathevi, R., Bakhtiar, Y., Mahadimenakbar, D., Maryati, M., Jones, D. T., & Bignell, D. E. (2002). A comparison of Termite (Insecta: Isoptera) assemblages in six primary Forest stands in Sabah, Malaysia. *Malayan Nature Journal, 56*, 225–237.

Inger, R. F., & Stuebing, R. B. (1989). *Frogs of Sabah*. Kota Kinabalu: Sabah Parks Publication. 129 pp.

Johnson, R. A. (1979). Configuration of the digestive tube as an aid to identification of worker Termitidae (Isoptera). *Systematic Entomology, 4*, 31–38.

Jones, J. A. (1990). Termites, soil fertility and carbon cycling in dry tropical Africa: a hypothesis. *Journal of Tropical Ecology, 6*, 291–305.

Jones, D. T. (2000). Termite assemblages in two distinct montane forest types at 1000 m elevation in Maliau Basin, Sabah. *Journal of Tropical Ecology, 16*, 271–286.

Jones, D. T., & Eggleton, P. (2000). Sampling termite species assemblages in tropical forests: Testing a rapid biodiversity assessment protocol. *Journal of Applied Ecology, 37*, 191–203.

Jones, D. T., & Gathorne-Hardy, F. (1995). Foraging activity of the processional termite *Hospitalitermes hospitalis* (Termitidae: Nasutitermitinae) in the rain forest of Brunei, north-west Borneo. *Insectes Sociaux, 42*, 359–369.

Jones, D. T., & Prasetyo, A. H. (2002). A survey of the termites (Insecta: Isoptera) of Tabalong District, South Kalimantan, Indonesia. *The Raffles Bulletin of Zoology, 50*, 117–128.

Jones, D. T., Tan, J., & Bakhtiar, Y. (1998). The termites (Insecta: Isoptera) of the Maliau Basin, Sabah. In M. Maryati, S. Waidi, & A. Anton (Eds.), *Monograph of the Maliau Basin scientific expedition biological monograph, no 1* (pp. 95–112). Kota Kinabalu: Universiti Malaysia Sabah.

Jones, D. T., Susilo, F. X., Bignell, D. E., Hardiwinoto, S., Gillison, A. N., & Eggleton, P. (2003). Termite assemblage collapse along a land-use intensification gradient in lowland central Sumatra, Indonesia. *Journal of Applied Ecology, 40*, 380–391.

Kamaruzaman, J. (1998, August). Forest resource inventory assessment in Gunung Rara Forest Reserve, Sabah, using stratified field sampling. In *Integrated tools proceedings boise* (pp. 86–93). Idaho, USA.

Krishna, K., Grimaldi, D. A., Krishna, V., & Engel, M. S. (2013). Treatise on the Isoptera of the world. *Bulletin of the American Museum of Natural History, 377*, 2704.

Langewald, J., Mitchell, J. D., Maniania, N. K., & Kooyman, C. (2003). Microbial control of termite in Africa. In P. C. Neuenschwander, C. Borgemeister, & J. Langewald (Eds.), *Biological control IPM system in Africa* (pp. 227–242). Wallingford: CABI Publishing.

Lavelle, P., Bignell, D., Lepage, M., Wolters, V., Roger, P., Ineson, P., Heal, O. W., & Dhillion, S. (1997). Soil function in a changing world: the role of invertebrate ecosystem engineers. *European Journal of Soil Biology, 33*, 159–193.

Luke, S. H., Fayle, T. M., Eggleton, P., Turner, E., & Davies, R. G. (2014). Functional structure of ant and termite assemblages in old growth forest, logged forest and oil palm plantation in Malaysian Borneo. *Biodiversity and Conservation, 23*, 2817–2832.

Malaysian Palm Oil Board. (2006). Malaysian oil palm statistic 2005, 25th edn. Economic and industry development division. Kelana Jaya Malaysian palm oil board

Marsh, C. W., & Greer, A. G. (1992). Forest land-use in Sabah, Malaysia: An introduction to Danum Valley. *Philosophical Transactions of the Royal Society of London. Series B, Biological Sciences, 335*, 331–339.

Melling, L. (1999). Sustainable agriculture development on peatland. In *Workshop on working towards integrated peatland management for sustainable development* (pp. 17–18).

Payne, J., Francis, C. M., & Philips, K. (1985). *A field guide to the mammals of Borneo*. Kota Kinabalu: The Sabah Society with WWF Malaysia. 332 pp.

Sabah Forestry Department. (1989). *Forestry in Sabah*. Sandakan: Forestry Department of Sabah. 164 pp.

Soepadmo, E. (1995). Background to the tree flora of Sabah and Sarawak project. In E. Soepadmo & K. M. Wong (Eds.), *Tree flora of Sabah and Sarawak* (Vol. 1, pp. XIII–XXIX). Kuala Lumpur: Forest Research Institute Malaysia.

Speight, M. R., & Wylie, F. R. (2001). *Insect pests in tropical forestry* (p. 307). Wallingford: CABI.

Stuebing, R. B. (1991). Checklist of snakes of Borneo. *The Raffles Bulletin of Zoology, 39*, 323–362.

Sudharto, P. S., Sipayung, A., & Desmeir de Chenon, R. (1991, September 9–14). Termites – A new problem on oil palm plantation in Indonesia. In *Proceedings of the PORIM international palm oil conference: Progress, prospect & challenges towards the 21st century* (pp. 407–417), Kuala Lumpur.

Thapa, R. S. (1981). Termites of Sabah. *Sabah Forest Record, 12*, 1–374.

Tho, Y. P. (1982). Gap formation by the termite *Microcerotermes dubius* in lowland forest of Peninsular Malaysia. *Malaysian Forester, 45*, 184–192.

Tho, Y. P. (1992). *Termites of peninsular Malaysia, Malayan Forest Record No 36*. Kepong/Kuala Lumpur: Forest Research Institute Malaysia. 224 pp.

Wardle, D. A. (1987). Control of termites in nurseries and young plantations in Africa: Established practices and alternative courses of action. *Commonw For Rev, 66*, 77–89.

Whitmore, T. C. (1998). *An introduction to tropical rain forests* (2nd ed.). Oxford/New York: Oxford University Press. 296 pp.

Wong, M. K., & Homathevi, R. (2014). The spatial distribution pattern of soil-dwelling termites in primary forest and oil palm plantation in Sabah, Malaysia. In *The 20th world congress of soil science proceedings* (pp. 310–311)

Wood, B. J. (1968). *Pest of oil palms in Malaysia and their control*. Kuala Lumpur: Incorporated Society of Planter. 204 pp.

Wood, T. G. (1996). The agricultural importance of termites in the tropics. *Agricultural Zoology Reviews, 7*, 117–155.

Wood, T. G., & Johnson, R. A. (1986). The biology, physiology and ecology of termites. In S. B. Vinson (Ed.), *The economic impact and control of social insects* (pp. 1–68). New York: Praeger publication.

Wood, G. H. S., & Meijer, W. (1964). *Dipterocarps of Sabah (North Borneo), Sabah Forest Record No. 5*. Sandakan: Forest Department. 344 pp.

Wood, T. G., Johnson, R. A., Bacchus, S., Shittu, M. O., & Anderson, J. M. (1982). Abundance and distribution of termites (Isoptera) in a Riparian forest in the Southern Guinea savanna vegetation zone of Nigeria. *Biotropica, 14*, 25–39.

Yohannes, G. (2006). *Evaluation of termite resistant plant attributes for their bioactivities against macrotermes termite*. M.Sc. thesis, Biology Department, Addis Ababa University, 62 pp.

Zulkefli, M. (2007). Termite population and the effect of different water table on *Coptotermes curvignathus* in peat soil (pp. 1–73). Institute for Tropical Biology and Conservation, Universiti Malaysia Sabah.

Chapter 5
Integrated Termite Management in the Context of Indoor and Outdoor Pest Situation

G.K. Mahapatro and Debajyoti Chatterjee

Contents

G.K. Mahapatro (✉) • D. Chatterjee
Division of Entomology, ICAR – Indian Agricultural Research Institute,
New Delhi 110012, India
e-mail: gagan_gk@rediffmail.com

M.A. Khan, W. Ahmad (eds.), *Termites and Sustainable Management*,
Sustainability in Plant and Crop Protection,
https://doi.org/10.1007/978-3-319-68726-1_5

Abstract With their well-known role indoor and outdoor, termites attract attention in plant protection. Herein, attempt is made to collect, compile and collate the information on termites, viz. categorization of termites as per feeding habit, role as outdoor pest particularly in agri-horticultural systems, their status as indoor pests, economic loss to buildings and constructions, suitable steps for sustainable management and various available options in indoor and outdoor scenario such as chemical, cultural, mound treatment and green management approach. Special emphasis is given on 3B technologies (borate, baiting and barrier) along with indigenous traditional knowledge (ITK) for termite management. We strongly advocate concerted need for formulating situation-specific package, judicious utilization of chemicals inclusive of frontier technologies and ITKs in adopting smart strategies under the aegis of integrated termite management (ITM).

Keywords Barrier • Baiting • Borate • Biocontrol • ITK

5.1 Introduction

Termites are the most important component of the living world accounting for about 10% of animal biomass in tropics and subtropics (Donovan et al. 2007). They are referred as true soil engineers owing to their immense contribution in soil-building activity (Black and Okwakol 1997; Donovan et al. 2001a; Dawes 2010). Despite this fact, they are often convicted as culprit for destruction of crops and human-made structures, especially those manufactured by wood or wooden materials. Destruction by termites includes crop losses, damage to various structures, standing trees and sometimes devouring forests (Ravan et al. 2015). Cite-worthy examples of termite attack and effects could be seen in India, where villages were abandoned by their inhabitants (Mahapatro et al. 2014).

Starting their evolution probably from the upper Jurassic times, the termites reasonably diversified by the Cretaceous era (Barbara et al. 2000). Latest literature reveals records of about 3138 species (fossil and living) worldwide (Mahapatro et al. 2015). Generic endemism of termites showed a vibrant spectrum of distribution, skewed more towards tropics than the temperate zones. Local species richness and regional generic richness were found greatly predisposed by respective climate, altitude and habitat, showing greater diversity of termites in the lowland tropical forests and decreasing towards the temperate areas (Eggleton 2000). Generic endemism showed highest in the case of Neotropics (78%) followed by Afrotropical (77%), Australasia (64%), Oriental (56%), Nearctic (29%), Palaearctic (21%), Malagasy (20%), Papuan (18%) and virtually to zero per cent in Oceania region (Eggleton 2000). This gives a clear picture of how termites have spread their diversity worldwide.

5.2 Classification of Termites as per Feeding Habit

Termites are best known for their extreme capacity of feeding on cellulose materials, due to which they acquired the major, present pest status. Besides taxonomic classification, it is therefore obvious to gather knowledge about their feeding habits and feeding groups of termites. Donovan et al. (2001b) depicted a fabulous framework of termite feeding groups, based on gut content analysis. Termites were grouped into four feeding groups; amongst which, Group I contained lower termites feeding on dead wood and grass; Group II contained Termitidae with an array of feeding pattern on dead wood, grass, leaf litter and micro-epiphytes [within Group II, subgroup II-F consisted termites which feed on grass, dead wood and leaf litter with the help of mandatory fungal mutualism (Termitomyces) in the fungal garden]. Group III included Termitidae devouring on the organic-rich upper soil layer, and Group IV included true soil feeders consuming on mineral soil. This feeding habit-based classification of termites could help largely to distinguish the bioecology of a species belonging to concerned feeding type, which in turn is useful to formulate appropriate management strategies.

5.3 Introduction to Integrated Termite Management

Integrated management of termites was addressed since the beginning of the twentieth century (Snyder 1927, 1935; Brown et al. 1934; Hartnack 1943). In order to develop a sustainable model of integrated termite management (ITM), its principal components are to be coherently active, i.e. proper communication between stakeholders, adequate biological information about the target termites and systematic inspection of affected area, implying correct action plan and following up the entire process meticulously (Forschler 2011). The idea of managing termites under an IPM view was initially raised by Su and Scheffrahn (1998). Their rationale assumed that structural or landscape alteration is not a feasible intervention for termite management, but whole house chemical barriers could be able to provide holistic protection. Forschler et al. (2007) suggested ITM using an IPM view should be an ongoing programme, with emphasis given in identification of site factors which could be altered to reduce termites' activity. ITM should be designed with full knowledge of the target species and target site, not only by killing the termite population but also driving them away from the target site by influencing a change in their behaviour (Forschler 2011). Success of ITM programme largely depends upon a holistic knowledge-based systematic action plan. This derives from intense inspections to deal, in a holistic perspective, with the information on the bioecology of termites. So far the major research and development of ITM focused on indoor management of termites, but as such scanty or no scientific data are available on ITM for management of termites as outdoor pest.

5.4 Termite as Outdoor Pest

5.4.1 In Agricultural and Horticultural Ecosystem

The comprehensive Indian termite fauna shares a small proportion that presently accounts for 261 species (Mahapatro et al. 2014), against the total 3138 world species (including living and fossil species). Despite their well-known role as pests, termites attracted less attention of researcher in India. Out of seven families known from India, the Termitidae (higher termite family) comprises most of the species that are a threat to agriculture. Mostly, *Microtermes* and *Odontotermes* species ravage in field situations. *Heterotermes indicola* and *Coptotermes heimi* have some agro-horticultural importance but limited to protected cultivation (except *H. indicola* in bitter gourd). Major crops affected by termites in the open field are cereals (upland rice, irrigated rice in rice-wheat system, maize, millets and barley), pulses (soybean, pigeon pea, chickpea, mung), oilseeds (groundnut), vegetables (brinjal, potato, root crops), ornamentals (rose, chrysanthemum, marigold), tree crops (mango, banana, citrus, coconut), etc.

Various species build large mounds under/above ground often containing hundreds of thousands of individuals. Termites construct shallow subterranean foraging galleries radiating from the nest for distance of up to 50 m. The main galleries ramify to a network of smaller galleries from which foraging workers exploit potential food resources over extensive areas. Termites forage directly on underground plant material. Seedlings are either cut just below or above the soil surface. In the latter case, termites gain access from soil-covered galleries impinging on the base of the plant. Usually, the seedlings are completely severed, resulting in lowered stands (e.g. wheat, maize).

However, damages to maturing plants are largely caused by species that have entirely subterranean nests consisting of a diffuse network of galleries and chambers. These species enter and consume the root system, which directly kills the plant or indirectly lowers yields through decreased water and nutrient translocation. The attack to the root system can also lead to increased susceptibility to pathogens, or lodging of mature plants. When the grain in lodged plants touches the ground, soil fungi such as *Aspergillus* spp. may invade it (e.g. groundnut). Termite damage to stored products generally results in invasion by *Aspergillus* spp., which causes indirect yield losses and contaminates productions with aflatoxins.

5.4.2 In Lawn

Yellow patches in the lawn are the general symptom of termite infestations. Sometimes, dog urine or other factors may also cause similar symptoms. Lawn grasses/turf can be uprooted to see the live termites to confirm. Undoubtedly, insecticide is the rapid and most effective method of control. In lawn periphery, termite-infested trees can be treated as mentioned earlier (trunk and/or basin treatment). Western nations follow drilling and using a foam-based termiticide or installing bait stations. The objective is to use least amount of chemicals. Slow-acting methods such as bait systems or biological control may be effective. Biological control methods involve the introduction of insects (ants) or entomopathogenic nematodes (EPNs) that prey on termites. While both ants and EPNs are natural predators for termites, they do not offer as effective prevention or control as bait or termiticide treatment systems. Few plants with insecticidal properties, such as sweet basil, turmeric, *Aloe vera*, *Calotropis*, etc., may help in protecting the lawn organically. In addition, vegetation and mulch should be kept at least 6 inches away from boundary/walls of the lawn to reduce food, shelter and moisture sources for termites (Mahapatro 2016).

5.5 Termite as Indoor Pest

5.5.1 Economic Loss Due to Termite Attack on Buildings and Constructions

Year after year humankind suffers substantial losses due to termite attacks on various buildings and constructions, which in other way retard the advancement of civilization. Initial report of termite occurrence goes back to the eighteenth century at the time of publication of 'Systema Naturae' by the legendary Carolus Linnaeus, who enlisted three species of termites in his book, including *Odontotermes*. Owing to their huge capacity of infesting buildings and constructions, Macrotermitinae and Rhinotermitinae became well-known pests in urban and rural areas (Sornnuwat 1996; Kirton and Azmi 2005). In a latest literature, it was estimated that the world suffered an annual loss of 40 billion USD, owing to the attack of termites on various constructions (Liston 2015). As documented by the United Nation Environment Program (UNEP) in finding alternatives to persistent organic pollutants (POPs) for termite management, African termite fauna was found to be more than 1000 species. Comparatively, a lower record was found for North America, estimating less than 50 species including indoor and outdoor pests and general termite fauna, even though their damage reached a toll of estimated 1.5 billion USD per year in the southern part of the USA (Ibrahim and Adebote 2012). South American termite fauna is a little richer, with around 400 species reported. European termite fauna was found to be meagre as there is almost no/less report of termite as agricultural pest from this part of the world. Australian termite fauna was found to be richer, estimating more than 360 species known from this region as by the UNEP report. An economic loss of more than 100 million Australian dollars per year was estimated by Scholz et al. (2010), due to termite attack on structures.

A vivid scientific report on Asian termites and their impact as structural pest is available which clearly shows the impact of these tiny giants on the human life and economy of various countries. As many as 13 species of termites were reported from

Pakistan as structural pests, which were found to be the causative agent of great economic losses every year (Manzoor and Mir 2010). As many as 16 species were found associated with structural damage in Iran (Ravan 2010). China reported a huge loss due to attacks by more than 435 species of termites, as reported by UNEP, including agricultural and structural pests and general termite fauna of the country. According to Zhong and Liu (2002), 80–90% damage of buildings were reported from China causing economic harm up to 1700–2000 million RMB per year. About 90% of households situated south to the Yangtze river were attacked by termites (GEI 2005; MRP 2010), with around 95% of buildings in Hong Kong infested by *Coptotermes formosanus* (Tai and Chen 2002). In Thailand the damage due to termites was severe with as many as 13 species associated with structural damage (Sornnuwat 1996) In Malaysia, five common genera of subterranean termites were reported to cause massive economic damage, ranging up to 8–12 million USD during the years 2000 and 2003, respectively (Lee 2002; Yeoh and Lee 2007). In a striking report from Singapore and Malaysia, seven genera of termites were found to attack various buildings and constructions in urban and semi-urban areas causing huge economic losses (Lee et al. 2007). In Taiwan, most important historical constructions were reported to be affected by *Coptotermes* sp. resulting in a great economic loss, estimated over 3–4 million USD per year (Li et al. 2011). Far East, Japan, was found to be seriously economically hit by termite attacks, causing damage of about 1 billion USD (Takahashi and Yoshimura 2002). In India, termites affect the economy of estimating a damage of around 280 million Indian rupees (Rajagopal 2002). Latest incident reported losses of almost 10 million Indian rupees (approx. 225,000 USD) which these insects chewed within an Indian bank (BBC News 26th April, 2011). The worst ever instance of termite damage in India concerned three villages which were totally abandoned by residents due to severe attacks. Long back in the 1950s, villages in Sri Hargobindpur of Punjab state, and recently Jhopadiya village of Rajasthan and Lambari village of Uttarakhand in India, were found to be attacked so severely by *Heterotermes indicola*, turning the entire victimized villages into ghost towns (Mahapatro and Kumar 2013; Mahapatro et al. 2014).

5.5.2 Structural Pest for Other Constructions

Besides buildings built for residential and other purposes, termite attack occurs in all sorts of human-made constructions where wood and/or cellulose-containing materials are used as a structural component. Devastation included banks, libraries, schools, temples and ancient heritages, museums, dams and dykes and bridges. Even communication devices such as network cables, etc. were not spared. Everywhere the impact of damage was found equally outsized. In tropical countries, drywood and subterranean termites were found responsible for attacking many archives, libraries and museums where books, furnitures and such items were displayed as chief exhibits (Pinniger 2012). Attacks by a few genera of subterranean termites, namely, *Reticulitermes*, *Coptotermes* and *Macrotermes*, were even found through soil and trees. In India, attack of termites in libraries and their counter measurement were reported by Roonwal (1979). Cultural heritages of many countries around the world were also reported to come under termite attack, leading to major economic and cultural losses. The giant foot of these tiny terminators could not spare national parks, landscapes and avenue trees too. Formosan subterranean termites were reported to attack the old oak trees of the beautiful avenues of New Orleans in the USA, leading to their destruction (Gilberg and Su 2012). From different corners of the world, many incidences of termite attack on dams and dykes were reported. In Vietnam, 64 species belonging to four families, mostly of Termitidae, were recorded as destructive at dam sites (Tuyen 2006). From India, Hirakud dam of Odisha on river Mahanadi was also reported to be infested by termites (Hoon 1960).

5.6 Steps Towards Sustainable Management of Termite

5.6.1 Correct Taxonomy for Pest Management

Success of pest management and formulation of sustainable pest control strategy largely depends upon correct identification of the targeted pest species, a task often challenging. As an example, solving the paradox on the identification of *Coptotermes gestroi* in the Southeast Asian region, Kirton (2005) concluded that (1) the species referred to as *C. havilandi* in parts of North and South America and the Caribbean, (2) the species referred to as *C. gestroi* and *C. havilandi* in Thailand and (3) the species wrongly identified as *C. travians* in Peninsular Malaysia are in fact one and the same species and should be known as *C. gestroi*, the Asian subterranean termite. The species once wrongly referred to as *C. havilandi* in Peninsular Malaysia should be called *C. travians*.

Uncertainty about correct identification of two more important species such as *C. vastator* in the Philippines and *C. heimi* in India was also reported by Kirton (2005). This essentially indicates the need of a trustworthy taxonomy in case of pest management. In Delhi and NCR, *Heterotermes indicola* was described as a serious structural pest by Mahapatro and Kumar (2013), though the pest was recorded long before in the early 1900s, pointing the lack of concerted attempts in taxonomic studies.

5.6.2 Management (Common Methods) in Indoor and Outdoor Attack

5.6.2.1 Chemical Approach

With enhancement of technology and population explosion, supply and demand of food stood as a challenge to humankind. Eventually, crop protection emerged as a major issue in the arena of agricultural production. Application of more effective remedies to reduce crop losses by insect attack and mitigation of food security could lead towards the invention of new pest control methods. The use of chemicals in pest management took its place with advancement of agrichemical techniques. Despite their harmful effect, chemical pesticides are an established practice in agriculture. With time and situation, changes have taken place in the use of pesticides, their concentration and sustainable use. Awareness of their dose, limit, application time and hazards was brought to the notice of people for greater interest, to protect life on earth. In India, Insecticide Act, 1968 has specified crop-wise use of chemical pesticides and their used pattern. Similarly, such efforts were seen in many countries abroad. Under leadership by the United Nations Environment Program, members of the UNEP/FAO/Global IPM Facility Expert Group on Termite Biology and Management (established in 2000) recommended some chemicals in support of international activities on persistent organic pollutants (POPs), in which as many as seven chemicals were classified as non-repellents and 11 were grouped into repellents (http://www.unep.org/chemicalsandwaste/Portals/9/Pesticides/Alternatives-termite-fulldocument.pdf). Repellents included bifenthrin, which was found as a sodium ion channel inhibitor. Non-repellent chemicals included chlorpyrifos (liquid, granular and RTU forms), imidacloprid and fipronil (both in liquid and gel form) as per application site and target population (Table 5.1). Effective termiticides like endosulfan, lindane, etc. are now banned in India (Mahapatro and Madhumita 2013, 2014).

5.6.2.2 Cultural Practice

This includes crop rotation, mulch management, soil management and soil tillage, high density sowing, etc. Termites mostly target diseased and stressed plants, and healthy plants are seldom affected even though termites are present nearby. Removal of debris and mulches from the field reduces food supply of termites, thereby controlling their population. Higher seed rate is recommended in the termite prone field so as to avoid a major crop loss. Crop rotation, specially including a fallow period, is recommended to reduce harmful effect on soil fertility. Intercropping of maize with groundnut or soybean was found to reduce termite activity and increased predatory ant activity (Mahapatro 2014). Pre-planting tillage destroys the subsoil tunnels of termites, thus restricting their foraging activities and also the damage to crops. The border/bund on the field perimeter is often overlooked and untreated resulting in termite reservoirs. Cleaning and cultivation of the bund/border of field can help reducing termite attack in the field.

Table 5.1 Insecticides approved by Insecticide Act, 1968 (India)

		Formulation	
Sl. No.	Insecticide/Termiticide	Agriculture	Structural use (indoor)
1.	Chlorpyrifos	20% EC	50% EC
2.	Imidacloprid	17.8% SL	30.5% SL
3.	Bifenthrin	10.0% EC	2.5% EC

5.6.2.3 Mound Treatment

Destruction of termitaria in and around the field and treatment of the spot with termiticides at the village/community level is a promising method of control. De-queening may kill the entire colony, but it has practical problems. Rapid and complete extermination of mound could be achieved by treating the mound by pouring in, through funnel, water suspension of termiticides. Table 5.2 presents a height-liquid ratio for management of sub-cylindrical mounds of *Odontotermes obesus*.

5.6.2.4 Green Management Approach

No doubt, indiscriminate use of pesticides and insecticides has resulted in excessive environmental hazards. Besides, food security and food availability also are topics of concern; thus, it is evident we need to give more emphasis on crop security. This opens the discussion on green management of termite and is a towering need of present pest control methods. Controlling pests using phyto-based products and biological agents was attempted since long. However, in spite of accumulating ample amount of scientific literature and effort, biological control of termites could not meet substantial commercial success in the last 50 years (Chouvenc et al. 2011).

Green and clean technological interventions are to be made in termite management in the best plausible way. Under National Fellow (ICAR) project, the integration of frontier science with indigenous knowledge is attained in an innovative way, to offer an ecotechnology basis. In the study area under North East Plain Zone, the stubble management, if implemented successfully, can protect the crop (40% of total area in India) from termites with such a no or low-input strategy and a sustainable agriculture.

This strategy named *push-pull* used in maize is a combination of behaviour-modifying stimuli to manipulate the distribution and abundance of pest and beneficial insects. It is a novel conservation agriculture system innovated and validated for 3 consecutive years (2011–2013). It was demonstrated to farmers at ATIC (Indian Agricultural Research Institute, New Delhi) during 2013–2015. The technology suggests leaving the maize stubbles after harvesting and sowing three rows of pre-treated wheat seeds in between maize stubbles (spacing between maize rows 75 cm). Termites are attracted and congregated in the stubbles, but are not damaging wheat rows. It is a very interesting case of harmony between farmers and termites. Decomposition of stubbles also adds soil fertility to the field (unpublished result).

Table 5.2 Height-liquid ratio for termite mound treatment

Sl. No.	Height of mound	Total quantity of termiticide solution
1	3 ft. (ca. 91 cm)	1 gal. (4.5 L)
2	4 ft. (ca. 122 cm)	5 gal. (23 L)
3	5 ft. (ca. 152 cm)	10 gal. (45 L)
4	6 ft. (ca. 183 cm)	5 gal. (82 L)

5.6.2.5 Nematodes and Pathogens as Biocontrol Agents

Successful application of EPNs in pest control resulted from the collective effects of various factors, which elicited the interest of science towards the use of nematodes in insect pest control, and showed success against many insect pests and availability of many nematode species in various IPM programmes, for inoculative and inundative release (Bedding 1998; Kaya et al. 1993). In Sri Lanka, a dampwood termite *Glyptotermes dilatatus* colony with several individuals in tea plantation was successfully reported to be managed with *Heterorhabditis* sp. The best achievement in this aspect was reported from South Pacific islands where another genus of dampwood termite, *Neotermes* sp., was controlled by EPNs in the unbranched trunks of coconut palms (Lenz 2005). Leaving such isolated success stories, major achievement in termite control using nematode is still waiting for its way to emerge. In Australia, the use of *Heterorhabditis* sp. against residual population of subterranean termite *Coptotermes* sp. proved better in its initial stage, but could not proceed to much success due to temperature accounting 30 °C near the nest centre, which turned lethal for the nematodes. Field experiments for elimination of diffuse nesting subterranean termite *Reticulitermes flavipes* using two types of commercially available nematodes failed (Lenz 2005). In the domain of present research on termite control with the eco-friendly technologies, administering commercial formulation of EPNs with soil treatment was found effective.

More than 700 species of fungi were listed as insect pathogen (Milner 2000), but examination with termite was vastly centric to two species, namely, *Beauveria bassiana* and *Metarhizium anisopliae* (Lenz 2005). Commercial bio-termiticide Bioblast™ used against subterranean termites in the USA and Japan was also reported to be relied upon *M. anisopliae.* Two isolates of these fungi isolated from India proved efficacy on the arboreal termite *Odontotermes* sp., which could prove better if used with baits for termite control (Balachander et al. 2009).

5.6.2.6 Phyto-Based Products

Using plant-derived materials for various purpose of crop protection was followed by humankind since the dawn of agricultural practices, in various forms and nature which varied from place to place. Scientific analysis revealed that various phytochemicals and secondary metabolites, e.g. alkaloid, terpenoids, unsaturated isobutylamide, etc., are toxic to termites. Handful information regarding management with phytoproducts was available from different places of the world. Pretreatment of wood and wooden materials, required for house-building purposes and for other structural construction by anti-termite herbal products, showed better outputs in termite control in India (Verma et al. 2009). Extracts of *Cannabis sativus*, *Datura alba*, *Curcuma amada*, *Ricinus communis* and *Asafoetida* showed better results. Despite their lower efficacy than chemical insecticides, they were found safe and environment friendly and, hence, more recommendable in green management. Plant extracts will certainly find a better place in wood protection against termites and its use as wood preservatives (http://www.kadinst.hku.hk/sdconf10/Papers_PDF/p571.pdf). A citable example of green management of termites drew our attention in an age-old temple in Kerala, India. Here a mixture of locally available phyto-based products and oils (total eight ingredients) for managing termite attack was successfully applied on the wooden sculptures and timber in service in the premises. Thus, this technology was honoured with the UNESCO Asia-Pacific Heritage Award for Cultural Heritage Conservation for the year 2015 (Times of India, TNN, 15th September, 2015).

5.7 Technologies for Termite Management (3B)

Building defence against termites is a challenging issue in pest management. Various integrative approaches were taken and some became landmarks. In the age of advanced science, application of borate, bait and barriers (3B) was found to be a turning point in termite control.

5.7.1 Borate

Borate refers to the compound comprised of boron and oxygen that naturally exists as sodium or calcium borate. Due to its high toxicity, it was found to be strikingly effective against termites (the Ohio Department of Agriculture – Pesticide Regulation – Certification and Training Section, 2007). In the north-east India, bamboos treated with borate were found to sustain termite attack (Anonymous 2016). Various other experiments left evidence of the use of boron as a good termiticide too (Rust and Venturina 2009). In India, Forest Research Institute at Dehradun suggested application of borate in the form of BCCA (borated copper chrome arsenic) and ACZB (ammoniacal copper zinc borate) for wood preservation against termites. A demerit of this method includes leaching of chemical into soil that limits the use of this practice only up to above ground.

5.7.2 Baiting

Using baits in pest control is a well-practiced technique. Chitin synthesis inhibitors (CSI) of various forms were reported to be applied as active ingredients and were used through food matrix or as dust. In certain applications, termites were reported to be successfully managed using this technique in the treated area. This practice proved good results against lower termites such as *Reticulitermes* and *Coptotermes*. Triumph against *Reticulitermes santonensis* was reported in Paris using baiting technique (source: UNEP report). Ngee et al. (2004) divided bait toxicants in three groups as (1) metabolic inhibitors, (2) biological control agents and (3) insect growth regulators. Application of baits including active toxic ingredients into feeding diet of termites, present near the attacked or unaffected structure and eventual insect death. Slow-acting poisons proved more advantage as the sublethal effect allows ample time to be spread amongst the large number of individuals in a colony. Fast poisons could even kill individuals rapidly, but the accumulation of dead individuals in the bait-affected place could leave signs of danger to the other nestmates, which may in turn avoid bait stations (Rawat 2013). In India, baiting technique for management of termite both indoor and outdoor condition has yet to arrive in the market. Research on chlorfluazuron 1% based bait system is under experimentation at Central Building Research Institute, Roorkee, India (Rawat 2013). Research and development on bait techniques in termite control have huge popularity and prospect in India and other developing countries as well, if it is made easily available to the end users.

5.7.3 Barrier

Another promising technology to defend against termite includes the use of various barriers according to the nature of the attack. Barriers could be of different types, based on the constituting materials. Setting up transparent sheet just below damp-proof course (DPC) level of new buildings is found to be a protective measure against termite and dampness, crushed stones and gravels. Sand could act as a physical barrier against subterranean termites. Also particulate material could act as good physical barrier against termite attack in constructions, when positioned in the crawling area (Rawat 2013). Small-sized mesh prepared by stainless steel was reported as better horizontal obstacle for many species (Lenz and Runko 1994). In Australia and Hawaii, 'Termi-Mesh' could hold promising barrier in termite control (Grace et al. 1997).

For sustainable management of termites in indoor condition, barrier chemicals were grouped into two categories, viz. repellent and non-repellent (see par 6.2.1). Striking differences between these two categories of chemical barrier lies in the termite response towards them. In case of non-repellent chemicals, toxicity could pass onto the other colony members giving better population control. On the other hand, while using repellent chemicals, termite could detect and stay away from the treated area, which results into no toxicity passing to other colony mates (source: UNEP). Preventing termite attack in a newly built construction or chasing out termites from existing buildings includes many processes of chemical barrier imposition. In India, government-initiated procedures of termite treatment were standardized in *Indian Standard* code of practice for anti-termite measures in buildings (IS 6313 – part 2 and 3), for pre- and post-constructional stages of buildings and other constructions. Direct applications of termiticides, recommended doses, etc. were specified as per the status of constructions. Few striking innovative approaches of termite control through chemical barrier, such as 'rodding', 'reticulation' and 'perimeter treatment', were elaborated with recommended termiticides and their dose in IS 6313 (2015).

For outdoor management of termites, barrier technologies are poorly known yet. Restricted applications of some outdoor techniques proved efficacy to keep termite away from target spots. Arrangement of sand around the horticulture and gardening pots, putting sand bed around the tree trunk and creating trench filled with sand around the outer periphery of a lawn are useful in combating termites in horticultural crops. Physical barriers include sand or stainless barriers, which proved effective in new constructions. Painting of earthen pots with red earth treated with termiticides or lime could also act as physical barrier in gardens. Painting tree trunk with mixture of lime and termiticide can prevent attack of termite on avenue trees and horticultural fruit trees in orchards. Painting earthen pots with lime and termiticide mixture even protects potted plants in nursery and gardens.

5.7.4 Termatrac (T3i)

Recent past development in termite management through radar-based tracking is a milestone achievement. In Australia, a device called Termatrac (T3i) was launched in 1999 for the detection and management of termites in buildings and structures. This instrument works on three techniques amalgamated in one, viz. termite track radar, laser-guided remote thermal sensor and moisture sensor (http://www.termatrac.com/products/termatrac-t3i.). This device was developed mostly for drywood termite detection. Like other wood-destroying insect pests, termites too maintain high temperature and humidity in the nest. The laser-guided remote thermal sensor and moisture sensor help in detecting the exact location of the nest and entry point of the nest. Termite detection radar decodes the level of activity and direction from where termites travelled to get a particular point. This makes the job of a termite control technician easy, to implement a proper management method. In developing countries like India, this device was marketed by Pest Control India, international distributor in this country (http://www.termatrac.com/about-termatrac/international-distributors/).

5.8 Indigenous Traditional Knowledge (ITK) for Termite Management

From the daybreak of the agricultural process discovery by the human civilization, the conflict between man and insects secured an important position in the human life. Great ancient literature acknowledged the presence of termite and their damage. In India, termites were known since Vedic era. Primordial Sanskrit literature referred to them as 'Kashthaharika' (wood feeder), describing their nature of existence in the human life. Development of pest control strategies took birth since those old days as human civilization started suffering from insects due to crop losses. Methods of pest control had seen changes in various eras as per the best effort could the human minds devote in this matter. Today with advanced science and technology, the matter has become easier in case of pest control, but often at the cost of environmental safety. ITKs were the age-old traditional practices of managing pest problems in crop protection using local knowledge about the pest, its habit habitat and the remedial habitat using locally available materials and plant-based products. This knowledge was transferred from generation to generation as the heritage of a community. Worldwide, these ITKs hold immense importance nowadays for green management of termites as they neither include the use of any sort of hazardous chemical nor involve any such process. The maximum use of such ITKs is seen in the areas which are backwards and lack scientific development. Examples from India include the use of oil waste as surface application on bamboos, in the state of Gujarat that provided the plants an anti-termite protection. Many ITKs were documented from coconut cultivation of Kerala, of which some were validated by subsequent scientific studies (Mahapatro and Kumar 2015). The buildings of the city of Jodhpur, Rajasthan states, were coloured with blue, which believed to protect them from termites. Eating termites as food and medicine was reported from many tribes living in Chhattisgarh (Mahapatro 2015). Application of ITK for termite management in upland rice cultivation was also documented in India (Mahapatro and Sreedevi 2014). Many more such instances from different areas of the world have been reported, from time to time. Most important matter here is the scientific validation of such ITKs, for their better and rational implementation.

5.9 Conclusion

Termites as pests have been associated with crops since immemorial times. We suggest useful approaches are the validation of relevant ITKs, verifying the true efficacy of biocontrol agents before integrating them, or the judicious utilization of chemicals, regulating the termite population with smart strategies under the aegis of IPM.

Acknowledgements Authors are thankful to the funding agency ICAR for the sanctioned National Fellow Project, 'Environmentally Sustainable Termite Control: Integrative and Inclusive Approach of Frontier and Indigenous Technologies'. Logistic supports provided by Director, IARI (New Delhi), are gratefully acknowledged.

References

Anonymous. (2016). *Preserving bamboo in buildings*. Pest net publication. http://www.pestnet.org/SummariesofMessages/Pests/PestManagement/Chemicalcontrol/Syntheticpesticides/Insecticides/Preservationofbambooinbuildings,Sikkim,India.aspx. Accessed on 30 June 2016.

Balachander, M., Remadevi, O. K., Sasidharan TO, & Bai, S. N. (2009). Infectivity of *Metarhizium anisopliae* (Deuteromycotina: Hyphomycetes) isolates to the arboreal termite *Odontotermes* sp. (Isoptera: Termitidae). *International Journal of Tropical Insect Science, 29*, 202–207.

Barbara, L. T., David, A. G., & Kumar, K. (2000). Early fossils history of the termites. In Y. Abe, D. E. Bignell, & T. Higashi (Eds.), *Termites: Evolution, sociality, symbiosis, ecology* (pp. 77–93). Dordrecht: Springer.

Bedding, R. A. (1998). Future possibilities for using entomopathogenic nematodes. *Japanese Journal of Nematology, 28*, 46–60.

Black, H. I. J., & Okwakol, M. J. N. (1997). Agricultural intensification, soil biodiversity and agroecosystem function in the tropics: The role of termites. *Applied Soil Ecology, 6*, 37–53.

Brown, A. A., Herms, W. B., Horner, A. C., Kelly, J. W., Kofoid, C. A., Light, S. F., & Randall, M. (1934). General recommendations for the control of termite damage. In C. A. Kofid (Ed.), *Termites and termite control* (pp. 539–551). Berkeley: University of California Press.

Chouvenc, T., NY, S., & Grace, J. K. (2011). Fifty years of attempted biological control of termites – Analysis of a failure. *Biological Control, 59*, 69–82.

Dawes, T. Z. (2010). Reestablishment of ecological functioning by mulching and termite invasion in a degraded soil in an Australian Savanna. *Soil Biology and Biochemistry, 42*, 1825–1834.

Donovan, S. E., Eggleton, P., & Bignell, D. E. (2001a). Gut content analysis and a new feeding group classification of termites. *Ecological Entomology, 26*, 356–366.

Donovan, S. E., Eggleton, P., Dubbin, W. E., Batchelder, M., & Dibong, L. (2001b). The effect of a soil-feeding termite, *Cubitermes fungifaber* (Isoptera: Termitidae) on soil properties: Termites may be an important source of soil microhabitat heterogeneity in tropical forests. *Pedobiologia, 45*, 1–11.

Donovan, S. E., Griffiths, G. J. K., Homathevi, R., & Winder, L. (2007). The spatial pattern of soil-dwelling termites in primary and logged forest in Sabah, Malaysia. *Ecological Entomology, 32*, 1–10.

Eggleton, P. (2000). Global pattern of termite diversity. In T. Abe, D. E. Bignell, & M. Higashi (Eds.), *Termites: Evolution, sociality, symbiosis, ecology* (pp. 25–51). Dordrecht: Springer.

Forschler, B. (2011). Sustainable termite management using an integrated pest management approach. In P. Dhang (Ed.), *Urban pest management: An environmental perspective* (pp. 133–144). Oxfordshire: CAB International.

Forschler, B. T., Jones, S. C., Suiter, D., Gold, R., Henderson, G., Kard, B., Baker, P., Jackson, D., & Howell, H. (2007). Subterranean termite management: Still an ongoing process. *Pest Control Magazine, 75*, 88–95.

GEI. (2005). *Demonstration project of alternative to chlordane and mirex for termite control in China*. Guangdong Entomological Institute. Beijing. From the World Bank http://web.worldbank.org/external/projects

Gilberg, M., & Su, N. Y. (2012). New termite baiting technologies for the preservation of cultural resources: Results of field trials in the national park system. *Park Science, 21*, 16–22.

Grace, J. K., Yates, III J. R., Tome, C. H. M., & Oshiro, R. J. (1997). Use of a stainless steel mesh to exclude Formosan subterranean termites. The international research group on wood preservation, IRG/WP97-10199 (http://www.ctahr.hawaii.edu/gracek/pdfs/117.pdf)

Hartnack, H. (1943). *Unbidden house guest Part I, Part II*. Tacoma: Hartnack Publishing.

Hoon, R. C. (1960, October 4–12). The incidence of white ants (termites) in the region of the Hirakud Dam project. In *Proceedings of international symposium on termites in the humid tropic*. New Delhi.

Ibrahim, B. U., & Adebote, D. A. (2012). Appraisal of the economic activities of termites: A review. *Bayero Journal of Pure and Applied Sciences, 5*, 84–89.

IS 6313. (2015). *Indian standard code of practice for anti-termite measures in buildings part 2 and 3 pre and post-constructional chemical treatment measures (third revision)*. New Delhi: Bureau of Indian Standards.

Kaya, H. K., Bedding, R. A., & Akhurst, R. J. (1993). An overview of insect-parasitic and entomopathogenic nematodes. In R. A. Bedding, R. Akhurst, & H. Kaya (Eds.), *Nematodes and the biological control of insect pests* (pp. 1–10). Melbourne: CSIRO Publishing.

Kirton, L. G. (2005). The importance of accurate termite taxonomy in the broader perspective of termite management. In C. Y. Lee & W. H. Robinson (Eds.), *Proceedings of the fifth inter-*

national conference on urban pests. Printed by Perniagaan Ph'ng @ P&Y Design Network, Malaysia.

Kirton, L. G., & Azmi, M. (2005). Patterns in the relative incidence of subterranean termite species infesting buildings in peninsular Malaysia. *Sociobiology, 46*, 1–15.

Lee, C. Y. (2002). Subterranean termite pests and their control in the urban environment in Malaysia. *Sociobiology, 40*, 3–9.

Lee, C. Y., Vongkaluang, C., & Lenz, M. (2007). Challenges to subterranean termite management for multi-genera faunas in Southeast Asia and Australia. *Sociobiology, 50*, 213–221.

Lenz, M. (2005). Biological control in termite management: The potential of nematodes and fungal pathogens. In C. Y. Lee & W. H. Robinson (Eds.), *Proceedings of the fifth international conference on urban pests*. Printed by Perniagaan Ph'ng @ P&Y Design Network, Malaysia.

Lenz, M., & Runko, S. (1994). Protection of buildings, other structures and materials in ground contact from attack by subterranean termites (Isoptera) with a physical barrier-a fine mesh of high grade stainless steel. *Sociobiology, 24*, 1–16.

Li, H. F., NY, S., WJ, W., & Hsu, E. L. (2011). Termite pests and their control in Taiwan. *Sociobiology, 57*, 575–586.

Liston, B. (2015). *UPDATE 1-Termite species mix in Florida, reproducing at breakneck rate*. Reuter publication. http://www.reuters.com/article/usa-florida-termites-idUSL2N0WS01120150326. Accessed on 1 June 2016.

Mahapatro, G. K. (2014). Rodents and termites: Non insect pests. In V. V. Ramamurthy, C. Srivastava, & K. Shankarganesh (Eds.), *A training manual on crop protection* (pp. 253–259). New Delhi: Indian Agricultural Research Institute.

Mahapatro, G. K. (2015). Ethno-entomology: Termites as food and medicine. *Current Biotica, 9*, 5–8.

Mahapatro, G. K. (2016). *Termite management in horticultural system* (p. 10). New Delhi: Published by ICAR National Fellow Project, ICAR-IARI.

Mahapatro, G. K., & Kumar, S. (2013). First report of a commonly prevalent termite *Heterotermes indicola* from Delhi region. *The Indian Journal of Agricltural Sciences, 83*, 459–462.

Mahapatro, G. K., & Kumar, S. (2015). Review on the incidence and management of coconut termites. *Indian Journal of Entomology, 77*, 152–159.

Mahapatro, G. K., & Madhumita, P. (2013). The case for banning endosulfan. *Current Science, 104*, 1476–1479.

Mahapatro, G. K., & Madhumita, P. (2014). Endosulfan issue: Science verses conscience. *Current Science, 106*, 152–155.

Mahapatro, G. K., & Sreedevi, K. (2014). Indigenous approaches for the management of termite and white grub in upland rice. *Current Biotica, 8*, 97–108.

Mahapatro, G. K., Kumar, S., & Kumar, A. K. (2014). *A unique case of termite devastated desert village near Kota, International conference on entomology, February 21–23, 2014*. Patiala: Punjabi University.

Mahapatro, G. K., Sreedevi, K., & Kumar, S. (2015). Krishna Kumar (1928–2014). *Current Science, 108*, 2277–2278.

Manzoor, F., & Mir, N. (2010). Survey of termite infested houses, indigenous building materials and construction techniques in Pakistan. *Pakistan Journal of Zoology, 42*, 693–696.

Milner, R. J. (2000). Current status of *Metarhizium* as a mycoinsecticide in Australia. *Biocontrol News and Information, 21*, 47N–50N.

MRP (2010). *Maxwell Robinson Phelps termite report*. Maxwell, R and Phelps. http://www.pestcontrolperth.com/wpcontent/uploads/2010/06/Maxwell-Robinson-Phelps-MRP-Termite-Report.pdf. Accessed on 26 Feb 2016.

Ngee, P. S., Yoshimura, T., & Lee, C. Y. (2004). Foraging populations and control strategies of subterranean termites in the urban environment with special reference to baiting. *Japanese Journal of Environmental Entomology Zoology, 15*, 197–215.

Pinniger, D. B. (2012). *Managing pests in paper-based collections*. London: Preservation Advisory Centre, The British Library.

Rajagopal, D. (2002). Economically important termite species in India. *Sociobiology, 40*, 33–46.
Ravan, S. (2010). Ecological distribution and feeding preferences of Iran termites. *African Journal of Plant Science, 4*, 360–367.
Ravan, S., Imtiaz, A. K., Farkhanda, M., & Khan, Z. (2015). Feeding habitats and wood preferences of termites in Iran. *Journal of Entomology and Zoology Studies, 3*, 20–23.
Rawat, B. S. (2013, January 27–28). Current scenario of termite management in buildings in India. In *Souvenir of PMA conference cum AGM 2013 at Chandigarh*.
Roonwal, M. L. (1979). *Termite life and termite control in tropical south Asia* (p. 177). Jodhpur: Scientific Publishers.
Rust, M. K., & Venturina, J. (2009). *Final report 2009, structural pest control board grant no: 084-4231-7*. http://pestboard.ca.gov/howdoi/research/2009_drywood_rpt.pdf. University of California. Accessed on 15 Feb 2016.
Scholz, G., Militz, H., Gascon Garrido, P., IbizaPalacios, M. S., Oliver Villanueva, J. V., Peters, B. C., & Fitzgerald, C. J. (2010). Improved termite resistance of wood by wax impregnation. *International Biodeterioration and Biodegradation, 64*, 688–693.
Snyder, T. E. (1927). Termites modify building codes. *Journal of Economic Entomology, 20*, 316–321.
Snyder, T. E. (1935). *Our enemy the termite*. Ithaca: Comstock Publishing.
Sornnuwat, Y. (1996). *Studies on damage of constructions caused by subterranean termites and its control in Thailand*. Wood Research No 83.
Su, N. Y., & Scheffrahn, R. F. (1998). A review of subterranean termite control practices and prospects for integrated pest management programme. *Integrated Pest Management Reviews, 3*, 1–13.
Tai, O. S., & Chen, F. (2002). *A course of termite control*. Guangdong: Science Press.
Takahashi, M., & Yoshimura, T. (2002). Recent development in the control of Japanese subterranean termites. *Sociobiology, 40*, 13–24.
Tuyen, V. V. (2006, March 6–7). Termite nest in dam site foundation before construction. In *Proceedings of The third conference of Pacific Rim Termite Research Group*, Guangzhou, China.
Verma, M., Sharma, S., & Prasad, R. (2009). Biological alternatives for termite control: A review. *International Biodeterioration and Biodegradation, 63*, 959–972.
Yeoh, B. H., & Lee, C. Y. (2007). Tunneling activity, wood consumption and survivorship of *Coptotermes gestroi*, *Coptotermes curvignathus* and *Coptotermes kalshoveni* (Isoptera: Rhinotermitidae) in the laboratory. *Sociobiology, 50*, 1087–1096.
Zhong, J. H., & Liu, L. L. (2002). Termite fauna in China and their economic importance. *Sociobiology, 40*, 25–32.

Chapter 6
Eco-Friendly Termite Management in Tropical Conditions

Monica Verma, Sharad Verma, and Satyawati Sharma

Contents

Abstract Being important pests of temperate and tropical regions, termites are responsible for massive losses to agricultural crops and wooden structures. Chemical control is the most popular and effective method of management. However, the deleterious effect of continuous usage of chemical termiticides is of serious concern, and researchers throughout the world are actually searching for alternative approaches. This chapter encompasses the various non-chemical strategies developed so far. Physical methods of control are also discussed. Focus has been given on biological means of management with major emphasis drawn on fungi, bacteria, nematodes and plant-derived natural products (botanicals). Botanical pesticides are sustainable, biodegradable and easily available, with no effect on nontarget species, and do not cause

M. Verma (✉) • S. Verma • S. Sharma
Center for Rural Development and Technology, Indian Institute of Technology Delhi, Hauz Khas, New Delhi 110016, India
e-mail: monica24_iitd@yahoo.com

M.A. Khan, W. Ahmad (eds.), *Termites and Sustainable Management*, Sustainability in Plant and Crop Protection,
https://doi.org/10.1007/978-3-319-68726-1_6

pest resistance. Furthermore, the bioactive components present in the botanicals can be isolated, characterized, formulated and used as commercial termiticides.

Keywords Tropical • Fungi • Bacteria • Nematodes • Botanicals • Bioactive components

6.1 Introduction

Termites evolved about 220 million years ago, as shown by fossil records, and are considered as a primitive group of winged insects having close similarity with the cockroaches (Blattoidea). They are soft-bodied, pale in colour, polymorphic, cellulose-eating, long-lived social insects (Collins 1988; Thorne and Carpenter 1992; Meyer 2005). Termites live in large colonies depend entirely on wood, either living or dead, or the woody tissue of plants, intact or partially decayed.

Termites are widely distributed in the tropics (both in humid and dry zones) as well as in some temperate regions. The ideal living conditions (temperature and humidity) in tropical environments render them active all the year round. They are highly diverse and abundant in the rain forests of Africa, South America, and Southeast Asia (Collins 1988; Bignell and Eggleton 1998). There are over 3000 designated species of termites, out of them about 185 are considered to be pests. In India, there are about 300 species of termites known so far, and about 35 species have been reportedly damaging agricultural crops and buildings (Lewis 1997; Rajagopal 2002; Verma et al. 2009).

Economically, termites cause enormous damage to wooden structures in buildings, agricultural and forest crops, stored timbers, books and records, specifically the products made of cellulose which incurred huge losses throughout the world. Termites destroy wooden constructions, because the main structural polymer of wood, lignocellulose, is well digested by their hindgut protozoa or bacteria (Breznak 1982; Breznak and Brune 1994). Members of the families Hodotermitidae (*Anacanthotermes* and *Hodotermes*), Kalotermitidae (*Neotermes*), Rhinotermitidae (*Coptotermes*, *Heterotermes* and *Psammotermes*) and Termitidae (*Amitermes*, *Ancistrotermes*, *Cornitermes*, *Macrotermes*, *Microcerotermes*, *Microtermes*, *Odontotermes*, *Procornitermes* and *Syntermes*) are responsible for great losses in agriculture (UNEP 2000). It is also known that termites damage a variety of materials ranging from paper fabrics to even non-cellulosic materials such as asbestos, asphalt bitumen, lead and metal foils (Bultman et al. 1979). They impart both positive and negative impact on the environment and population. By breaking down cellulose, they help to recycle soil nutrients, thus serving an important role in the ecosystem. Some termite species in tropical regions grow fungi within their nests which develop into large mushrooms and are entirely cultured and cultivated by termites but also eaten in some communities of Africa (Sileshi et al. 2009).

These deleterious insects can be controlled by the application of synthetic termiticides. The use of chemical termiticides remains the primary method to prevent termite attack on wooden structures and agricultural crops. Unlimited input of pesticides resulted in insecticide resistance in pests; safety risks for humans, due to their long persistence in soil; and entry of toxic residual chemicals in food chains, coupled with bioaccumulation and biomagnifications, contamination of groundwater and decrease in biodiversity (Sindhu et al. 2011). Chemicals like methyl bromide, chlorofluorocarbons, etc. are established culprits of depletion of the ozone layer (Cox et al. 1995; UNEP 1992; UNEP 2000). The increased awareness about the negative health hazards, adverse side effects and environmental pollution caused by the incessant use of chemical pesticides shifted the focus on biological alternatives for termite control.

There is a renewed concern in the usage of biological instead of chemical pesticides. Biopesticides are derived from animals, plants and microorganisms such as bacteria and viruses. They are biodegradable and do not leave problematic residues. They are less harmful than chemical pesticides, affecting only the target pests and their close relatives. Chemical pesticides instead often destroy friendly insects, birds and mammals (Devlin and Zettel 1999). Research on the active ingredients, pesticide preparations, application rates and environmental impact of botanical pesticides are a prerequisite for sustainable agriculture (Buss and Park-Brown 2002). Also, as consumers' demand for organically produced food increases, scientific research on the use of botanical pesticides is now gaining momentum (Nas 2004).

Due to the increased demand of organic and good quality food, the use of biopesticides all the over world is increasing. In India this approach showed a vast potential, as the most commonly used biopesticides include *Bacillus thuringiensis* (Bt) and neem. However, their adoption by farmers in India needs education and training for maximizing gains. This shows a lacuna in overall biopesticide research and formulation, as India is a rich source of biodiversity and thousands of plants showed a biopesticidal potential, which could be exploited for pest control. Therefore, the research in this area should be emphasized and accelerated so that novel pest control agents could be explored, tested, formulated and registered as commercial biopesticides.

Botanical pesticides are ingredients derived from plants suitable for insect control. That may provide potential alternatives to currently used insect control agents because they include a rich source of bioactive chemicals. Over 2000 plants belonging to some 60 plant families are known to exhibit insecticidal activities (Dev and Koul 1997). These bioactive chemicals include insecticides, antifeedants, insect growth regulators (IGRs), juvenile hormones, ecdysones, repellents, attractants, arrestants, etc. Plants are hence viewed an important alternative source for chemical pesticides (Kannaiyan 1999). This chapter describes the deleterious effects of chemical insecticides and the critical need of environmentally friendly approaches of termite control, including the use of bacteria, fungi and nematodes, with special reference to plant-derived insecticides.

6.2 Economic Losses by Termites

Through their vital role in the tropical ecosystems by decomposing dead wood and other plant material rich in cellulose (Wood and Sands 1978; Abe 1995), termites recycle woody and other plant material and aerate the soil with their tunnelling efforts.

They are abundant in tropical and subtropical environments where they help in breaking down and recycling one third of the annual production of dead wood. Termites become economic pests when they start destroying wood and wooden human products such as homes, building materials, forests and other commercial items (Meyer 2005).

Termites cause over 3 billion dollars' worth of damage to wooden structures annually throughout the USA (Lewis 1997). Control and repair costs due to Formosan subterranean termites in New Orleans were estimated at 300 million dollars annually (Suszkiw 1998). They caused losses of about RMB 1700–2000 million in China (Zhong and Liug 2002). An amount of 800 million US dollars in Japan and more than 100 million (Australian $) in Australia are lost to termite damage (www.chem.unep.ch/pops/termites).

Severe losses in different regions of India have been recorded on highly susceptible crops such as wheat and sugarcane, maize, groundnut, sunflower, tea, cotton and others. Members of genus *Odontotermes* caused 5–50% plant mortality and up to 46% pod damage in groundnut (Rajagopal 2002), 30.6% yield reduction in sugarcane and 19% in potato, pulses and wheat (Verma et al. 2001). They are responsible for the loss of 15–25% of maize yield and about 1478 million rupees (Joshi et al. 2005), which account for about 8% of total production loss.

6.3 Biology of Termites

Termites inhabit in underground colonies, above ground mounds, mud channels, gallery, tubes and tunnels (Haverty et al. 2005; Diehl et al. 2005). Their colonies are polymorphic and characterized with a division of labour with different castes (reproductive, workers, soldiers and immature individuals) (Thompson et al. 2004; Husseneder and Simms 2008). Figure 6.1 shows termite workers foraging for food.

Mainly, two types of reproductive (primary and secondary) occur in a termite colony. King and queen are the primary reproductives, whereas secondary reproductive includes neotenic, nymphoid or ergatoid. Primary reproductive performs the reproduction task by producing new individuals. Its role is the egg production and distribution by colonizing flights. The queen lays about 3000 eggs a day through his distended abdomen (Thompson 2000) and may live up to 25 years (Myles 2005). The eggs are yellowish white in colour requiring 50–60 days of incubation. Generally, termite colonies have only one pair of primary reproductive. After their death, they are replaced by numerous supplementary reproductives which are with or without wing pads and are slightly larger and more pigmented than workers. The workers and the soldiers are sterile and wingless and usually lack eyes (Myles 2005). A fully grown termite colony is acquired in approximately 4–5 years including 60,000 to 2,000,000 workers. Out of total individuals, the majority of termites are workers, whereas soldiers comprise approx. 1–2% of the total population (Husseneder et al. 2005). Workers and soldiers mature in a year and live up to 3–5 years (Myles 2005). They are pale cream in colour and 6 mm in length. Soldiers have enlarged heads, almost half their body length, with prominent black jaws. Workers perform most of

Fig. 6.1 Termite workers forage for food

the tasks such as mound building, maintaining, cleaning and repairing, taking care of eggs and feeding and grooming the immature soldiers and other individuals. Immature individuals may develop into any caste (worker, soldiers or reproductive) depending on the requirement of a colony (Pearce 1997). Colony protection and guarding from intruders is the main function of soldiers (Higashi et al. 1991; Boomsma et al. 2005). The mass nuptial flight of winged reproductives (Alates) emerges in April–May and is the first indication of termite infestations (Philip 2004). Alates shed their wings after a brief flight and females search for nesting sites with males following closely behind. After finding a moist crevice with wooden materials, they form the royal chamber and lay the eggs (Su and Scheffrahn 2000).

6.4 Termite Management by Physical, Chemical and Biological Means

6.4.1 Physical Methods

Physical barriers are materials that are installed around the foundation of buildings or houses to restrict the entry of termites. There are two kinds of barriers: toxic and non-toxic. Toxic physical barriers are the chemical termiticides incorporated in the soil around the structure. Soil treatment with chlorfenapyr as physical barrier treatment is very effective, as it is non-repellent and has delayed toxicity (Rust and Saran 2006).

Non-toxic physical barriers include substances such as sand or gravel aggregates, metal mesh or sheeting that prevent termites to enter the structure acting as physical/

mechanical barrier that termites are unable to penetrate. Gravel-made barrier, the BTB (basaltic termite barrier) is highly used in Hawaii for preconstruction treatment of buildings (Grace et al. 1996). In recent years, nongraded stone products such as copper shields, solid steel sheets, polymer sheets and stainless steel mesh were popular in Australia, Texas, Hawaii and other parts of the USA, as physical barriers (Grace et al. 1996; Potter 2004; Baker 2005). Such a stainless steel wire mesh named Termimesh® was developed and patented in Australia. It is a non-corrosive flexible steel mesh of size 0.66 × 0.45 mm which can be installed within the stem walls of the foundations. It is exceptionally efficient in preventing termite entry as shown by various field trials (Lenz and Runko 1994; Grace et al. 1996; Kard 1999).

Physical methods also involve the use of physical treatments such as heat, freezing, electricity, microwaves and electromagnetic waves. In heat treatment, the building is covered with nylon tarps, and hot air is blown inside to raise the temperature to 45 °C for 35 min or 50 °C for 1 h (Myles 2005). High temperature kills the termites. Woodrow and Grace (1998) used high-temperature treatments to control *Cryptotermes brevis.* For freezing treatments, liquid nitrogen is pumped into the infected area to kill the termites with low temperature. The electrical treatment includes applying the termite-infested wooden board an electrical shock of low current (-0.5 amps), high voltage (90,000 volts) and high frequency (60,000 cycles), by using Electro-Gun® (Myles 2005). The lethal effects of the electrical shock treatments on western drywood termite, *Incisitermes minor* (Hagen), were studied by Lewis and Haverty (2000). In microwave treatment, heat generated by the microwaves kills termites (UNEP 2000). Electromagnetic waves can also be used as a non-chemical approach to control termites. This technology is odourless, noiseless, environmentally friendly and easy to apply. Dibaa et al. (2013) studied electromagnetic waves as non-destructive method to control subterranean termites *Coptotermes curvignathus* Holmgren and *Coptotermes formosanus* Shiraki. They used electromagnetic waves at frequencies 30–300 Hz, 0.3–3 KHz and 3–300 KHz with variation of irradiation time exposure (15, 30, 45 and 60 min) for *C. curvignathus*. For *C. formosanus*, 300 and 500 KHz with irradiation time exposures of 30 and 60 min were applied. The results revealed that the average value of termite mortality ranged between 25.45 and 82.27%, and the average value of termite filter paper consumption ranged between 8.89 and 39.44% for *C. curvignathus* (Dibaa et al. 2013).

6.4.2 Chemical Control

Chemical termiticides provide the most effective and widely used solutions so far. Various termiticides with active constituents, bifenthrin, chlorfenapyr, cypermethrin, fipronil, imidacloprid, permethrin, spinosad, disodium octaborate tetrahydrate, calcium arsenate and chlorpyrifos, are registered for termite control across the world. Insecticides such as chlorpyrifos, bifenthrin, imidacloprid, endosulfan and lindane are currently being used for control of termites in stored wood, as well as for crops (Su et al. 1999).

Soil treatment with bifenthrin, cypermethrin and permethrin formulations was most effective against termites as reported by Smith and Rust (1990). Cypermethrin (100 ppm), fenvalerate (500 ppm) and imidacloprid (1000 ppm) were effective against subterranean termites at high concentrations (Kuriachan and Gold 1998). Minor damage was observed in cotton, maize, rice, sorghum and sugarcane due to three species of termites: *Trinervitermes trinervius*, *Odontotermes smeathmani* and *Amitermea evuncifer* following application of the insecticides chlorpyrifos, lindane or thiamethoxam (Bhanot and Singal 2007). Ahmed et al. (2006a) studied the effectiveness of imidacloprid, monomehypo and chlorpyrifos against subterranean termites in sugarcane and found that chlorpyrifos was most effective of all. Seed treatment with chlorpyrifos, endosulfan, formothion and monocrotophos has been standardized in wheat, barley and gram (Bhanot et al. 1991a, b; Bhanot et al. 1995). Southern yellow pine and radiata pine treated with chromated copper arsenate as wood preservative prevented attacks of Formosan subterranean termites, *Coptotermes formosanus* (Grace 1998).

For control of subterranean termites, soil termiticide injection is applied. In this procedure, the foundation wall/slab is drilled, and termiticide was injected below the slab and in the soil in contact with the foundation. For control of drywood termite infestation, chemical fumigation is the best method. It includes the use of chemical fumigant, usually methyl bromide, sulfuryl fluoride and carbon dioxide. The whole building is covered with a tent, and the fumigant is pumped in the building, later removing the tent to vent off the gas. The chemical absorbent materials are removed from the building to be fumigated (Myles 2005). Active ingredients in various fumigants are carbon dioxide (asphyxiant), methyl bromide, phosphine and sulfuryl fluoride (metabolic poison). Methyl bromide is a frequently used fumigant. However, it is a very toxic and harmful gas. Its impact on the atmospheric ozone layer, leftover odour in household materials, and long aeration times needed for fumigated structures has limited the use of this fumigant (UNEP 2000). Guan et al. (2011) gave a novel and environmentally friendly technique for termite control by photo-immobilizing bifenthrin-embedded chitosan on the surface of wood. They combined the effectiveness of bifenthrin against termites and photo-immobilization technique of chitosan by developing chitosan as a carrier to insert bifenthrin and immobilize it by ultraviolet light on the surface of wood. Immobilized bifenthrin is hence protected from free and noncontrolled releasing, and has a long-term stability, with high efficiency against termites, at a dose of 2.5 mg/cm^2.

Biological resistance of particleboard panels was studied against *Anacanthotermes vagans*. Wood particles were treated with copper and zinc salicylate and then exposed to the termite. Results showed minimum weight loss recorded with copper-salicylate-treated specimens (4.7%), while control showed maximum weight losses (16.3%). This confirmed significant protection of copper salicylate against termites which can therefore be recommended to the industry to provide protection against *A. vagans* (Bayatkashkoli et al. 2016).

Chemical control is an effective strategy to control termite population, but their excessive use causes toxic side effects and pesticide residue in soil and water, whereas insurgence of a pest resistance is deleterious to humans and the environment. Alternative approach to replace and reduce chemical insecticides is an essential part of current management practices, globally. Researchers are now developing novel and environmentally friendly methods of control. Plant-derived natural products, entomopathogenic fungi, nematodes and bacteria are some of the alternative methods for termite control. The use of botanicals as termiticide is discussed in detail along with other biological alternatives.

6.4.3 Biological Control

6.4.3.1 Fungi

Successful management through entomopathogenic fungi is reported by several authors. Fungal control is gaining significance due to certain advantages like easier mass production techniques, infecting almost all types of insects and simpler and inexpensive mode of application. Almost 750 species of fungi have been documented to be pathogenic, but only a few are currently being used as entomopathogenic agents against insect pests (Fukatzu et al. 1997). Numerous fungi of genera *Termitaria*, *Neotermus*, *Metirolella*, *Laboulbenia*, *Antennopsis*, *Metarhizium*, *Beauveria*, *Leboulbeniopsis* and *Coreomyceptosis* are parasitic to termites (Staples and Milner 1996). The pathogenicity of a fungus towards insects is dependent upon a complex relationship between the ability of the fungus to germinate on the cuticle, and penetrate the cuticle, and the ability of the insect's immune system to prevent fungus growth and infection. Isolates pathogenic for one particular host species may not show the same growth characteristics and pathogenicity in another insect species (Huxham et al. 1989).

Metarhizium anisopliae is a biocontrol agent that requires special application and handling technique. *Metarhizium* causes a "green muscardine" disease to its insect hosts, because of the green colour of its conidial cells. It has wide host range including ticks, thrips, white grubs and different species of termites such as *Reticulitermes* sp., *Odontotermes* sp. and *Coptotermes formosanus* (Bahiense et al. 2006; Dong et al. 2009). Many strains of *M. anisopliae* have been isolated from termites and are reported as effective myco-incesticides for the management of subterranean species. A fungal isolate, C4-B, identified as *M. anisopliae* (Metchnikoff) caused rapid mortality in Formosan subterranean termite alates (Wright et al. 2005). A new virulent *M. anisopliae* variety (*M. anisopliae* var. dcjhyium, DQ288247) was evaluated against *Odontotermes formosanus* and for which it was found to be highly infectious and virulent (Dong et al. 2007). Isolates Ma2 and Ma13 of *M. anisopliae* were found to be pathogenic to workers of *Odontotermes* sp., under laboratory conditions (Balachander et al. 2009).

The effectiveness of the *M. anisopliae* strain ARSEF 6911 was tested in the laboratory and field against *Microtermes obesi* and *O. obesus*. The conidial suspensions induced mortality in laboratory tests. Balachander et al. (2013) conducted an experiment on termite management using fungal (*M. anisopliae*) conidia with a mixture of attractants such as sugarcane bagasse, sawdust and cardboard powder. They obtained increased termite (*O. obesus*) mortality from 50 to 98% (workers) and 16 to 78% (soldiers) with IWST-Ma13 among five tested isolates after 10 days. Augmented foraging activity (from 23 to 58%) was observed for workers and soldiers with same isolate, when conidia were mixed with attractants. They also recorded higher wood protection and reduced mud galleries covering the tree bark in field experiments.

Ravindran et al. (2015) isolated four fungal isolates of *M. anisopliae* and examined their sporulation characteristics and virulence against *C. formosanus*. They found significant termite mortality on the fourth day of inoculation (1×10^7 conidia/ml suspension) of *Metarhizium* isolates. Conidia development on the cadaver surface and complete mycelial growth were also observed, proving the potential of *M. anisopliae* for termite control. Samsuddin et al. (2016) tested pathogenicity of ten isolates of *M. anisopliae* against *Coptotermes curvignathus*. They studied the termite mortality due to fungal transmission, along with fungal progression rate and conidia sporulation. The highest mortality (97%) and shortest median lethal time (LT_{50} = 1.5 days) were obtained with 1×10^7 conidia/ml suspension of PR1 isolate, whereas the highest rate (88%) of mycelia formation and conidia sporulation (80%) was obtained with TFFH3 and PKLG isolates of *M. anisopliae*, respectively. Same isolates were recorded with 93% mortality and LT_{50} above 2 days. They recommended PR1 isolate to be used as potent biopesticide based on performance in infecting *C. curvignathus*.

Yii et al. (2016) investigated synergistic efficacy of termiticide fipronil with *M. anisopliae* in combination, and separately, against *C. curvignathus*. They observed higher termite mortality and reduced lethal effect time for fungus-insecticide bait formulation, due to the compatibility between them. A reduced LT_{50} value of 6.46 days was found, in comparison to control, with combination of sublethal dose of 0.05 mg a.i./l fipronil and fungal conidia concentrations of 10^7 conidia/g bait followed by 4.89 days of LT_{50} at 10^8 conidia/g bait. The highest mortality (99.66%) was recorded with 0.01 mg a.i./l fipronil $+10^9$ conidia/g. They advocated that the sublethal fipronil in the formulated bait weakened the termites and reduced their defence mechanism, which enabled a progress of the fungal infection.

Beauveria bassiana, the white muscardine fungus, causes white (later yellowish or occasionally reddish) muscardine disease (Zimmermann 2007). The genus *Beauveria* has about 49 species, out of which approximately 22 are known as entomopathogenic (Kirk 2003). *Beauveria bassiana* (Balsamo) Vuillemin is highly pathogenic to many insect species in both temperate and tropical regions (Stranes et al. 1993). Isolates of *B. bassiana* and *M. anisopliae* caused complete mortality of *C. formosanus* within 15 days (Delate et al. 1995). *Metarhizium anisopliae* benefits, over *B. bassiana*, by the social behaviour and high production of fungal biomass (Sun et al. 2002). It is, therefore, widely used in microbial management of termites.

Sileshi et al. (2013) studied the antitermitic evaluation of *M. anisopliae* and *B. bassiana* in lab conditions. The concentrations 1×10^5 to 1×10^9 conidia ml^{-1} were used for two isolates of each *M. anisopliae* (PPRC-2 and MM) and *B. bassiana*

(PPRC-56 and 9609), applied by direct spraying of spore suspensions on *Macrotermes* sp. At the fourth day after treatment, 60 to 100% mortalities were recorded with MM (1×10^5) and PPRC-2 (1×10^9) of *M. anisopliae.* Similarly, mortality of termites varied at the same time from 25% for *B. bassiana* isolate 9609 (low concentration) to 95% for PPRC-56 (highest concentration). After 7 days, complete mortality was observed at all the concentrations and isolates. Their study suggested that *M. anisopliae* and *B. bassiana* could be used as effective eco-friendly myco-insecticides, with better efficacy of *M. anisopliae.*

Rana and Kachhawa (2014) used *B. bassiana* and *Paecilomyces fumosoroseus* as biocontrol agents along with *M. anisopliae*, for *in vivo* suppression of termites in a maize field in 2012 and 2013. The fungal pathogens were applied at the rate of 5×10^{13} spores/ha of farmland with farmyard manure (FYM) mixed thoroughly in soil prior to sowing. Promising results were obtained with all the bio-agents used, however, the fungi enriched with FYM treatment showed better results than the direct application of fungi in the field. Soil comprising FYM applied with *M. anisopliae* showed highest germination of maize (91% in 2012 and 95% in 2013) and maximum yields (39.37 q/ha in 2012 and 39.40 q/ha in 2013) with minimum plant mortality after 90 days (3.6% in 2012 and 4.6% in 2013), followed by *P. fumosoroseus* and *B. bassiana.* Some fungi-based commercial products are also available in various countries, such as the fungal termiticide Bio-Blast™ (made of *M. anisopliae*), extensively produced in the USA by EcoScience Co. for household applications (Ramanujam et al. 2014). Entomopathogenic fungi are either used alone or in combination with some baits or attractants. The virulence of the fungal entomopathogens depends on processes such as adhesion, germination, differentiation and penetration. They are affected by a range of intrinsic and external factors, which ultimately determine the fungus pathogenicity (Shahid et al. 2012). Lectins, enzymes and hydrophobic and electrostatic forces also play important roles in fungal propagation (Boucias et al. 1998).

Despite numerous reports on effective termite control by entomopathogenic fungi, the success rate greatly depends on the viability and virulence of fungal inoculum (conidia), infection and transmission, conidia dispersal, enzyme requirement during adhesion and penetration, competition for resources by other opportunistic fast-growing fungi and disease outbreak in the termite population (Yii et al. 2016; Chouvenc et al. 2012; Shahid et al. 2012).

The cryptic life habit of termites also plays a major role in limiting their establishment. Yanagawa et al. (2008) investigated the defence mechanism of termites against entomopathogenic fungi. They reported that *Coptotermes formosanus* removes and discards foreign organisms, such as fungal conidia, from the body surface of its nest-mates by a mutual grooming behaviour to protect themselves. They further proved that the antennae play an important role in detection and removal of conidia from their cuticle, as termites are able to distinguish between treated fungi (Yanagawa et al. 2009). In addition, Hussain and Tian (2013) proved failure of entomopathogenic fungi for termite management in field trials. Disease transmission through "trap and treat" method was found unfeasible due to allogrooming and a large number of individuals in a colony. They found that little inoculums cannot be spread among the nest-mates of a termite colony comprising millions of individuals. Considering the above facts, it can be concluded that the use of entomopathogenic fungi is a sustainable alternative for termite control under controlled conditions. It may serve as an effective method at the field level by (1) improving the ability of fungal adherence and penetration through the host integuments; (2) using an appropriate dose and amount of fungal inocula, depending on the colony size and number of termites in a mound; (3) selecting the favourable conditions for fungi to grow inside the mound; and (4) identifying the optimal environmental conditions, inside and outside the mound.

6.4.3.2 Bacteria

Bacteria may also be utilized in non-chemical termite management. *Bacillus thuringiensis*, *Serratia marcescens*, *Acinetobacter calcoaceticus*, *Aeromonas caviae*, *Alcaligenes latus*, *Arthrobacter* sp., *Bacillus* sp., *Chromobacterium* sp., *Corynebacterium urealyticum*, *Enterobacter gergoviae*, *Micrococcus*, *Neisseria* and *Rhizobium radiobacter* are responsible for termite mortality (Khan et al. 1977a, b, 1985; Osbrink et al. 2001; Devi et al. 2007; Singh 2007).

Pathogenic *Pseudomonas aeruginosa* (Schroeter) caused the mortality of *Microcerotermes championi*, *Heterotermes indicola* and *Coptotermes heimi* (Rhinotermitidae) in the range 25–52% at 7-day postinoculation and up to 84–100% at 25-day postinoculation, in the laboratory (Khan et al. 1992).

Serratia marcescens isolate T8 killed 24% *C. formosanus* after 2 days and 99% after 19 days of assay (Connick et al. 2001). *Bacillus thuringiensis* (Bt) is the most extensively used microbial agent for Lepidoptera, Coleoptera and Diptera. The insecticidal proteins of *B. thuringiensis* have highly specific insect gut toxins and safe towards nontarget organisms (Lacey and Goettel 1995). Two local strains of *B. thuringiensis* impart 100% workers' mortality for *M. championi* and *H. indicola*, within 13 days in laboratory tests (Khan et al. 1977a).

Khan et al. (1985) reported that *H. indicola*, *M. championi* and *Bifiditermes beesoni* (Gardner) were highly susceptible to infection by a Bt commercial preparation (Thuricide-HP, concentrate), showing 100% mortality within 6 days of contact. Few rhizobacterial species produce and emit hydrogen cyanide (HCN) into the rhizosphere which can be toxic to subterranean faunas. These rhizobacteria could be selectively introduced into termite mounds for control. Three species of HCN-producing rhizobacteria, *Rhizobium radiobacter*, *Alcaligenes latus* and *Aeromonas caviae*, were found to be effective in killing the *O. obesus* termites under in vitro conditions (Devi et al. 2007). Cyanide poisoning caused by HCN-producing bacteria such as *P. fluorescens* can kill termites by blocking respiration through inhibition of cytochrome c oxidase, rather than infection (Devi and Kothamasi 2009).

6.4.3.3 Nematodes

Entomopathogenic nematodes (phylum Nematoda) are important biocontrol agents for many insect populations. They parasitize species in the orders Hemiptera, Diptera, Hymenoptera, Lepidoptera, Orthoptera, Coleoptera, Thysanoptera, Siphonaptera as well as Isoptera (Nickle and Welch 1984). Two families, Steinernematidae and Heterorhabditidae, are obligate insect parasites (Poinar 1979). The infective juvenile stage of the nematode infects the host and causes septicaemia and death by releasing symbiotic bacteria into the insect's haemocoel (Kaya and Gaugler 1993).

Infective stage of *Steinernema carpocapsae* (Weiser) (Steinernematidae) provided 96% mortality of *C. formosanus* within 7 days of exposure, in laboratory experiments (Fujii 1975). High mortality of *R. flavipes* (Kollar) termites with nematodes was observed in laboratory tests (Trudeau 1989). Weeks and Baker (2004) reported the efficacy of two species of entomopathogenic nematode, *Heterorhabditis bacteriophora* (Poinar) and *Steinernema carpocapsae* (Weiser). The latter proved to be more potent in causing termite mortality than *H. bacteriophora*, which could be directly linked to the survivability of both species. Yu et al. (2006) reported that *Steinernema riobrave*, *S. carpocapsae*, *S. feltiae* (Filipjev), and *Heterohabditis bacteriophora* efficiently infected and killed three termite species, *Heterotermes aureus* (Synder), *Gnathamitermes perplexus* (Banks) and *R. flavipes* (Kollar), in laboratory sand assays. However, *S. riobrave* and *S. feltiae* caused low mortality in *R. virginicus* (Banks) under similar conditions.

In laboratory bioassays, Yu et al. (2008) observed that *S. riobrave* Cabanillas, Poinar and Raulston, *S. carpocapsae*, *S. feltiae* and *H. bacteriophora* infected and killed the subterranean termite, *Heterotermes aureus* (Snyder). *Steinernema carpocapsae*, *S. riobrave* and *H. bacteriophora* successfully reproduced in *H. aureus*, and infective juveniles exited the termite cadavers successfully. Yu et al. (2010) conducted studies on entomopathogenic nematodes against termites and found a novel strain of *S. riobrave* which possesses superior virulence to subterranean termites. Murugan and Vasugi (2011) conducted laboratory studies on the bioactivities of entomopathogenic nematodes and neem seed kernel extract (NSKE) against workers of *R. flavipes*.

Entomopathogenic nematodes have a number of characteristics that make them appropriate for biological control and commercial production as microbial biopesticides. They have a broad host range among soil-dwelling pests; are easy to produce, store and apply; are safe to vertebrates, plants and other nontarget organisms; and are responsive to genetic selection (Kaya and Gaugler 1993; Kaya et al. 1993).

6.4.3.4 Botanicals

Plants may provide potential alternatives to currently used insect control agents because they constitute a rich source of bioactive chemicals. Over 2000 plants belonging to some 60 families are known to exhibit insecticidal activities (Dev and

Koul 1997). These chemicals include insecticides, antifeedants, insect growth regulators (IGRs), juvenile hormones, ecdysones, repellents, attractants, arrestants, etc. Plants are hence thought to be an important alternative source of chemical pesticides (Kannaiyan 1999). Plant-derived natural products and biocontrol agents are promising replacements to chemicals for pest control. The alternative control strategies were initiated in 1935, where cellulose blocks were treated with citronellic acid, to test against termites (Trikojus 1935).

Different plant parts (leaves, stem, root, bark, flower, fruit, resin, etc.) and their extracts were tested by researchers throughout the world to exploit their pest control efficiency. These botanicals are locally available and low cost and found to be active in termite control. Laboratory and field studies have recognized a wide range of plants with toxic and repellent or antifeedant properties against termites, and many have been considered for use as insecticides (Stoll 1986; Gerrits and VanLatum 1988; Harborne 1988). These plant-based insecticides are biodegradable and safe to humans, the environment and nontarget species.

Essential oils are volatile oils that have strong aromatic components that give distinctive odours, flavours or scents. These are the by-products of plant metabolism and are usually referred to as volatile plant secondary metabolites. They are found in glandular hairs or secretory cavities of the plant cell wall and are present as droplets of fluid in the leaves, stems, bark, flowers, roots and fruits, in different species. They often are complex mixtures of monoterpenes, phenols and sesquiterpenes. Plant essential oils display a broad spectrum of activity against insects and plant pathogenic fungi, ranging from insecticidal, antifeedant, repellent, deterrents of oviposition, growth regulators and vector antagonists (Koul et al. 2008). Antitermitic activity of various plant essential oils is enlisted in Table 6.1.

The insecticidal activity of essential oils was evaluated as early as 1972 (Nakashima and Shimizu 1972). Some plant essential oils not only repel insects but also have contact and fumigant insecticidal activities against specific pests (Isman 2000). The essential oil from the fruits of *Myristica fragrans*, and its major constituents, was evaluated against *Microcerotermes beesoni*. The LC_{50} value of fruit essential oil is 28.6 mg/g, and the compound myristicin caused 100% mortality at a dosage of 5 mg/g after 14 days (Pal et al. 2011). Different parts of a number of plants like leaf, flower, fruit and root contain indeed some bioactive components, and many can be utilized as termite control agents (Table 6.2).

Laboratory assays were conducted for barrier and repellent activity of *Jatropha curcas* oil against the Philippine milk termite, *Coptotermes vastator* Light. The oil was found to be antifeedant, with a reduction in tunnelling activity observed along with increased termite mortality (Acda 2009). Kareru et al. (2010) produced a surface coating based on *Thevetia peruviana* (Pers.) K. Schum seed oil, with antitermite properties. The paint significantly repelled and protected wood from subterranean termite attacks (*Microtermes* sp.). Ding and Hu (2010) studied the repellent activity of flowers, leaves, stems and roots of *Lantana camara* L. The addition of chipped fresh lantana leaves and stems into soil caused substantial reduction in tunnelling and repelled *C. formosanus* termite species from penetrating the barrier and thus prevented infestation of a piece of wood on the other side of the

Table 6.1 Essential oils as termite control agents

Essential oil	Bioactive constituent	Termite sp.	Mode of action	References
Chamaecyparis obtusa var. *formosana* *Calocedrus macrolepis* var. *formosana* and *Cryptomeria japonica* (heartwood, sapwood and leaves)	–	*C. formosanus*	Repellent and toxic	Cheng et al. (2007)
Citrus (peel)	*d*-limonene	*C. formosanus*	Toxic	Raina et al. (2007)
Trachyspermum ammi, *Pimenta dioica*, *Carum carvi*, *Anethum graveolens*, *Pelargonium graveolens* and *Litsea cubeba*	Thymol and carvacrol, eugenol, citronellol, geraniol, verbenol, 2-phenylethanol, nerol	*Reticulitermes speratus* Kolbe	Fumigant and toxic	Seo et al. (2009)
Mentha arvensis, *Cymbopogon citratus*, *Carum copticum*	–	*O. obesus*	Toxic	Gupta et al. (2011)
Myristica fragrans (fruits)	α-Pinene, Sabinene, β-pinene, myrcene, limonene, terpine-4-ol, safrole and myristicin	*Microcerotermes beesoni*	Toxic	Pal et al. (2011)
Cymbopogon citratus, *Eucalyptus globulus*, *Syzygium aromaticum*, *Origanum vulgare*, *Rosmarinus officinalis*, *Cinnamomum verum* and *Thymus vulgaris*	–	*O. assamensis Holmgren*	Toxic	Pandey et al. (2012)
Chamaemelum nobile, *Santolina chamaecyparissus*, *Ormenis multicaulis* and *Eriocephalus punctulatus*	Trans-pinocarveol, caryophyllene oxide, sabinene hydrate, and santolina alcohol	*Reticulitermes speratus* Kolbe	Toxic	Seo et al. (2014)
Tagetes erecta and *Citrus sinensis*	–	*Odontotermes obesus*	Repellancy	Verma et al. (2016)

Table 6.2 Plant extracts (leaves, roots, stem, flower, fruit, seed, seed oil) as termite control agents

Botanical	Bioactive constituent	Termite sp.	Mode of action	References
Lantana camara var. *aculeate* (leaves)	Triterpenoid, 22 β-acetoxylantic acid	*O. obesus*	Toxic	Verma et al. (2005); Verma and Verma (2006)
Withania somnifera, *Croton tiglium*, *Hygrophila auriculata* (seed and leaf)	–	*Microtermes obesi*	Toxic	Ahmed et al. (2006b)
Echinops ritro L., *E. spinosissimus* subsp. *Spinosissimus* Turra, *E. albicaulis* Kar. and Kir., *E. transiliensis* Golosh	Thiophenes, 2,2′:5′,2″-Terthiophene and 5′-(3-buten-1-ynyl)-2,2′-bithiophene	*C. formosanus*	Toxic	Fokialakis et al. (2006)
Piper nigrum (seed)	Guineensine	*C. formosanus*	Toxic	Meepagala et al. (2006)
Sophora flavescens Aiton	Alkaloids, matrine and oxymatrine	*C. formosanus*	Antifeedant and acute residual toxicity	Mao and Henderson (2007)
Ligularia macrophylla	*Eremophilanes*	*C. formosanus*	Toxic	Cantrell et al. (2007)
Ocimum canum, *O. gratissimum*, *Zanthoxylum zanthoxyloides*, *Sporobolus pyramidalis*, *Allium sativum* (whole plant)	–	*Macrotermes* sp.	Repellent, Antifeedant	Owusu et al. (2008)
Euphorbia kansui (roots)	Diterpenes	*R. speratus*	Toxic	Shi et al. (2008)
Anacardium occidentale (nut shell liquid)	Anacardic acid, cardanol, methyl anacardate	*C. vastator*	Toxic and repellent	Boongaling et al. (2008)
Jatropha curcas (seed oil)	–	*Microcerotermes beesoni*	Toxicity	Singh and Kumar (2008)

(continued)

Table 6.2 (continued)

Botanical	Bioactive constituent	Termite sp.	Mode of action	References
Jatropha curcas (seed oil)	–	*C. vastator*	Toxic, Repellent and antifeedant	Acda (2009)
L. camara, *Monodora myristica*, *Euphorbia Lateriflora* (leaves)		*M. michaelseni*	Toxic and antifeedant	Ogunsina et al. (2009)
Protium javanicum (leaves)	Scopoletin	*C. formosanus*	Toxic	Adfa et al. (2010)
Capparis decidua (whole plant)		*O. obesus*	Toxic	Upadhyay et al. (2010)
Aleurites moluccana (seed oil)	–	*C. formosanus*	Antifeedant	Nakayama and Osbrink (2010)
Lantana camara (leaves, stem and flowers)	–	*C. formosanus*	Repellent	Ding and Hu (2010)
Thevetia peruviana (Pers.) K. Schum oil	–	*Microtermes* sp.	Repellent	Kareru et al. (2010)
Opuntia ficus indica cladodes, *Moringa oleifera* (seeds)	Lectin	*N. corniger*	Toxic	Paiva et al. (2011)
Ocimum sanctum (inflorescences, leaves, stem and root)	–	*Heterotermes indicola*	Toxic and repellent	Manzoor et al. (2011a)
Curcuma longa, *Nerium indicum*, *Melia azedarach*	–	*Heterotermes indicola*	Toxic and repellent	Manzoor et al. (2011b)
Pongamia pinnata, *Jatropha curcas* (seed oil)	Karanjin, phorbol esters	*O. obesus*	Toxic	Verma et al. (2011)
L. camara (leaves)	–	*R. virginicus*	Toxic, antifeedant	Yuan and Hu (2011)
L. camara (leaves)		*R. flavipes*	Repellent, antifeedant and toxic	Yuan and Hu (2012)

Andrographis lineata, *A. paniculata*, *Argemone mexicana* L., *Aristolochia bracteolata* Lam., *Datura metel* L., *Eclipta prostrata* L., *Sesbania grandiflora* L., *Tagetes erecta* L. (leaves)	–	*C. formosanus*	Toxic	Elango et al. (2012)
Capparis decidua (stem, root, flower and fruit extracts)	Heneicosylhexadecanoate, triacontanol, and 2-carboxy-1, 1-dimethylpyrrolidine) and one novel compound 6-(1-hydroxy-non-3-enyl)-tetrahydropyran-2 one	*Odontotermes obesus*	Toxic and repellent	Upadhyay et al. (2012)
Maesa lanceolata, *Chenopodium* spp., *Croton macrostachyus, Tagetes minuta*, *Datura stramonium*, *Vernonia amygdalina*, *Phytolacca dodecandra*, *Nicotiana tobacum*, *Schinus molle* and *Ficus vasta* (leaves), *A. indica* (seeds)	–	–	Toxic	Ibrahim and Demisse (2013)
Jatropha curcas (root)	Terpenes and fatty acids	*Microtermes obesi*	Toxic	Verma et al. (2013)
Coconut (shell oil)	–	*Odontotermes horni*, *O. obesus*, *O. redemanni* and *Microtermes obesi*		Shiny and Remadevi (2014)
Camellia oleifera Abel	Saponin	*Reticulitermes flavipes* (e.g., *santonensis*)	Toxic	Hu et al. (2015)

barrier. Leaves, stems and flowers appeared more repellency than roots. They suggested the use of fresh-cut lantana leaves, stems and flowers as additives to garden mulches to control termites. The termiticidal activity of *Opuntia ficus indica* and *Moringa oleifera* seeds was studied by Paiva et al. (2011). The lectin of *O. ficus indica* was found to be more toxic against *Nasutitermes corniger* workers than that isolated from *M. urundeuva* bark, heartwood and leaf. Purified *O. ficus indica* lectin showed stronger termiticidal activity against workers (LC_{50} of 0.116 mg ml/1) than soldiers. Antitermitic activity of *Curcuma longa*, *Nerium indicum* and *Melia azedarach* was evaluated against the subterranean termite *Heterotermes indicola* (Wasmann). All the extracts showed repellency and extracts were toxic in a no-choice test. Soil barrier tests also showed that the *C. longa* extract was the most efficient in soil treatments (Manzoor et al. 2011b).

Upadhyay et al. (2012) investigated the antitermitic activity of *Capparis decidua* stem, root, flower, fruit extracts and pure compounds against *Odontotermes obesus*. Crude stem extracts were very effective against termites. Three pure compounds were isolated from stem extract (heneicosylhexadecanoate, triacontanol and 2-carboxy-1, 1-dimethylpyrrolidine), whereas one novel compound 6-(1-hydroxy-non-3-enyl)-tetrahydropyran-2 one) was isolated from flower extracts. All had very low LD_{50} values in worker termites.

Ibrahim and Demisse (2013) conducted a field study to estimate the efficacy of 11 plants against termites on hot pepper at Bako, Western Ethiopia. Lowest damage caused by termites was observed with *Maesa lanceolata* and *Azadirachta indica*, which was equivalent to the chemical diazinon 60EC (2500 ml/ha). The study confirmed that these two botanicals can be effective in integrated termite management practice.

Apart of its repellent activity (see above), solvent extracts of various parts of *Jatropha curcas* were tested against *Microtermes obesi*. Verma et al. (2013) tested the methanolic extract of *J. curcas* root and found it to be the most promising against termites. GCMS analysis of *J. curcas* roots revealed an array of active constituents with terpenes and fatty acids as dominant compounds.

Seo et al. (2014) examined the fumigant toxicity of *Chamaemelum nobile*, *Santolina chamaecyparissus*, *Ormenis multicaulis* and *Eriocephalus punctulatus* and their constituents against the Japanese termite *Reticulitermes speratus* Kolbe. After 2 days of treatment, *Chamaemelum nobile* exhibited the strongest fumigant toxicity, followed by those from *Santolina chamaecyparissus*, *Ormenis multicaulis* and *Eriocephalus punctulatus*. Different concentrations of *Azadirachta indica*, *Jatropha curcas*, *Maesa lanceolata*, *Chenopodium ambrosioides* and *Vernonia hymenolepis* were tested on *Macrotermes* sp. All the extracts were able to cause mortality on *Macrotermes* with a lethal time (LT_{50}) lower than the untreated control. Among all extracts, *J. curcas* was observed to be the most potent followed by seed extracts of *A. indica* and leaf extracts of *M. lanceolata* (Addisu et al. 2014).

Hu et al. (2015) studied the activity of *Camellia oleifera* Abel. against *R. flavipes*. The results showed that shells are not repulsive but toxic by ingestion, due to the presence of a saponin which can prevent termite attacks. Coconut shell oil was also tested as a potential termiticide during 18 months of field trial. The control stakes were destroyed entirely vs 34.2% damage in the coconut shell oil brush-coated stakes. High phenol content or the presence of high oxygenated fractions might be the reason of the termiticidal activity (Shiny and Remadevi 2014).

Some timbers are resistant to termite attacks, due to the presence of active components as part of their natural defence systems. Researchers exploited this potential of wood by extracting and testing the active components. Table 6.3 shows the different tree species and the active components responsible for termite control. Wooden blocks impregnated with liquid carbon dioxide and ethanol extracts of eastern red cedar (*Juniperus virginiana* L.) tested for resistance against *R. flavipes* showed noteworthy resistance to termite damage, compared to untreated controls. These extracts may provide a natural source of safe natural wood preservatives (Eller et al. 2010). Similarly, Franca et al. (2016) investigated the natural resistance of heartwood and sapwood of two African mahogany species, *Khaya ivorensis* and *K. senegalensis*, which showed resistance to subterranean termites.

Until now, numerous plants and their parts have been identified as potent termiticidal agents. However, their efficacy is not comparable to chemical termiticides. To overcome this limitation, the active components present in botanicals have to be isolated and tested for their termiticidal ability. Further, these bioactive components should be combined in various ratios to assess their synergistic potential. Synergism may increase their bioactivity many folds, as compared to the single compounds. Although the botanical pesticides are low in action as compared to chemical pesticides, they offer a safe perspective to retain healthy soil, water and air for future generations. Continuous efforts should hence be applied in the search for novel and efficient natural insecticides.

Table 6.3 Wood extracts as termite control agents

Botanical	Bioactive constituent	Termite sp.	Mode of action	References
Artocarpus heterophyllus	Artocarpin	*C. formosanus* and *R. speratus*	Toxic	Shibutani et al. (2006)
Thuja plicata, *Chamaecyparis nootkatensis*	–	*C. formosanus*	Antifeedant	Taylor et al. (2006)
Bagassa guianensis, *Erisma uncinatum*	–	*Nasutitermes* sp.	Toxic	Peres et al. (2006)
Myracrodruon urundeuva	Lectin	*N. corniger*	Repellent and toxic	Sa´ et al. (2008)
Bowdichia virgilioides	–	*N. corniger*	Repellent	Santana et al. (2010)
Juniperus virginiana	Cedrol and α-cedrene	*R. flavipes*	Toxic	Eller et al. (2010)
Sextonia rubra	Rubrynolide	–	Toxic	Rodrigues et al. (2011)
Khaya ivorensis, *Khaya senegalensis*	–	*R. flavipes*	Antifeedant	Franca et al. (2016)

6.5 Conclusion

Termites are an enormous hazard to agricultural crops and wooden structures, due to the pronounced economic losses induced throughout the tropic and temperate regions. Chemical termiticides are the most common and effective methods of control. Nevertheless, their detrimental consequences are a matter of concern for researchers and people throughout the world. The undesirable side effects and destruction of nontarget species by these chemicals generate an urgent need to search for ecologically safer alternatives. Various non-chemical alternatives may be applied for termite control, including fungi, bacteria and nematodes, with plant-derived termiticidal botanicals. Their utilization is sustainable, offering many advantages. These are biodegradability, easily and locally availability, no effect on nontarget species and lack of induction of pest resistance. These botanicals and their bioactive components can be isolated, characterized, formulated and tested in field conditions for use as commercial termiticides.

References

Abe, T. (1995). *The termite-symbionts system: how does it work and has it evolved as a superefficient decomposer in tropical terrestrial ecosystems?* Kyoto: Center for Ecological Research, Kyoto University.

Acda, M. N. (2009). Toxicity, tunneling and feeding behavior of the termite, *Coptotermes vastator,* in sand treated with oil of the physic nut, *Jatropha curcas*. *Journal of Insect Science, 9*, 64.

Addisu, S., Mohamed, D., & Waktole, S. (2014). Efficacy of botanical extracts against termites, *Macrotermes* sp., (Isoptera: Termitidae) under laboratory conditions. *International Journal of Agricultural Research, 9*, 607–603.

Adfa, M., Yoshimura, T., Komura, K., & Koketsu, M. (2010). Antitermite activities of coumarin derivatives and scopoletin from *Protium javanicum* Burm. f. *Journal of Chemical Ecology, 36*, 720–726.

Ahmed, S., Mustafa, T., Riaz, M. A., & Hussain, A. (2006a). Efficacy of insecticides against subterranean termites in sugarcane. *International Journal of Agricultural and Biology, 8*, 508–510.

Ahmed, S., Riaz, M. A., & Shahid, M. (2006b). Response of *Microtermes obesi* (Isoptera: Termitidae) and its gut bacteria towards some plant extracts. *Journal of Food, Agriculture and Environment, 4*, 317–320.

Bahiense, T. C., Fernandes, E. K. K., & Bittencourt, V. R. E. P. (2006). Compatibility of the fungus *M. anisopliae* and deltamethrin to control a resistant isolate of *Boophilus microplus* tick. *Veterinary Parasitology, 141*, 319–324.

Baker, P. (2005). *Termite management for homeowners – cooperative extension*. The University of Arizona Cooperative Extension http://extension.arizona.edu/sites/extension.arizona.edu/files/pubs/az1356.pdf

Balachander, M., Remadevi, O. K., Sasidharan TO, & Sapna Bai, N. (2009). Infectivity of *Metarhizium anisopliae* (Deuteromycotina: Hyphomycetes) isolates to the arboreal termite *Odontotermes* sp. (Isoptera: Termitidae). *International Journal of Tropical Insect Science, 29*, 202–207.

Balachander, M., Remadevi, O. K., & Sasidharan TO. (2013). Dissemination of *Metarhizium anisopliae* infection among the population of *Odontotermes obesus* (Isoptera: Termitidae)

by augmenting the fungal conidia with attractants. *Journal of Asia-Pacific Entomology, 16*, 199–208.
Bayatkashkoli, A., Taghiyari, H. R., Kameshki, B., Ravan, S., & Shamsian, M. (2016). Effects of zinc and copper salicylate on biological resistance of particleboard against *Anacanthotermes vagans* termite. *International Biodeterioration & Biodegradation, 115*, 26–30.
Bhanot, J. P., Batra, G. R., & Verma, A. N. (1991a). Effect of seed treatment with different insecticides on germination, termite damage and grain yield of barley. *Journal of Insect Science, 4*, 61–63.
Bhanot, J. P., Verma, A. N., & Batra, G. R. (1991b). Effect of seed treatment with different insecticides on germination, termite (*Microtermes obesi*) damage and grain yield of wheat (*Triticum aestivum*). *Indian Journal of Agriculture Science, 61*, 689–691.
Bhanot, J. P., & Singal, S. K. (2007). Management of termites in field crops. In J. P. Bhanot, H. R. Rohilla, & V. K. Kalra (Eds.), *Recent trends in biology and management of polyphagous pests of agricultural importance* (pp. 166–172). Hisar: CCS Haryana Agricultural University.
Bhanot, J. P., Sharma, A. K., Batra, G. R., & Verma, A. N. (1995). Influence of different levels of irrigation and fertilizer on termite damage and yield of gram crop raised from aldrin treated and untreated seed. *Journal of Insect Science, 7*, 115–116.
Bignell, D. E., & Eggleton, P. (1998). Termites. In P. Calow (Ed.), *Encyclopedia of ecology and environmental management* (pp. 744–746). Oxford: Blackwell Scientific.
Boomsma, J. J., Baer, B., & Heinze, J. (2005). The evolution of male traits in social insects. *Annual Review of Entomology, 50*, 395–420.
Boongaling, E. G., Acda, M. N., Hernandez, H. P., & Rosario, E. J. (2008). Fractionation, derivatization and termiticidal activity of cashewnut (*Anacardium occidentale* L.) shell liquid against the Philippine milk termite (*Coptotermes vastator* Light). *The Philippine Agricultural Scientist, 91*, 408–415.
Boucias, D. G., Farmerie, W. G., & Pendland, J. C. (1998). Cloning and sequencing the cDNA of the insecticidal toxin, Hirsutellin A. *Journal of Invertebrate Pathology, 72*, 258–261.
Breznak, J. A. (1982). Intestinal microbiota of termites and other xylophagous insects. *Annual Review of Microbiology, 36*, 323–343.
Breznak, J. A., & Brune, A. (1994). Role of microorganisms in the digestion of lignocelluloses by termites. *Annual Review of Entomology, 39*, 453–487.
Bultman, J. D., Beal, R. H., & Ampong, F. F. K. (1979). Natural resistance of some tropical African woods to *Coptotermes formosanus* Shiraki. *Forest Products Journal, 29*, 46–51.
Buss, E. A., & Park-Brown, S. G. (2002). *Natural products for insect pest management*. London: Chapman Hall.
Cantrell, C. L., Duke, S. O., Fronczek, F. R., Osbrink, W. L. A., Mamonov, L. K., Vassilyev, J. I., Wedge, D. E., & Dayan, F. E. (2007). Phytotoxic eremophilanes from *Ligularia macrophylla*. *Journal of Agricultural and Food Chemistry, 55*, 10656–10663.
Cheng, S. S., Chang, H. T., CL, W., & Chang, S. T. (2007). Anti-termitic activities of essential oils from coniferous trees against *Coptotermes formosanus*. *Bioresource Technology, 98*, 456–459.
Chouvenc, T., Efstathion, C. A., Elliott, M. L., & Su, N. Y. (2012). Resource competition between two fungal parasites in subterranean termites. *Naturwissenschaften, 99*, 949–958.
Collins, N. M. (1988). Termites. In E. Cranbrook (Ed.), *Key environments of Malaysia*. Oxford: Pergamon.
Connick, W. J., Jr., Osbrink, W. L. A., Wright, M. S., Williams, K. S., Daigle, D. J., Boykin, W. L., & Lax, A. R. (2001). Increased mortality of *Coptotermes formosanus* (Isoptera: Rhinotermitidae) exposed to eicosanoid biosynthesis inhibitors and *Serratia marcescens* (Eubacteriales: Enterobacteriaceae). *Environmental Entomology, 30*, 449–455.
Cox, R. A., Rattigan, O. V., & Jones, R. L. (1995). Laboratory studies of BrO reactions of interest for the atmospheric ozone balance. In A. R. Bandy (Ed.), *The chemistry of the atmosphere: Oxidants and oxidation in the earth's atmosphere* (pp. 47–64). Cambridge: Royal Society of Chemistry.

Delate, K. M., Grace, J. K., & Tome, C. H. M. (1995). Potential use of pathogenic fungi in baits to control the Formosan subterranean termite (Isoptera, Rhinotermitidae). *Journal of Applied Entomology, 119*, 429–433.

Devlin, J. F., & Zettel, T. (1999). *Ecoagriculture: Initiatives in eastern and Southern Africa*. Harare: Weaver Press.

Dev, S., & Koul, O. (1997). Higher plants. In *Insecticides of natural origin* (p. 365). Amsterdam: Harwood Academic Press.

Devi, K. K., & Kothamasi, D. (2009). *Pseudomonas fluorescens* CHA0 can kill subterranean termite *Odontotermes obesus* by inhibiting cytochrome oxidase of the termite respiratory chain. *FEMS Microbiology Letters, 300*, 195–200.

Devi, K. K., Seth, N., Kothamasi, S., & Kothamasi, D. (2007). Hydrogen cyanide-producing rhizobacteria kill subterranean termite *Odontotermes obesus* (Rambur) by cyanide poisoning under in vitro conditions. *Current Microbiology, 54*, 74–78.

Dibaa, F., Hadary, F., Panjaitan, S. D., & Yoshimura, T. (2013). Electromagnetic waves as nondestructive method to control subterranean termites Coptotermes curvignathus Holmgren and Coptotermes formosanus Shiraki. The 3rd International conference on sustainable future for human security SUSTAIN 2012. *Procedia Environmental Sciences* 17:150–159.

Diehl, E., Junqueira, L. K., & Berti-Filho, E. (2005). Ant and termite mound coinhabitants in the wetlands of Santo Antonio da Patrulha, Rio Grande do Sul, Brazil. *Brazilian Journal of Biology, 65*, 431–437.

Ding, W., & Hu, X. P. (2010). Antitermitic effect of the *Lantana camara* plant on subterranean termites (Isoptera: Rhinotermitidae). *Insect Sci., 17*, 427–433.

Dong, C., Zhang, J., Chen, W., Huang, H., & Hu, Y. (2007). Characterization of a newly discovered China variety of *Metarhizium anisopliae* (*M. anisopliae* var. dcjhyium) for virulence to termites, isoenzyme, and phylogenic analysis. *Microbiological Research, 162*(1), 53–61.

Dong, C., Zhang, J., Huang, H., Chen, W., & Hu, Y. (2009). Pathogenicity of new China variety of *Metarhizium anisopliae* (*M. anisopliae* var. dcjhyium) to subterranean termite, *Odontotermes formosanus*. *Microbiological Research, 164*, 27–35.

Elango, G., Abdul Rahuman, A., Kamaraj, C., Bagavan, A., Abduz Zahir, A., Santhoshkumar, T., Marimuthu, S., Velayutham, K., Jayaseelan, C., Vishnu Kirthi, A., & Rajakumar, G. (2012). Efficacy of medicinal plant extracts against Formosan subterranean termite, *Coptotermes formosanus*. *Industrial Crops and Products, 36*, 524–530.

Eller, F. J., Clausen, C. A., Green, F., & Taylor, S. L. (2010). Critical fluid extraction of *Juniperus virginiana* L. and bioactivity of extracts against subterranean termites and wood-rot fungi. *Industrial Crops and Products, 32*, 481–485.

Fokialakis, N., Osbrink, W. L. A., Mamonov, L. K., Gemejieva, N. G., Mims, A. B., Skaltsounis, A. L., Lax, A. R., & Cantrell, C. L. (2006). Antifeedant and toxicity effects of thiophenes from four *Echinops* species against the Formosan subterranean termite, *Coptotermes formosanus*. *Pest Management Science, 62*, 832–838.

Franca, T. S. F. A., Franca, F. J. N., Arango, R. A., Woodward, B. M., & Arantes, M. D. C. (2016). Natural resistance of plantation grown African mahogany (*Khaya ivorensis* and *Khaya senegalensis*) from Brazil to wood-rot fungi and subterranean termites. *International Biodeterioration & Biodegradation, 107*, 88–91.

Fujii, J. K. (1975). *Effect of an entomogenous nematode Neoaplectana carpocapsae Weiser, on the Formosan subterranean termite, Coptotermes formosanus Shiraki, with ecological and biological studies on C. formosanus*. Ph.D dissertation, University of Hawaii, Honolulu, Hawaii, USA, pp. 163.

Fukatzu, T., Sato, H., & Kuriyama, H. (1997). Isolation, inoculation to insect host, and molecular phylogeny of an entomogenous fungus *Paecilomyces tenuipes*. *Journal of Invertebrate Pathology, 70*, 203–208.

Gerrits, R., & VanLatum, E. B. J. (1988). *Plant-derived pesticides in developing countries: Possibilities and research needs* (p. 104). The Hague: Netherlands' Ministry of Housing, Physical Planning and Environment.

Grace, J. K. (1998). Resistance of pine treated with chromated copper arsenate to the Formosan subterranean termite. *Forest Products Journal, 48*, 79–82.

Grace, J. K., Yates, J. R., III, Tome, C. H. M., & Ohino, R. J. (1996). Termite-resistant construction: Uses of a stainless steel. Mesh to exclude *Coptotermes formasanus* (Isoptera: Rhinotermitidae). *Sociobiology, 28*, 365–372.

Guan, Y. Q., Chen, J., Tang, J., Yang, L., & Liu, J. M. (2011). Immobilizing bifenthrin on wood for termite control. *International Biodeterioration & Biodegradation, 65*, 389–395.

Gupta, A., Sharma, S., & Naik, S. N. (2011). Biopesticidal value of selected essential oils against pathogenic fungus, termites, and nematodes. *International Biodeterioration & Biodegradation, 65*, 703–707.

Harborne, J. B. (1988). *Introduction to ecological biochemistry* (3rd ed.p. 356). London: Academic.

Haverty, M. I., Woodrow, R. J., Nelson, L. J., & Grace, J. K. (2005). Identification of termite species by the hydrocarbons in their feces. *Journal of Chemical Ecology, 31*, 2119–2151.

Higashi, M., Yamamura, N., Abe, T., & Burns, T. P. (1991). Why don't all termite species have a sterile worker caste? *Proceedings of the Biological Sciences, 246*, 25–29.

Hu, J., Chang, S., Peng, K., Hu, K., & Thevenon, M. F. (2015). Bio-susceptibility of shells of *Camellia oleifera* Abel. fruits to fungi and termites. *International Biodeterioration & Biodegradation, 104*, 219–223.

Hussain, A., & Tian, M. Y. (2013). Germination pattern and inoculum transfer of entomopathogenic fungi and their role in disease resistance among *Coptotermes formosanus* (Isoptera: Rhinotermitidae). *International Journal of Agriculture and Biology, 12*, 319–324.

Husseneder, C., Messenger, M. T., Su, N. Y., Grace, J. K., & Vargo, E. L. (2005). Colony social organization and population genetic structure of an introduced population of Formosan subterranean termite from New Orleans, Louisiana. *Journal of Economic Entomology, 98*, 1421–1434.

Husseneder, C., & Simms, D. M. (2008). Size and heterozygosity influence partner selection in the Formosan subterranean termite. *Behavioral Ecology, 19*, 764–773.

Huxham, L. M., Samueli, K. D. Z., Hele, J. B., & McCorkindale, N. J. (1989). *In vivo* and *in vitro* assays for pathogenicity of wild type and mutant strains of *Metarhizium anisopliae* for three insect species. *Journal of Invertebrate Pathology, 53*, 143–151.

Ibrahim, A., & Demisse, G. (2013). Evaluation of some botanicals against termites' damage on hot pepper at Bako, western Ethiopia. *International Journal of Agricultural Policy and Research, 1*, 48–52.

Isman, M. B. (2000). Plant essential oils for pest and disease management. *Crop Protection, 19*, 10603–10608.

Joshi, P. K., Singh, N. P., Singh, N. N., Gerpacio, R. V., & Pingali, P. L. (2005). *Maize in India: Production systems, constraints and research priorities* (p. 22). Mexico: D.F. CIMMYT.

Kannaiyan, S. (1999). *Botanicals in pest control.* Inaugural address in the training programme on botanicals in pest management, December 1–10, 1999. Coimbatore: Tamilnadu Agriculture University.

Kard, B. M. (1999). Mesh may fit in as a termite barrier. *Pest Control, 67*, 50–53.

Kareru, P. G., Keriko, J. M., Kenji, G. M., & Gachanja, A. N. (2010). Anti-termite and antimicrobial properties of paint made from *Thevetia peruviana* (Pers.) Schum. oil extract. *African Journal of Pharmacy and Pharmacology, 4*, 087–089.

Kaya, H. K., & Gaugler, R. (1993). Entomopathogenic nematodes. *Annual Review of Entomology, 38*, 181–206.

Kaya, H. K., Bedding, R. A., & Akhurst, R. J. (1993). An overview of insect-parasitic and entomopathogenic nematodes. In R. A. Bedding, R. Akhurst, & H. Kaya (Eds.), *Nematodes and the biological control of insect pest* (pp. 1–10). East Melbourne: CSIRO.

Khan, K. I., Fazal, Q., & Jafri, R. H. (1977a). Pathogenicity of locally discovered *Bacillus thuringiensis* strain to the termites: *Heterotermes indicola* (Wassman) and *Microcerotermes championi* (Snyder). *Pakistan Journal of Scientific Research, 29*, 12–13.

Khan, K. I., Fazal, Q., Jafri, R. H., & Ahmad, M. (1977b). Susceptibility of various species of termites to a pathogen, Serratia marcescens. *Pakistan Journal of Scientific Research, 29*, 46–47.

Khan, K. I., Jafri, R. H., & Ahmad, M. (1985). The pathogenicity and development of *Bacillus thuringiensis* in termites. *Pakistan Journal of Zoology, 17*, 201–209.

Khan, K. I., Jafri, R. H., Ahmad, M., & Khan, K. M. S. (1992). The pathogenicity of *Pseudomonas aeruginosa* against termites. *Pakistan Journal of Zoology, 24*, 243–245.

Kirk, P. M. (2003). *Indexfungorum*. Available at http://www.indexfungorum.org (28/10/2009).

Koul, O., Walia, S., & Dhaliwal, G. S. (2008). Essential oils as green pesticides: Potential and constraints. *Biopesticides International, 4*, 63–84.

Kuriachan, I., & Gold, R. E. (1998). Evaluation of the ability of *Reticulitermes flavipes Kollar*, a subterranean termite (Isoptera: Rhinotermitidae) to differentiate between termiticide treated and untreated soils in laboratory tests. *Sociobiology, 32*, 151–166.

Lacey, L. A., & Goettel, M. (1995). Current developments in microbial control of insect pests and prospects for the early 21st century. *Entomophaga, 40*, 3–27.

Lenz, M., & Runko, S. (1994). Protection of buildings, other structures and materials in ground contact from attack by subterranean termites (Isoptera) with a physical barrier-a fine mesh of high grade stainless steel. *Sociobiology, 24*, 1–16.

Lewis, V. R. (1997). Alternative control strategies for termites. *Journal of Agricultural Entomology, 14*, 291–307.

Lewis, V. R., & Haverty, M. I. (2000). Lethal effects of electrical shock treatments to the western drywood termite (isopteran: Kalotermitidae) and resulting damage to wooden test boards. *Sociobiology, 37*, 163–183.

Manzoor, F., Beena, W., Malik, S., Naz, N., Naz, S., & Syed, W. H. (2011a). Preliminary evaluation of *Ocimum sanctum* as toxicant and repellent against termite, *Heterotermes indicola* (Wasmann) (Isoptera: Rhinotermitidae). *Pakistan Journal of Science, 63*, 59–62.

Manzoor, F., Pervez, M., Adeyemi, M. M. H., & Malik, S. A. (2011b). Effects of three plant extracts on the repellency, toxicity and tunneling of subterranean termite *Heterotermes Indicola* (Wasmann). *Journal of Applied Environmental and Biological Sciences, 1*, 107–114.

Mao, L., & Henderson, G. (2007). Antifeedant activity and acute and residual toxicity of alkaloids from *Sophora flavescens* (Leguminosae) against Formosan subterranean termites (Isoptera: Rhinotermitidae). *Journal of Chemical Ecology, 100*, 866–870.

Meepagala, K. M., Osbrink, W., Ballew, A. Lax, A. R., & Duke, S. O. (2006). *Natural product based amides against Formosan subterranean termites (Coptotermes formosanus)*. Abstracts of papers, 232nd ACS National Meeting, San Francisco, CA, United States Sept 10–14.

Meyer, J. R. (2005). *Isoptera*. Department of Entomology. NC State University. Online at. http://www.cals.ncsu.edu/course/ent425/compendium/termites.html

Murugan, K., & Vasugi, C. (2011). Combined effect of *Azadirachta indica* and the entomopathogenic nematode *Steinernema glaseri* against subterranean termite, *Reticulitermes flavipes*. *Journal of Entomological and Acarological Research, 43*, 253–259.

Myles, T. G. (2005). *Termite biology, Urban Entomology Programme*. http://www.utoronto.ca/forest/termite/termite.htm

Nakashima, Y., & Shimizu, K. (1972). Antitermitic activity of *Thujopsis dolabrata* var *Hondai*. III. Components with a termiticidal activity. *Miyazaki Daigaku Nogakubu Kenkyu Hokoku, 19*, 251–259.

Nakayama, F. S., & Osbrink, W. L. (2010). Evaluation of kukui oil (*Aleurites moluccana*) for controlling termites. *Industrial Crops and Products, 31*, 312–315.

Nas, M. N. (2004). In vitro studies on some natural beverages as botanical pesticides against *Erwinia amylovora* and *Curtobacterium flaccumfaciensis* subsp. *poinsettiae*. *Turkish Journal of Agriculture, 28*, 57–61.

Nickle, W. R., & Welch, H. E. (1984). History, development and importance of insect nematology. In W. R. Nickle (Ed.), *Plant and insect nematodes* (pp. 627–653). New York: Dekker.

Ogunsina, O. O., Oladimeji, M. O., & Faboro, E. O. (2009). Mortality and anti-feedants evaluation of hexane and ethanol extracts of *Lantana camara* (Verbenaceae), African nutmeg (*Monodora myristica* (Gaerth) Dunal) and Enuopiri (*Euphorbia Laterifloria*, Schum and Thonner) against subterranean termite workers (*Macroterme michaelseni*). *Toxicological and Environmental Chemistry, 91*, 971–977.

Osbrink, L. A. W., Williams, K. S., Connick, W. J., Jr., Wright, M. S., & Lax, A. R. (2001). Virulence of bacteria associated with the Formosan subterranean termite (Isoptera: Rhinotermitidae) in New Orleans, LA. *Environmental Entomology, 30*, 443–448.

Owusu, E. O., Akutse, K. S., & Afreh Nuamah, K. (2008). Effect of some traditional plant components on the control of termites, *Macrotermes* sp. (Isoptera: Termitidae). *African Journal of Science and Technology, 9*, 2–89.

Paiva, P. M. G., Santana, G. S. M., Souza, I. F. A. C., Albuquerque, L. P., Agra Neto, A. C., Albuquerque, A. C., Luz, L. A., Napoleão, T. H., & Coelho, L. C. C. B. (2011). Effect of lectins from *Opuntia ficus-indica* Cladodes and *Moringa oleifera* seeds on survival of *Nasutitermes corniger*. *International Biodeterioration & Biodegradation, 65*, 982–989.

Pal, M., Verma, R. K., Krishan, S., & Tewari, S. K. (2011). Anti-termite activity of essential oil and its components from *Myristica fragrans* against *Microcerotermes beesoni*. *Journal of Applied Sciences and Environmental Management, 15*, 559–561.

Pandey, A., Chattopadhyay, P., Banerjee, S., Pakshirajan, K., & Singh, L. (2012). Antitermitic activity of plant essential oils and their major constituents against termite *Odontotermes assamensis* Holmgren (Isoptera: Termitidae) of north East India. *International Biodeterioration & Biodegradation, 75*, 63–67.

Pearce, M. J. (1997). *Laboratory culture and experimental techniques using termites*. Chatham: Natural Resources Institute.

Peres Filho, O., Dorval, A., Duda, M. J., & Moura, R. G. (2006). *Nasutitermes* sp. (Isoptera, Termitidae) response to extracts from four Brazilian woods. Scientia Forestalis/Forest. *Sciences, 71*, 51–54.

Philip, H. (2004). *Biology and control of the subterranean termite*. Pest Management Factsheet 98–01. http://www.agf.gov.bc.ca/cropprot/termite.htm

Poinar, G. O., Jr. (1979). *The natural history of nematodes*. Englewood Cliffs: Prentice Hall.

Potter, M. F. (2004). Termites. In D. Moreland (Ed.), *Handbook of pest control: The behavior, life history, and control of household pests* (pp. 216–361). Cleveland: GIE Media.

Raina, A., Blan, J., Doolittle, M., Lax, A., Boopathy, R., & Folkins, M. (2007). Effect of orange oil extract on the Formosan subterranean termite (Isoptera: Rhinotermitidae). *Journal of Economic Entomology, 100*, 880–885.

Rajagopal, D. (2002). Economically important termite species in India. *Sociobiology, 41*, 33–46.

Ramanujam, B., Rangeshwaran, R., Sivakmar, G., Mohan, M., & Yandigeri, M. S. (2014). Management of insect pests by microorganisms. *Proceedings of the Indian National Science Academy, 80*, 455–471.

Rana, B. S., & Kachhawa, D. (2014). Study of bio-efficacy of entomopahogenic fungi for suppression of termite incidence in maize. *International Journal of Plant Protection, 7*, 77–381.

Ravindran, K., Qiu, D., & Sivaramakrishnan, S. (2015). Sporulation characteristics and virulence of *Metarhizium anisopliae* against subterranean termites (*Coptotermes formosanus*). *International Journal of Microbiological Research, 6*, 01–04.

Rodrigues, A. M., Amusant, N., Beauchêne, J., Eparvier, V., Leménager, N., Baudasse, C., Espindola, L. S., & Stien, D. (2011). The termiticidal activity of *Sextonia rubra* (Mez) van der Werff (Lauraceae) extract and its active constituent rubrynolide. *Pest Management Science, 67*, 1420–1423.

Rust, M. K., & Saran, R. K. (2006). Toxicity, repellency, and transfer of chlorfenapyr against western subterranean termites (Isoptera: Rhinotermitidae). *Journal of Economic Entomology, 99*, 864–872.

Sa, R. A., Napoleao, T. H., Santos, N. D. L., Gomes, F. S., Albuquerque, A. C., Xavier, H. S., Coelho, L. C. B. B., Bieber, L. W., & Paiva, P. M. G. (2008). Induction of mortality on *Nasutitermes corniger* (Isoptera, Termitidae) by *Myracrodruon urundeuva* heartwood lectin. *International Biodeterioration & Biodegradation, 62*, 460–464.

Samsuddin, A. S., Sajap, A. S., & Mohamed, R. (2016). *Metarhizium anisopliae* of peninsular Malaysia origin poses high pathogenicity toward *Coptotermes curvignathus*, a major wood and tree pest. *The Malaysian forester, 78*, 41–48.
Santana, A. L. B. D., Maranhao, C. A., Santos, J. C., Cunha, F. M., Conceiçao, G. M., Bieber, L. W., & Nascimento, M. S. (2010). Antitermitic activity of extractives from three Brazilian hardwoods against *Nasutitermes corniger*. *International Biodeterioration & Biodegradation, 64*, 7–12.
Seo, S. H., Kim, J., Kang, J., Koh, S.-H., Ahn, Y.-J., Kang, K.-S., & Park, K. (2014). Fumigant toxicity and acetylcholinesterase inhibitory activity of 4 Asteraceae plant essential oils and their constituents against Japanese termite (*Reticulitermes speratus* Kolbe). *Pesticide Biochemistry and Physiology, 113*, 55–61.
Seo, S. M., Kim, J., Lee, S. G., Shin, C. H., Shin, S. C., & Park, I. K. (2009). Fumigant antitermitic activity of plant essential oils and components from ajowan (*Trachyspermum ammi*), allspice (*Pimenta dioica*), caraway (*Carum carvi*), dill (*Anethum graveolens*), geranium (*Pelargonium graveolens*), and litsea (*Litsea cubeba*) oils against Japanese termite (*Reticulitermes speratus Kolbe*). *Journal of Agriculture and Food Chemistry, 57*, 6596–6602.
Shahid, A. A., Rao, A. Q., Bakhsh, A., & Husnain, T. (2012). Entomopathogenic fungi as biological controllers: New insights into their virulence and pathogenicity. *Archives of Biological Sciences, 64*, 21–42.
Shi, J., Li, Z., Izumi, M., Baba, N., & Nakajima, S. (2008). Termiticidal activity of diterpenes from the roots of *Euphorbia kansui*. *Zeitschrift für Naturforschung. Section C, 63*, 51–58.
Shibutani, S., Yusuf, S., & Doi, S. (2006). Anti-termite (Isoptera) component from *Artocarpus heterophyllus* heartwood. *Sociobiology, 47*, 711–720.
Shiny, K. S., & Remadevi, O. K. (2014). Evaluation of termiticidal activity of coconut shell oil and its comparison to commercial wood preservatives. *European Journal of Wood and Wood Products, 72*, 139–141.
Sileshi, A., Sori, W., & Dawd, M. (2013). Laboratory evaluation of entomopathogenic fungi *Metarhizium anisopliae* and *Beauveria bassiana* against termite, *Macrotermes* (Isoptera: Termitidae). *Asian Journal of Plant Sciences, 12*, 1–10.
Sileshi, G. W., Nyeko, P., Nkunika, P. O. Y., Sekematte, B. M., Akinnifesi, F. K., & Ajayi, O. C. (2009). Integrating ethno-ecological and scientific knowledge of termites for sustainable termite management and human welfare in Africa. *Ecology and Society, 14*(1), 48.
Sindhu, S. S., Rakshiya, Y. S., & Verma, M. K. (2011). Biological control of termites by antagonistic soil microorganisms. In A. Singh, N. Parmar, & R. C. Kuhad (Eds.), *Bioaugmentation, biostimulation and biocontrol, soil biology* (pp. 261–309). Berlin\Heidelberg: Springer-Verlag.
Singh, Y. (2007). *Isolation and identification of bacteria having pathogenic interactions with termites (Isoptera)*. M. Sc. thesis, CCS Haryana Agricultural University, Hisar, pp. 104.
Singh, N., & Kumar, S. (2008). Antitermite activity of *Jatropha curcas* Linn. biochemicals. *Journal of Applied Sciences and Environmental Management, 12*, 67–69.
Smith, J. L., & Rust, M. K. (1990). Tunneling response and mortality of the western subterranean termite (Isoptera: Rhinotermitidae) to soil treated with termiticides. *Journal of Economic Entomology, 83*, 1395–1401.
Staples, J., & Milner, R. (1996). Metarhizium anisopliae as a mycotermiticide-laboratory behavioural bioassay. In: Wildey, K. B. (Ed.), *Proceedings of the 2nd international conference on insect pests in the urban environment, Edinburgh, 7–10 July 1996, Exeter organizing committee of the international conference on insect pests in the urban environment*, pp. 493.
Stoll, G. (1986). *Natural crop protection*. Josef Margraf, Langen, West Germany, pp. 188.
Stranes, R. L., Liu, L., & Marrane, P. G. (1993). History, use and future of microbial insecticides. *American Entomologist, 39*, 83–91.
Su, N. Y., Ban, P. M., & Scheffrahn, R. H. (1999). Longevity and efficacy of pyrethroid and organophosphate termiticides in field degradation studies using miniature slabs. *Journal of Economic Entomology, 92*, 890–898.

Su, N. Y., & Scheffrahn, R. H. (2000). *Formosan subterranean termite, Coptotermes formosanus Shiraki (Insecta: Isoptera: Rhinotermitidae)*. EENY121 Department of Entomology and Nematology, UF/IFAS Extension. University of Florida.

Sun, J., Fuxa, J. R., & Henderson, G. (2002). Sporulation of *Metarhizium anisopliae* and *Beauveria bassiana* on *Coptotermes formosanus* in vitro. *Journal of Invertebrate Pathology, 84*, 78–85.

Suszkiw, J. (1998). The Formosan termite: A formidable foe. *Agribiological Research, 46*, 4–9.

Taylor, A. M., Gartner, B. L., Morell, J. J., & Tsunoda, K. (2006). Effect of heartwood extractive fractions of *Thuja plicata* and *Chamaecyparis nootkatensis* on wood degradation by termites or fungi. *Journal of Wood Science, 52*, 147–153.

Thompson, G. (2000). *Termites*. Tropical Topics News Letter No. 64, Tropical Savanna, Australia.

Thompson, G. J., Kitade, O., Lo, N., & Crozier, R. H. (2004). On the origin of termite workers: Weighing up the phylogenetic evidence. *Journal of Evolutionary Biology, 17*, 217–220.

Thorne, B. L., & Carpenter, J. M. (1992). Phylogeny of the Dictyoptera. *Systematic Entomology, 17*, 253–268.

Trikojus, V. M. (1935). Some synthetic and natural antitermitic substances. *Australian Chemical Institute Proceedings, 2*, 171–176.

Trudeau, D. (1989). *Selection of entomophilic nematodes for control of the eastern subterranean termite, Reticultermes flavipes (Kollar) (Isoptera: Rhinotermitidae)*. Master's thesis, University of Toronto, Toronto, Ontario, Canada, pp. 93.

UNEP (United Nations Environment Programme). (1992). *Montreal protocol assessment supplement, Methyl Bromide: Its science, technology, and economics*. Synthesis Report of the Methyl Bromide Interim Scientific Assessment and Methyl Bromide Interim Technology and Economic Assessment.

UNEP. (2000). *Finding alternatives to Persistent Organic Pollutants (POPs) for termite management – Prepared by members of the UNEP/FAO/Global IPM Facility Expert Group on Termite Biology and Management*. http://www.chem.unep.ch/pops/termites/termite

UNEP/FAO/Global IPM Facility Expert Group on Termite Biology and Management. (2000). *Finding alternatives to persistent organic pollutants (POPs) for termite management*, online at www.chem.unep.ch/pops/termites/termite_ch4.htm

Upadhyay, R. K., Jaiswal, G., & Ahmad, S. (2010, September). Anti-termite efficacy of *Capparis decidua* and its combinatorial mixtures for the control of Indian white termite *Odontotermes obesus* (Isoptera: Odontotermitidae) in Indian soil. *Journal of Applied Science and Environmental Management, 14*, 101–105.

Upadhyay, R. K., Jaiswal, G., Ahmad, S., Khanna, L., & Jain, S. C. (2012). Antitermite activities of C. decidua extracts and pure compounds against Indian white termite Odontotermes obesus (Isoptera: Odontotermitidae). *Psyche, 2012*, Article ID 820245, 9 p.

Verma, R. K., & Verma, S. K. (2006). Phytochemical and termiticidal study of *Lantana camara* var. *aculeata* leaves. *Fitotrapia, 77*, 466–468.

Verma, S., Yadav, P. R., & Singh, R. (2001). Distribution of termites and yield loss in potato-based cropping sequence in western Uttar Pradesh. *Potato Journal, 28*, 119–120.

Verma, S. K., Verma, R. K., & Saxena, K. D. (2005). Termiticidal triterpenoid from leaves of *Lantana camara* var. *aculeate*. *Journal of the Institution of Chemists (India), 77*, 23–25.

Verma, M., Sharma, S., & Prasad, R. (2009). Biological alternatives for termite control: A review. *International Biodeterioration & Biodegradation, 63*, 959–972.

Verma, M., Pradhan, S., Sharma, S., Naik, S. N., & Prasad, R. (2011). Efficacy of karanjin and phorbol ester fraction against termites (*Odontotermes obesus*). *International Biodeterioration & Biodegradation, 65*, 877–882.

Verma, S., Verma, M., Sharma, S., & Malik, A. (2013). Determination of phytocomponents by GC-MS analysis of *Jatropha curcas* root and its termiticidal activity. *International Journal of Ecology and Environmental Sciences, 39*, 159–169.

Verma, S., Sharma, S., & Malik, A. (2016). Termiticidal and repellency efficacy of botanicals against *Odontotermes obesus*. *International Journal of Research in BioSciences, 5*, 52–59.

Weeks, B., & Baker, P. (2004). *Subterranean termite (Isoptera: Rhinotermitidae) mortality due to entomopathogenic nematodes (Nematoda: Steinernematidae, Heterorhabditidae).* University of Arizona, College of Agriculture Turfgrass and Ornamental Research Report, index at: http://cals.arizona.edu/pubs/crops/az1359/

Wood, T. G., & Sands, W. A. (1978). The role of termites in ecosystems. In M. V. Brian (Ed.), *Production ecology of ants and termites* (pp. 245–292). Cambridge: Cambridge University Press.

Woodrow, R. J., & Grace, J. K. (1998). Field studies on the use of high temperatures to control *Cryptotermes brevis* (Isoptera: Kalotermitidae). *Sociobiology, 32*, 27–49.

Wright, M. S., Raina, A. K., & Lax, A. R. (2005). A strain of the fungus *Metarhizium anisopliae* for controlling subterranean termites. *Journal of Economic Entomology, 98*, 1451–1458.

Yanagawa, A., Yokohari, F., & Shimizu, S. (2008). Defence mechanism of the termite, *Coptotermes formosanus* Shiraki, to entomopathogenic fungi. *Journal of Invertebrate Pathology, 97*, 165–170.

Yanagawa, A., Yokohari, F., & Shimizu, S. (2009). The role of antennae in removing entomopathogenic fungi from cuticle of the termite, *Coptotermes formosanus. Journal of Insect Science, 9*, 6.

Yii, J. E., Bong, C. F. J., King, J. H. P., & Kadir, J. (2016). Synergism of entomopathogenic fungus, *Metarhizium anisopliae* incorporated with Fipronil against oil palm pest subterranean termite, *Coptotermes curvignathus. Plant Protection Science, 52*, 35–44.

Yu, H., Gouge, D. H., & Baker, P. (2006). Parasitism of subterranean termites (Isoptera: Rhinotermitidae: Termitidae) by entomopathogenic nematodes (Rhabditida: Steinernematidae; Heterorhabditidae). *Journal of Economic Entomology, 99*, 1112–1119.

Yu, H., Gouge, D. H., Stock, S. P., & Baker, P. B. (2008). Development of entomopathogenic nematodes (Rhabditida: Steinernematidae; Heterorhabditidae) in the desert subterranean termite *Heterotermes aureus* (Isoptera: Rhinotermitidae). *Journal of Nematology, 40*, 311–317.

Yu, H., Gouge, D. H., & Shapiro Ilan, D. I. (2010). A novel strain of *Steinernema riobrave* (Rhabditida: Steinernematidae) possesses superior virulence to subterranean termites (Isoptera: Rhinotermitidae). *Journal of Nematology, 42*, 91–95.

Yuan, Z., & Hu, X. P. (2011). Evaluation of differential antitermitic activities of *Lantana camara* oven-dried tissues against *Reticulitermes virginicus* (Isoptera: Rhinotermitidae). *Insect Science., 18*, 671–681.

Yuan, Z., & Hu, X. P. (2012). Repellent, antifeedant, and toxic activities of *Lantana camara* leaf extract against *Reticulitermes flavipes* (Isoptera: Rhinotermitidae). *Journal of Economic Entomology, 105*, 2115–2121.

Zhong, J. H., & Liug, L. L. (2002). Termite fauna in China and their economic importance. *Sociobiology, 40*, 25–32.

Zimmermann, G. (2007). Review on safety of the entomopathogenic fungus *Beauveria bassiana* and *Beauveria brongniartii. Biocontrol Science and Technology, 17*, 553–596.

Chapter 7
Bio-intensive Integrated Management of Termites

K. Sahayaraj

Contents

Abstract Various viable options in termite management are termite baits (novaluron, hexaflumuron), the use of synthetic pesticides/insecticides/termiticides (bifenthrin, chlorpyrifos, cypermethrin, fenvalerate, Imidachloprid, permethrin, dexamethasone, ibuprofen, aldrin, Dieldrin, etc.), chemicals (boric acid, ibuprofen sodium salt), or botanicals (*Withania somnifera*, *Croton tiglium*, *Hygrophila auriculata*, *Trachyspermum ammi*, *Pimenta dioica*, *Carum carvi*, *Anethum graveolens*,

K. Sahayaraj (✉)
Crop Protection Research Centre, St. Xavier's College,
Palayamkottai, Tamil Nadu 627 002, India
e-mail: ksraj48@gmail.com

M.A. Khan, W. Ahmad (eds.), *Termites and Sustainable Management*,
Sustainability in Plant and Crop Protection,
https://doi.org/10.1007/978-3-319-68726-1_7

Pelargonium graveolens, *Litsea cubeba*, *Croton urucurana*, *Melia azedarach*, *Crotalaria burhia*, and *Anacardium occidentale*). Further, natural enemies include mammals, birds, insects (ants, reduviids), Araneae (spiders), microbes–bacteria [*Bacillus thuringiensis*], fungi [*Conidiobolus* sp., *Aspergillus flavus*, *Metarhizium anisopliae*, *Beauveria bassiana*, *B. pseudobassiana*, and *Isaria fumosorosea*), and nematodes (*Heterorhabditis* sp., *Steinernema* sp.). Other methods such as dug trench are also available for termite management. In addition, many commercial products are available in the market (e.g., Bioblast). No one has integrated more than two or three individual components for termite control that are environmentally safe and effective. Here we discuss how to utilize many integrated pest management components, to save crops and environment.

Keywords Predators • Parasitoids • Microbes • Nematodes • Botanicals

7.1 Introduction

Despite their importance in many ecosystems, termites are also serious pests of structures, houses, rangelands, tropical forestry, and agriculture in many countries. *Coptotermes heimi*, *Microcerotermes championi*, *Odontotermes obesus*, *O. guptai*, *Microtermes obesi*, and *Microtermes mycophagus* are species common in India. In Japan, *Reticulitermes speratus* is considered as pestiferous. In Malaysia, *Coptotermes curvignathus* is a serious pest of the oil palm, *Elaeis guineensis,* when grown on peat soils (Sajap et al. 2002). The two termite species *M. obesi* and *O. obesus* are commonly found in Pakistan. In this country *C. heimi*, *M. championi*, *O. obesus*, *O. guptai*, *M. obesi*, and *M. mycophagus* were also recorded from houses and garden trees in district Gujranwala, where they were found variably inhabiting different portions of garden trees (Azam et al. 2015). *Coptotermes formosanus* is present in subtropical areas in China (Ruan et al. 2015). More than 61 species of termites belonging to 25 genera and 4 families have also been recorded in eastern and southern regions of Africa (Assefa 1990).

7.1.1 Termites as Pests

Families such as Hodotermitidae (with genera *Anacanthotermes* and *Hodotermes*), Kalotermitidae (*Neotermes*), Rhinotermitidae (*Coptotermes*, *Heterotermes*, and *Psammotermes*), and Termitidae (*Amitermes*, *Ancistrotermes*, *Cornitermes*, *Macrotermes*, *Microcerotermes*, *Microtermes*, *Odontotermes*, *Procornitermes*, and *Syntermes*) cause great losses in agriculture (UNEP Report 2000). A variety of plants like yam (*Dioscorea* sp.) (Loko et al. 2016), African mahogany (*Khaya* sp.)

(Franca et al. 2016), oil palm (Mahapatro and Kumar 2015), coconuts, bamboo species (Niken et al. 2015), cassava (Atu 1993), field crops (Bigger 1966), groundnut (Reddy and Ghewande 1986), sugarcane (Mora et al. 1996), wheat (Verma 1980), and rice (Nwilene et al. 2008a) are attacked by various termite species.

Termites reported to attack citrus include *Reticulitermes lucifugus* in Israel; *Paraneotermes simplicicornis* in Texas; *Coptotermes* spp. in Malaya, Australia, and Surinam; *Microcerotermes diversus* in Arabia; and *Macrotermes* spp. in Nigeria. The eastern subterranean termites, *Reticulitermes yavipes* and *R. virginicus*, are pests of young citrus trees in Florida, killing them by girdling at or just below the soil line. In Africa, termites mostly attack eucalyptus seedlings and saplings (Atkinson et al. 1992). In Mali, termites damage about 10–30% of harvested kernels of groundnut (Umeh and Ivbijaro 1999). In India, they are responsible for the loss of 15–25% of maize yield (Joshi et al. 2005). The spectrum of damage is varied due to different climatic conditions, seasons, topography, etc. Considering the importance of termites in crops, it is essential to minimize their density using integrated control measures. Though an enormous amount of literature is available for physical and mechanical control, in this chapter we discuss only a number of eco-friendly products.

7.2 Eco-friendly Termite Control Measures

For many decades, chemical insecticides have remained popular for termite management, worldwide. However, with growing environmental concerns over pesticides, biological control measures using plants, microbes, predators, and parasitoids became important. Botanical pesticides possess many desirable properties such as insecticidal, repellent or deterrent activities, and insect growth regulation (IGR) against many agricultural pests. D- and l-tetrahydrocarvone and d- and l-dihydrocarvone are effective for the control of termites, fire ants, and carpenter ants (Bedoukian and Raina 2015). These components prevent the termites to approach crops, reducing their population near the agroecosystems and increasing the plants' resistance capacity toward termites. Furthermore, they are economically feasible. Commercial products available in the market for termite control are shown in Table 7.1.

7.3 Termite Baits

Baiting as a means of controlling active infestations of termites was initiated more than 45 years ago. However, research into termite baiting reemerged by the end of the 1980s. Termite baits comprise a cellulose matrix treated with a toxicant and placed in specific locations in the soil, close to where termites are active. Foraging

Table 7.1 Commercial products for termite control, available in the market

Active ingredient	Termites	Tested countries
Mirex	*Mas. darwiniensis, R. flavipes*	Australia, Canada, the USA
Deltamethrin, sulfluramid, avermectin, hydramethylnon	*C. formosanus*	The USA
Abamectin	*C. formosanus, R. flavipes, R. virginicus*	The USA
Zinc borate	*R. flavipes, R. virginicus*	The USA
Fipronil	*O. formosanus, R. hageni, R. flavipes*	The USA, China
	Microtermes mycophagus (Desneux)	Pakistan
Hexaflumuron (chitin synthesis inhibitor)	*C. acinaciformis, C. curvignathus, C. formosanus, C. gestroi, C. travians, Heterotermes* sp., *N. exitiosus, R. flavipes, R. lucifugus, R. speratus, R.* sp., *R. virginicus*	Australia, Cayman Islands, Italy, Japan, Malaysia, Puerto Rico, the USA
Bioblast™	Subterranean termites	The USA and Japan
Margosan-O	*Coptotermes formosanus*	–

termites aggregating in the bait can be exposed to the active ingredient in the food matrix or by treating them directly in the bait station. The suitability of materials such as wood, paper, cardboard, cork, or cellulose powder as a bait matrix has been investigated. Hydroquinone, catechol, resorcinol, phloroglucinol, 4-methoxyphenol, methoxyhydroquinone, toluhydroquinone, 1,4-dimethoxybenzene, 4-phenoxyphenol, phenylhydroquinone, polyphenylether, 4-benzyloxyphenol, and hydroquinone-bis-(2-hydroxyethyl)-ether are used and dissolved in distilled water or ethanol for this purposes.

There are two general types of bait products on the market for subterranean termite control. One type is the in-ground (IG) system, which is installed in the soil surrounding a structure to intercept foraging termites. The IG bait system is used for both preventive and remedial control, and it may take weeks and sometimes months for termites to find the IG stations (Su et al. 1995). Another type of bait system, often referred to as an aboveground (AG) bait, is intended for remedial control only and is designed to be placed over active termite infestations so that termites may come into stations to feed on baits (Su et al. 1997). Su (2015) formulated a fluid bait comprising cellulose and fine-ground phagostimulants impregnated with 0.5% hexaflumuron (AI, wt/wt) and mixed with 1% methylcellulose, to yield 10% dry weight. It was injected into simulated foraging galleries of *C. formosanus* and *R. virginicus* for a laboratory efficacy evaluation. Six weeks after the injection, mortalities for both species exceeded 90%, and all termites died within 8 weeks. At field level, 0.5% novaluron was applied as a bait to control *Reticulitermes* sp. and *C. formosanus* (Rhinotermitidae) (Keefer et al. 2015).

Though chemical-based baits are common for termite control, fungi-based bait was also evaluated. However, the presence of a massive amount of fungal matter triggered an avoidance behavior (Milner et al. 1998; Mburu et al. 2009) within the miniature termite colonies. Highest bait repellence was recorded in the treatments at 10^9 conidia/g bait, with slight repellence observed at 10^8 conidia/g bait and non-repellence at 10^7 conidia/g bait.

7.4 Botanicals

One of the most important biological methods for pest control is the use of botanicals or pesticide plants. Various parts of the plants such as seed, bark, leaf, fruit, root, wood, etc. have been utilized for pest control. Various solvent extracts of plants, their column chromatographic fractions, or bioactive principles have been recommended for pest management worldwide. Botanicals are eco-friendly options and alternative for synthetic pesticides, economically viable and easily available as a natural resource. They act as feeding deterrent, behavioral modulators, toxic agents, morphogenetic agents, and phagostimulants.

Various extract concentrations in water and methanol (100, 200 and 300 ppm) of *Melia azedarach* were tested against *M. obesi* and *O. obesus* by Qureshi et al. (2015). Results showed that water and methanol solvents were effective against *O. obesus*. Laboratory studies were carried out to assess the repellent efficacy of aqueous root extracts of *Crotalaria burhia* and powdered leaf dusts of *Anacardium occidentale* against *O. obesus*. It was found that *C. burhia* at 10% and 20% repelled 66.7% and 70% of *O. obesus*, whereas leaf dust of *A. occidentale* at 5% and 7.5% repelled 56.7% and 60% of termites (Bajya et al. 2015). Three different concentrations (20%, 10%, 5%) of various solutions consisting of garlic, neem, and tobacco were tested against *Heterotermis indicola* under laboratory conditions. High and low mortality was caused by garlic and tobacco concentrations (Ahmed et al. 2016). The most common plants utilized for different termite control applications are shown in Table 7.2.

7.5 Essential Oils

The biopesticidal potential of six plant-derived essential oils (mint [*Mentha arvensis*], ajwain [*Carum capticum*], lemongrass [*Cymbopogon citratus*], clove [*Eugenia caryophyllata*], cedarwood [*Cedrus deodara*], and eucalyptus [*Eucalyptus globulus*]) was evaluated against *O. obesus* (Gupta et al. 2011).The results showed that at 10% concentration of essential oils, all the termite workers were killed within 30 min. The authors further concluded that mint oil is superior to the other oils as it takes comparatively the least amount of time to kill the workers, at all concentrations tested. This may be due to the presence of the active component menthol present in the oil. Nootkatone (Fig. 7.1), a sesquiterpene ketone from vetiver (*Vetiveria zizanioides*) oil, showed strong termiticide repellent and toxicant activities against *C. formosanus*, as did other compounds (Zhu et al. 2001; Meepagala et al. 2006).

Table 7.2 Plants, their parts, bioactive principles, and activity against different termites

Plant	Parts used	Bioactive principle	Activity
Tagetes erecta	Leaf essential oil	(Z)-β-ocimene	Termiticidal
Lepidium meyenii	Aerial parts essential oil	3-Methoxyphenylacetonitrile, benzylthiocyanate	Feeding deterrent
	Leaf	Benzylthiocynate, 3-methoxyphenylacetonitrile, and b-ionon	Feeding deterrent
Xylopia aethiopica	Fruits and seeds	Diterpenes and amides	Antifeedant
Nepeta cataria	Essential oil	–	Barrier
Calocedrus formosana	Leaf essential oil	T-Muurolol	100% mortality
Melauleuca spp.	Leaf essential oil	g-Terpinene and terpinolene	Termicidal
Garlic oil	–	Diallyl trisulfide, diallyl disulfide, and diallyl sulfide	Antitermitic
Dipterocarpus kerrii	Resin	Sesquiterpenoids	Toxic
Adina racemosa Miq.	Bark	Benzoic acid	Toxic
Moneses uniflora	Aerial parts	Naphthoquinones, 2, 7-dimethyl-	Toxic
		1,4-naphthoquinone and 3-hydroxy-2,7-dimethyl-1,4-naphthoquinone	
Lantana camara var. *aculeate*	Leaves	Triterpenoid, 22 b-acetoxylantic acid	Toxic
Thujopsis dolabrata	Wood	b-Thujaplicin and carvacrol	Toxic
Chamaecyparis pisifera	Wood	Chamaecynone and isochamaecynone	Toxic
Cryptomeria japonica	Wood	b-Eudesmol and cedrol	Toxic
Chamaecyparis obtuse	Wood	Monoterpene, sesquiterpene, and sesquiterpene alcohol	Toxic
Thujopsis dolobrata	Wood	Thujopsene	Toxic
Cinnamomum osmophloeum	Leaf	Cinnamaldehyde	Toxic
Adina racemosa	Bark	Benzoic acid	Toxic
Dipterocarpus spp.	Resin	Sesquiterpenes	Antitermitic
Parthenium argentatum	Guayule resin	–	Antitermitic, repellant, antifeedant
Gray (Guayule)			
Aframomum meleguata	seeds	Gingerol [5-hydroxy-L-(4-hydroxy-3-methoxyphenyl) decan-3-one] and shogaol [1-(4-hydroxy-3-methoxyphenyl) dec-5-en-3-one]	Antifeedant

(continued)

Table 7.2 (continued)

Plant	Parts used	Bioactive principle	Activity
Detarium microcarpum	Leaves	Clerodane diterpenes, 3,13Eclerodien-15-oic acid, 4(18),13Eclerodien-15-oic acid, 18-oxo-3,13E–clerodien-15-oic acid and 2-oxo-3,13E–clerodien-15-oic acid	Antifeedant
Azadirachta indica	Seed	Limonoids	Antifeedant
Aframomum meleguata	Seed	Gingerol [5-hydroxy-L-(4-hydroxy- 3-methoxyphenyl) decan-3-one] and shogaol [1-(4-hydroxy-3- methoxyphenyl) dec-5-en-3-one]	Antifeedant
Vetiveria zizanoides	Root	Nootkatone (a sesquiterpene alcohol) and cedrene	Arrestants, feeding deterrent, repellent, and toxic
Lonchocarpus castilloi Standley	Wood	Flavonoids, castillen D and castillen E	Feeding deterrent activity
Juniperus procera	Wood	Cedrol and cedrene	Termiticidal ability
Diospyros sylvatica	Root	Quinines plumbagin, isodiospyrin, and microphyllone	Termiticidal ability

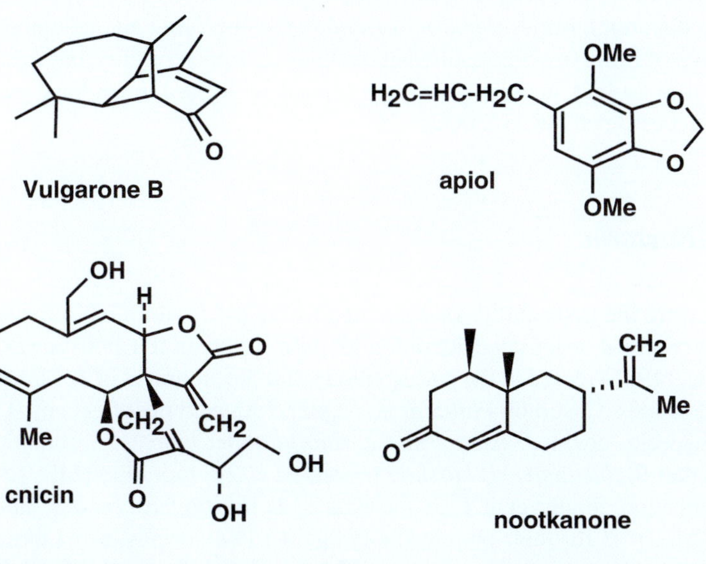

Fig. 7.1 Chemical structure of nootkatone and other products with termiticidal activity (Adapted from Meepagala et al. 2006)

Pandey et al. (2012) reported strong termiticidal activity of essential oils from lemongrass (*Cymbopogon citratus*), eucalyptus (*Eucalyptus globulus*), clove (*Syzygium aromaticum*), oregano (*Origanum vulgare*), rosemary (*Rosmarinus officinalis*), cinnamon (*Cinnamomum verum*), and thyme (*Thymus vulgaris*) against *Odontotermes assamensis* Holmgren. These authors further reported that phenol constituents exhibited the strongest termiticidal activity among the other major constituents. Furthermore, alcohol, acetate, and aldehyde groups were more toxic than hydrocarbons and ketones. Qureshi et al. (2015) noticed that extracts of *M. azedarach* reduced carbohydrate and lipid contents in *M. obesi* and *O. obesus*, leading to insect death. Change in these biochemical components may be due to the insecticidal stress caused by extracts which lowered the feeding or proper digestion of food and metabolism.

7.6 Microbes for Termite Control

The use of pathogens as biological control agents has long been considered a promising technology for termite control. Before 1960, few reports noted the pathogenic effect of microorganisms on termites (Chouvenc et al. 2011). In a preliminary field tests, Hanel and Watson (1983) utilized the fungus *Metarhizium anisopliae* to control *Nasutitermes exitiosus*. Hoe et al. (2009) successfully used three local *M. anisopliae* isolates (TA, MG, LR2) against *C. curvignathus*. They further reported that these isolates have potential to be developed as biopesticide. Lenz and Runko (1992) managed coconut termite, *Neotermes rainbowi*, using fungi and nematodes. Direct application of fungi to nests resulted in complete colony mortality, but studies where feeding sites or bait stations have been treated with the fungus have yet to show similar success levels (Rath 2000).

7.6.1 Bacteria

Bacteria were the first candidates evaluated for its use in termite biological control, but never received serious consideration for field application (Chouvenc et al. 2011). Khan et al. (1985) recorded the pathogenicity and development of *Bacillus thuringiensis* in termites. Castilhos-Fortes et al. (2002) evaluated the effects of *B. thuringiensis* subspecies against *Nasutitermes ehrhardti* under laboratory conditions. They reported that *B. thuringiensis kurstaki* registered <72% mortality at the seventh day after application. Singha et al. (2010) evaluated *B. thuringiensis* and *B. thuringiensis* subsp. *israelensis* for their pathogenicity against two species of tea termites, viz., *M. obesi* and *Microcerotermes beesoni*. They reported that *B. thuringiensis* strains caused >80% mortality in both termite species. However, *B. thuringiensis* subsp. *israelensis* was noticed to be the most virulent compared to other types. Osbrink et al. (2001) isolated *Serratia* from dead termites and reported that three of their isolates induced >85% mortality within 19 days in Petri dish assays. In another study, *Pseudomonas fluorescens* CHA0 was reported to kill *O. obesus* by inhibiting cytochrome c oxidase of the termite respiratory chain (Devi and Kothamasi 2009). In 2001, Osbrink and co-workers suggested bacteria like *Acinetobacter calcoaceticus*, *A. baumannii*, *Citrobacter freundii*, *Corynebacterium urealyticum*, *Enterobacter gergoviae*, and *Serratia marcescens* for the management of the Formosan subterranean termite in New Orleans. A list of bacteria used for termite control is given in Table 7.3.

7.6.2 Fungi

A number of entomopathogenic Hyphomycetes, including *Metarhizium anisopliae* and *Beauveria bassiana*, have demonstrated considerable potential for termite control. Different strains of entomopathogenic fungi are virulent against various insect life stages by infecting them through the cuticle (ectoparasites) or by entering into the body and producing toxins (endoparasites). Fungal products are widely used to control various agricultural pests: *M. anisopliae* is a well-known entomopathogenic fungus occurring naturally, reported to infect over 200 species of insect including termites. Milner et al. (1996) recommended *M. anisopliae* for termite management, whose strain (SRRC 2558) was evaluated against *Reticulitermes flavipes* and *C. formosanus*, in the laboratory. Results showed that this new strain was highly infectious against termites (Wang and Powell 2004).

Hussain et al. (2011) evaluated the efficacy of *M. anisopliae* strain ARSEF 6911 in the laboratory and field conditions against two sugarcane termites, *M. obesi* and *O. obesus*. They noticed that all conidial suspensions were able to induce

Table 7.3 Most common bacteria applied in termite control

Bacteria	Termite(s)	Country
Pseudomonas fluorescens	*Odontotermes obesus*	India
Aeromonas caviae		
Alcaligenes latus		
Xenorhabdus nematophila	*Coptotermes curvignathus*	Thailand
Photorhabdus luminescens	Subterranean termite	Pakistan
Serratia aureus, *Serratia marcescens*, *Bacillus megaterium*, *B. cereus*, *B. subtilis*, *Pseudomonas aeruginosa*, *Citrobacter freundii*	*Macrotermes bellicosus*	Nigeria
B. thuringiensis	*Coptotermes formosanus*	The USA (New Orleans)
B. thuringiensis	*Reticulitermes flavipes*, *R. virginicus*, *R. hesperus*, *Zootermopsis angusticollis*	–
B. thuringiensis subsp. *israelensis*	*Microtermes obesi*, *Microcerotermes beesoni*	India

mortality, and there were no significant differences in the LT_{50} values between species. Field results also revealed that tested treatments significantly reduced termite infestations, compared to the untreated control. Kramm et al. (1982) reported that conidia of *M. anisopliae* generally spread among individual termites by grooming. Maniania et al. (2002) reported that a granular formulation of *M. anisopliae* might be a useful option for the management of termites in the maize agroecosystem in Kenya.

7.6.3 *Virus*

Nuclear polyhedrosis virus (NPV) is considered an important microbial agent against termites. Initially, Al-Fazairy and Hassan (1988) recorded the infection of termites by *Spodoptera littoralis* NPV. Literature is, however, limited on viruses use in biological control (Chouvenc et al. 2011). This limited documentation on viruses may reflect the fact that termitologists are not typically skilled in virology and vice versa, as virologists generally focus their work on different biological models (human health, plant disease). Chouvenc and Su (2010) suggested that a good candidate for biological control of termites should have particular characteristics like capability to complete its life cycle and spread, before the host death.

7.6.4 *Nematode*

Nematodes belonging to the families Steinernematidae and Heterorhabditidae are obligate insect parasites. They can be utilized either alone or in combinations with other IPM components for termite management. In India, the efficacy of an entomopathogenic nematode-based product, Pusa Nemagel, was tested against subterranean termites infesting wheat (*Triticum aestivum*) and pearl millet (*Pennisetum glaucum*), as soil application at the time of sowing (Rathour et al. 2014). A single application of Pusa Nemagel was shown to reduce the termite incidence in wheat crop by 48–78%, as compared to control. This reduction in termite incidence resulted in a corresponding increase in the wheat yield by 22.2–43.3%. The use of specific nematode species against termites is listed in Table 7.4.

7.7 Predators and Parasitoids

Termites have a wide variety of predators, both opportunist and specialist and vertebrate and invertebrate. Attack occurs mainly upon alate reproductives or foraging workers outside or within the nest. Among the vertebrates, fish, anurans, lizards, snakes, birds, and mammals deserve a special mention. Insect and other arthropod

Table 7.4 Entomopathogenic nematode species used for termite control

Nematode	Termite species
Steinernema sp.	*Coptotermes formosanus*
Steinernema feltiae	*Reticulitermes tibialis*, *Heterotermes aureus*
S. riobrave	*H. aureus*, *Gnathamitermes perplexus*
S. carpocapsae	*H. aureus*, *Zootermopsis angusticollis*
Heterorhabditis bacteriophora	*H. aureus*
H. sonorensis	*Macrotermes bellicosus*, *Trinervitermes occidentalis*
Neoaplectana carpocapsae	*Coptotermes formosanus*
N. longicurvicauda	*Reticuldermes flavipes*
Oigolaimella attenuata	*Reticulitermes*

groups reportedly prey opportunistically on termites. All orders of entomophagous insects probably contain at least some species that feed on termites (Logan et al. 1990; Culliney and Grace 2000). Ants are the greatest enemies of termites in all regions of the world. In agriculture fields, ground beetles are among the termite predators (Lovei and Sunderland 1996). McMahan (1983) observed that reduviid predators are having adaptive features to feed on termites in large numbers. Although reduviids feed on a wide variety of arthropods (as generalist predators), they exhibit preference on termites, ants, and bees (Hwang and Weirauch 2012). The pantropical Salyavatinae (Hemiptera: Reduviidae) contains members with enigmatic morphology and specialized behavior for feeding on termites. All Salyavatinae (17 genera, 107 species) are possibly specialist termite predators. However, existing observations are limited only to seven species (Gordon and Weirauch 2016).

Many spiders are specific to feed on preys that are highly aggregated, including termites and ants. *Chrosiotes tonala* (Araneae: Theridiidae) (Eberhard 1991), *Diores* sp. (Jocque and Dippenaar-Schoeman 1992), *Mashonarus guttatus* (Wesolowska and Cumming 2002), *Ammoxenus pentheri* (Dippenaar-Schoeman et al. 1996), etc. are considered as specialized for termites. The spider *Ammoxenus amphalodes* is a monophagous true termite eater, capturing only *Hodotermes mossambicus*. However, its role as a biocontrol agent against termites is limited, due to an insufficient numerical response (Haddad et al. 2016).

7.8 IPM and Traditional Practices

7.8.1 IPM Practices

Ramakrishnan et al. (1999) demonstrated that the susceptibility of *R. flavipes* to *M. anisopliae* was improved in the presence of imidacloprid. The compatibility and synergy in efficacy of the termiticide fipronil with *M. anisopliae,* alone or in combination, against the subterranean termite *C. curvignathus* were investigated by Yii et al. (2016). Results showed that sublethal doses of fipronil were found relatively less detrimental to fungal growth in a compatibility test. The fungus–insecticide bait formulation showed the greatest synergistic effects that increased termite mortality as well as reduced the lethal time at a sublethal dose. The insecticidal stress caused by sublethal fipronil in the formulated bait may weaken defense mechanism in termites, which facilitates fungus infection. The observed synergism treatments showed a potential for development of an integrated fungus–insecticide control method. In another study, the combined treatment of *M. anisopliae* and diesel oil significantly reduced insect damage by attaining higher germination (> 55%) and lower bud damage (< 5.50%) at both sites, in both seasons. The results suggest that the application of *M. anisopliae* and diesel oil in combination might be a useful treatment option for the management of sugarcane termites (Hussain et al. 2011).

7.8.2 *Traditional Practices*

Generally most of the traditional methods of pest control are marginally used in developed countries, but these practices still play a significant role in pest control programs in developing countries. The use of plant materials in the management of insect pests, and especially in termite control, has been an old strategy in the African culture (Owusu et al. 2008). Traditionally, in Gushegu District of Northern Ghana, planting of elephant grass, burial of plant and animal materials, application of wood ash, and application of a mixture of salt and shea butter residues were practiced by farmers to reduce termite attack (Dokurugu et al. 2012). Nwilene et al. (2008b) assessed the effectiveness of traditional practices with two plant extracts (neem seed oil and neem powder) that showed that seed oil was more effective than the powder for termite control.

7.9 Conclusion

From the above review, we may conclude that alternatives to environmental adverse components of termite management can be applied by farmers by using locally available and eco-friendly means. These include botanical extracts or their bioactive principles, indigenous entomopathogenic microbes, nematodes, natural predators, and parasitoids. For effective management of termite pests, it is recommended that i) most toxic plants should be formulated and tested under laboratory and field conditions for commercialization, ii) traditional methodology followed in different parts of the world should be reviewed for researchers' knowledge, and iii) synergistic impact of pesticides and natural products can be evaluated to maximize management efficacy.

Acknowledgments The author gratefully acknowledges laboratory assistance and support by the management people of St. Xavier's College, Palayamkottai.

References

Ahmed, N., Huma, Z., Rehman, S. U., Ullah, M., & Ahmed, S. (2016). Effect of different plant extracts on termite species (Heterotermis indicola). *Journal of Bioresource Management, 3*, 9–16.

Al-Fazairy, A. A., & Hassan, F. A. (1988). Infection of termites by *Spodoptera littoralis* nuclear polyhedrosis virus. *International Journal of Tropical Insect Science, 9*, 37–39.

Assefa Gebre-Amlak. (1990). Maize stalkborer research in Ethiopia. In: Gebrekidan B (ed) Maize improvement, production and protection in Eastern and Southern Africa Regional. Proceedings of the 3rd eastern and southern Africa Regional Maize Workshop. 18–22 September 1989, Nairobi, Kenya, pp 239–252.

Atkinson, P. R., Nixon, K. M., & Shaw, M. J. P. (1992). On the susceptibility of eucalyptus species and clones to attack by *Macrotermes natalensis* Havilland (Isoptera: Termitidae). *Forest Ecology and Management, 48*, 15–30.

Atu, U. G. (1993). Cultural practices for the control of termite (Isoptera) damage to yams and cassava in South-eastern Nigeria. *International Journal of Pest Management, 39*, 462–466.

Azam, I., Sarwar, M. K., Waheed, I., Nadia, I., & Fareeha, A. (2015). Studies on population density and diversity of termites of district Gujranwala, Pakistan. *Journal of Entomology and Zoology Studies, 3*, 160–163.

Bajya, D., Manoharan, T., Sridharan, S., & Kuttalam, S. (2015). Repellent efficacy of *Crotalaria burhia* and *Anacardium occidentale* against *Odontotermes obesus* (Isoptera: Termitidae) under laboratory conditions. *Indian Journal of Agricultural Science, 85*, 1234–1236.

Bedoukian, R. H., & Raina, A. (2015). US Patent No. 9,021,738. Washington, DC: US Patent and Trademark Office.

Bigger, M. (1966). The biology and control of termites damaging field crops in Tanganyika. *Bulletin of Entomological Research, 56*, 417–444.

Castilhos-Fortes, R. D., Matsumura, A. T., Diehl, E., & Fiuza, L. M. (2002). Susceptibility of Nasutitermes ehrhardti (Isoptera: Termitidae) to Bacillus thuringiensis subspecies. *Brazilian Journal of Microbiology, 33*(3), 219–222.

Chouvenc, T., & Su, N. Y. (2010). Apparent synergy among defense mechanisms in subterranean termites (Rhinotermitidae) against epizootic events – The limits and potential for biological control. *Journal of Economic Entomology, 103*, 1327–1337.

Chouvenc, T., NY, S., & Grace, J. K. (2011). Fifty years of attempted biological control of termites–Analysis of a failure. *Biological Control, 59*, 69–82.

Culliney, T. W., & Grace, J. K. (2000). Prospects for the biological control of subterranean termites (Isoptera: Rhinotermitidae), with special reference to *Coptotermes formosanus*. *Bulletin of Entomological Research, 90*, 9–21.

Devi, K. K., & Kothamasi, D. (2009). Pseudomonas fluorescens CHA0 can kill subterranean termite Odontotermes obesus by inhibiting cytochrome c oxidase of the termite respiratory chain. *FEMS Microbiology Letters, 300*(2), 195–200.

Dippenaar-Schoeman, A. S., DeJager, M., & VandenBerg, A. (1996). Behaviour and biology of two species of termite-eating spiders, *Ammoxenus amphalodes* and *A. penlheri* (Araneae: Ammoxenidael, in South Africa). *African Plant Protection, 2*, 15–17.

Dokurugu, M., Bernard, B. N., & Irunuoh, A. O. (2012). Indigenous knowledge of termite control: A case study of five farming communities in Gushegu district of northern Ghan. *Journal of Entomology and Nematology, 4*, 58–64.

Eberhard, W. G. (1991). *Chrosiotes tonala* (Araneae, Theridiidae), a web-building spider, specializing on termites. *Psyche, 98*, 7–19.

Franca, T. S. F. A., França, F. J. N., Arango, R. A., Woodward, B. M., & Arantes, M. D. C. (2016). Natural resistance of plantation grown African mahogany (Khaya ivorensis and Khaya senegalensis) from Brazil to wood-rot fungi and subterranean termites. *International Biodeterioration and Biodegradation, 107*, 88–91.

Gordon, E. R., & Weirauch, C. (2016). Efficient capture of natural history data reveals prey conservatism of cryptic termite predators. *Molecular Phylogenetics and Evolution, 94*, 65–73.

Gupta, A., Sharma, S., & Naik, S. N. (2011). Biopesticidal value of selected essential oils against pathogenic fungus, termites, and nematodes. *International Biodeterioration and Biodegradation, 65*, 703–707.

Haddad, C. R., Brabec, M., Pekar, S., & Fourie, R. (2016). Seasonal population dynamics of a specialized termite-eating spider (Araneae: Ammoxenidae) and its prey (Isoptera: Hodotermitidae). *Pedobiologia, 59*, 105–110.

Hanel, H., & Watson, J. A. L. (1983). Preliminary field tests on the use of *Metarhizium anisopliae* for control of *Nasutitermes exitiosus* (Hill) (Isoptera: Termitidae). *Bulletin of Entomological Research, 73*, 305–313.

Hoe, P. K., Bong, C. F. J., Jugah, K., & Rajan, A. (2009). Evaluation of *Metarhizium anisopliae* (Deuteromycotina: Hyphomycete) isolates and their effects on subterranean termite *Coptotermes curvignathus* (Isoptera: Rhinotermitidae). *American Journal of Agricultural and Biological Sciences, 4*, 289–297.

Hussain, A., Ahmed, S., & Shahid, M. (2011). Laboratory and field evaluation of *Metarhizium anisopliae* var. anisopliae for controlling subterranean termites. *Neotropical Entomology, 40*, 244–250.

Hwang, W. S., & Weirauch, C. (2012). Evolutionary history of assassin bugs (Insecta: Hemiptera: Reduviidae): Insights from divergence dating and ancestral state reconstruction. *PloS One, 7*, e45523.

Jocque, R., & Dippenaar-Schoeman, A. S. (1992). Two new termite-eating *Diores* species (Araneae: Zodariidae) and some observations on unique prey immobilization. *Journal of Natural History, 26*, 1405–1412.

Joshi, P. K., Singh, N. P., Singh, N. N., Gerpacio, R. V., & Pingali, P. L. (2005). *Maize in India: Production systems, constraints, and research priorities* (p. 22). Mexico: D.F. CIMMYT.

Keefer, T. C., Puckett, R. T., Brown, K. S., & Gold, R. E. (2015). Field trials With 0.5% Novaluron insecticide applied as a bait to control subterranean termites (*Reticulitermes* sp. and *Coptotermes formosanus* [Isoptera: Rhinotermitidae]) on structures. *Journal of Economic Entomology, 108*, 2407–2413.

Khan, K., Jafri, R. H., & Ahmad, M. (1985). The pathogenicity and development of *Bacillus thuringiensis* in termites. *Pakistan Journal of Zoology, 17*, 201–209.

Kramm, K. R., West, D. R., & Rockenbach, P. G. (1982). Termite pathogens: transfer of the entomopathogen *Metarhizium anisopliae* between *Reticulitermis* sp. termites. *Journal of Invertebrate Pathology, 40*, 1–6.

Lenz, M., & Runko, S. (1992). Use of microorganisms to control colonies of the coconut termite *Neotermes rainbowi* (Hill) on Vaitupu, Tuvalu. Commonwealth scientific and industrial research Organisation, Division of Entomology, Termite Group Report No. 92/16, pp 47.

Logan, J. W. M., Cowie, R. H., & Wood, T. G. (1990). Termite (Isoptera) control in agriculture and forestry by non-chemical methods: a review. *Bulletin of Entomological Research, 80*, 309–330.

Loko, Y. L., Agre, P., Orobiyi, A., Dossou-Aminon, I., Roisin, Y., Tamo, M., & Dansi, A. (2016). Farmers' knowledge and perceptions of termites as pests of yam (Dioscorea spp.) in Central Benin. *International Journal of Pest Management, 62*, 75–84.

Lovei, G. L., & Sunderland, K. D. (1996). Ecology and behavior of ground beetles (Coleoptera: Carabidae). *Annual Review of Entomology, 41*, 231–256.

Mahapatro, G. K., & Kumar, S. (2015). Review on the incidence and management of coconut termites. *Indian Journal of Entomology, 77*, 152–159.

Maniania, N. K., Ekesi, S., & Songa, J. M. (2002). Managing termites in maize with the entomopathogenic fungus *Metarhizium anisopliae*. *International Journal of Tropical Insect Science, 22*, 41–46.

Mburu, D. M., Ochla, L., Maniania, N. K., Njagi, P. G. N., Gitonga, L. M., Ndungu, M. W., Wanjoya, A. K., & Hassanali, A. (2009). Relationship between virulence and repellency of entomopathogenic isolates of *Metarhizium anisopliae* and *Beauveria bassiana* to the termite *Macrotermes michaelseni*. *Journal of Insect Physiology, 55*, 774–780.

McMahan, E. A. (1983). Adaptations, feeding preferences, and biometrics of a termite-baiting assassin bug (Hemiptera: Reduviidae). *Annals of the Entomological Society of America, 76*, 483–486.

Meepagala, K., Osbrink, W., Sturtz, G., & Lax, A. (2006). Plant-derived natural products exhibiting activity against Formosan subterranean termites (*Coptotermes formosanus*). *Pest Management Science, 62*, 565–570.

Milner, R. J., Staples, J. A., Hartley, T. R., Lutton, G. G., & Watson, J. A. L. (1996). Occurrence of the entomogenous fungus, *Metarhizium anisopliae,* in nests and feeding sites of Australian termites. *Mycological Research, 102*, 216–220.

Milner, R. J., Staples, J. A., & Lutton, G. G. (1998). The selection of an isolate of the hyphomycete fungus *Metarhizium anisopliae*, for control of termites in Australia. *Biological Control, 11*, 240–247.

Mora, P., Rouland, C., & Renoux, J. (1996). Foraging and nesting damage caused by *Microtermes subhyalinus* (Isoptera: Termitidae) in a sugar cane plantation in the Central African Republic. *Bulletin of Entomological Research, 86*, 387–395.

Niken, S., Yoshimurab, T., Rokhmanc, F., & Mastur, Z. (2015). Potential for subterranean termite attack against five bamboo species in correlation with chemical components. *Procedia Environmental Sciences, 28*, 783–788.

Nwilene, F. E., Nwanze, K. F., & Youdeowei, A. (2008a). Impact of IPM on food and horticultural crops in Africa – A review. *Entomologia Experimentalis et Applicata, 128*, 355–363.

Nwilene, F. E., Agunbiade, T. A., Togola, M. A., Youm, O., Ajayi, O., Oikeh, S. O., & Falola, O. O. (2008b). Efficacy of traditional practices and botanicals for the control of termites on rice at Ikenne, southwest Nigeria. *International Journal of Tropical Insect Science, 28*, 37–44.

Osbrink, W. L., Williams, K. S., Connick, W. J., Jr., Wright, M. S., & Lax, A. R. (2001). Virulence of bacteria associated with the Formosan subterranean termite (Isoptera: Rhinotermitidae) in New Orleans, LA. *Environmental Entomology, 30*(2), 443–448.

Owusu, E. O., Akutse, K. S., & Afreh-Nuamah, K. (2008). Effect of some traditional plant components on the control of termites, Macrotermes spp (Isoptera: Termitidae). *African Journal of Science and Technology, 9*, 83–89.

Pandey, A., Chattopadhyay, P., Banerjee, S., Pakshirajan, K., & Singh, L. (2012). Antitermitic activity of plant essential oils and their major constituents against termite Odontotermes assamensis Holmgren (Isoptera: Termitidae) of North East India. *International Biodeterioration & Biodegradation, 75*, 63–67.

Qureshi, N. A., Ashraf, A., Afzal, M., Ullah, N., Iqbal, A., & Haleem, S. (2015). Toxic potential of *Melia azedarach* leaves extract against *Odontotermes obesus* and *Microtermes obesi*. *International Journal of Biosciences, 6*, 120–127.

Ramakrishnan, R., Suiter, D. R., Nakatsu, C. H., Humber, R. A., & Bennett, G. W. (1999). Imidacloprid-enchanced *Reticulitermes flavipes* (Isoptera: Rhinotermitidae) susceptibility to the entomopathogen *Metarhizium anisopliae*. *Journal of Economic Entomology, 92*, 1125–1132.

Rath, A. C. (2000). The use of entomopathogenic fungi for control of termites. *Biocontrol Science and Technology, 10*, 563–581.

Rathour, K. S., Ganguly, S., Das, T. K., Singh, P., Kumar, A., & Somvanshi, V. S. (2014). Biological management of subterranean termites (*Odontotermes obesus*) infesting wheat and pearl millet crops by entomopathogenic nematodes. *Indian Journal of Nematology, 44*, 97–100.

Reddy, P. S., & Ghewande, M. P. (1986). Major insect pests of groundnut and their management. *Pesticides, 20*, 52–56.

Ruan, G., Song, X., Hu, Y., Han, N., & Zhang, D. (2015). Foraging activities of *Coptotermes formosanus* in subtropical areas in China. *Journal of Economic Entomology, 108*, 701–706.

Sajap, A. S., Amin, M. J., & Ouimette, D. (2002). Above -ground baiting for controlling *Coptotermes* termites in Selangor, Malaysia. *Sociobiology, 39*, 345–352.

Singha, D., Singha, B., & Dutta, B. K. (2010). In vitro pathogenicity of Bacillus thuringiensis against tea termites. *Journal of Biological Control, 24*, 279–281.

Su, N. Y. (2015). A fluid bait for remedial control of subterranean termites. *Journal of Economic Entomology, 108*, 274–276.

Su, N. Y., Thoms, E. M., Ban, P. M., & Scheffrahn, R. H. (1995). A monitoring/baiting station to detect and eliminate foraging populations of subterranean termites (Isoptera: Rhinotermitidae) near structures. *Journal of Economic Entomology, 88*, 932–936.

Su, N. Y., Ban, P. M., & Scheffrahn, R. H. (1997). Remedial baiting with hexaflumuron in above-ground stations to control structure-infesting populations of the Formosan subterranean termite (Isoptera: Rhinotermitidae). *Journal of Economic Entomology, 90*, 809–817.

Umeh, V. C., & Ivbijaro, M. F. (1999). Effects of termite damage to maize of seed extracts of *Azadirachta indica* and *Piper guineense* in farmers' fields. *The Journal of Agricultural Science, 133*, 403–407.

UNEP/FAO/Global IPM Facility Expert Group on Termite Biology and Management. (2000). Finding alternatives to persistent organic pollutants (POPs) for termite management, online at www.chem.unep.ch/pops/termites/termite_ch4.htm

Verma, A. N. (1980). *Effect of number of irrigations on termite damage in wheat crop*. Haryana Agricultural University.

Wang, C., & Powell, J. E. (2004). Cellulose bait improves the effectiveness of *Metarhizium anisopliae* as a microbial control of termites (Isoptera: Rhinotermitidae). *Biological Control, 30*, 523–529.

Wesolowska, W., & Cumming, M. S. (2002). *Mashonarus guttatus*, gen and sp n., the second termitivorous jumping spider from Africa (Araneae: Salticidae). *Bulletin of the British Arachnological Society, 12*, 165–170.

Yii, J. E., Bong, C. F. J., King, J. H. P., & Kadir, J. (2016). Synergism of entomopathogenic fungus, *Metarhizium anisopliae* incorporated with fipronil against oil palm pest subterranean termite, *Coptotermes curvignathus*. *Plant Protection Science, 52*, 35–44.

Zhu, B. C., Henderson, G., Chen, F., Maistrello, L., & Laine, R. A. (2001). Nootkatone is a repellent for Formosan subterranean termite (*Coptotermes formosanus*). *Journal of Chemical Ecology, 27*, 523–531.

Chapter 8
Role of Botanicals in Termite Management

Syed Kamran Ahmad, Natalie Dale-Skey, and Md. Aslam Khan

Contents

Abstract Plant-derived pesticides (botanical) deliver a potential alternative to highly hazardous synthetic pesticides for insect pest control. They can be derived by leaves, floral system, fruits or seeds, wood, and/or roots. The active chemical compounds are extracted via drying, grinding, and mixing the plant parts in suitable solvents. Some of the well-known botanical pesticides are pyrethrin, rotenone, sab-

S.K. Ahmad (✉)
Centre for Environmental Research and Studies (CERS), Jazan University, Jazan, Saudi Arabia
e-mail: entosaif@rediffmail.com

N. Dale-Skey
The Natural History Museum, London, UK

M.A. Khan
Department of Biology, Faculty of Science, Jazan University, Jazan, Saudi Arabia

M.A. Khan, W. Ahmad (eds.), *Termites and Sustainable Management*, Sustainability in Plant and Crop Protection,
https://doi.org/10.1007/978-3-319-68726-1_8

adilla, nicotine, ryanodine, etc. Little attention has been paid to explore the use of botanicals against termites. Most of the studies are limited to the extraction of plant chemicals with water and methanol and to the application against termites to assess their killing potential. The botanical bioassays against termites seem incomplete because either some aerial parts like leaves, fruits, seeds, and stems or roots have been utilized alone. Similarly, oil extraction and plant crude extracts were not assessed for the same plants. Isolation and synthesis of active compound are also very rare. In this chapter, we review the properties of various plant parts and their potential role in termite management, highlighting the gaps concerning the available informations.

Keywords Termite • Botanical insecticides • Resin • Essential oils

8.1 Introduction

Insect pest management relies heavily on synthetic chemicals. Calendar-based and non-judicious use of these pesticides for the control of agricultural and urban insect pests has created severe environmental hazards. Excess use of synthetic pesticides resulted in phytotoxicity, mammalian toxicity, pesticides residues, pesticide resistance in target insect pests, insect outbreaks, and increased production costs (Elango et al. 2012). Plants are well known to have a cumulative defense mechanism and to produce secondary metabolites in order to survive in the ecosystem facing insect infestations. Since immemorial times, the botanical remedies were the only options used against biting arthropods (Birkinshaw and Colquhoun 1998). Unfortunately, major scientific efforts have been directed toward the production of pharmaceuticals, cosmetics, and medicinal stuffs from plants, rather than developing botanical pesticides (Nakayama and Osbrink 2010). The deleterious effects of plant extracts on insects are manifested in several ways, including suppression of calling behavior (Khan and Saxena 1986), growth retardation (Breuer and Schmidt 1995), toxicity (Hiremath et al. 1997), oviposition deterrence (Zhao et al. 1998), feeding inhibition (Wheeler and Isman 2001), and reduction of fecundity and fertility (Muthukrishnan and Pushpalatha 2001). Frequent studies have documented the natural resistance of certain plant species to insect attacks, but most of the studies are incomplete as they were conducted without exploring up to the active compound. Also, the majority of studies are laboratory based, lacking the know-how about environmental errors. Developmental steps toward botanical insecticides are shown in Fig. 8.1.

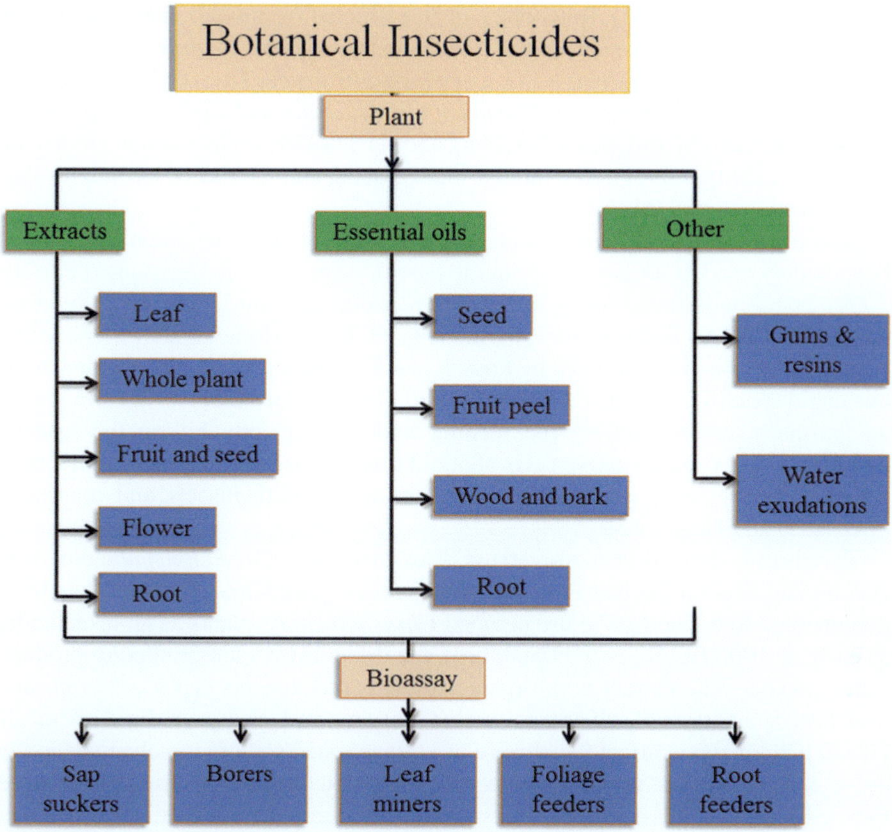

Fig. 8.1 Different plant parts for development of botanical insecticide

8.2 Popular Botanical Insecticides

Despite having a long history, the numbers of successfully isolated and commercialized botanical insecticides are countable on fingers. Rotenone, one of the oldest botanicals, is a colorless, crystalline isoflavone used as a broad-spectrum insecticide. The active chemical component was first isolated in 1895 by a French botanist, Emmanuel Geoffroy, who called it nicouline, from *Lonchocarpus nicou* (Ambrose and Harvey 1936). Later on, the Japanese chemist Nagai Nagayoshi isolated a pure crystalline compound from *Derris elliptica* in 1902 and named it "rotenone." Rotenone acts as a pesticide, an insecticide, and a fish killer (Fimrite 2007). It is a nonselective insecticide, effective against a range of insect pests like potato beetles, cucumber beetles, flea beetles, cabbage worms, raspberry and asparagus beetles, and various other arthropods. It is known to interfere with nicotinamide adenine dinucleotide (NAD) (a coenzyme found in all living cells) during the formation of ATP (adenosine triphosphate) which is the central unit of intracellular energy and metabolism (Hayes 1991). Rotenone can be extracted from seeds, stems, and roots of several tropical and subtropical plant species such as *Tephrosia virginiana*, *Pachyrhizus erosus*, *Deguelia utilis*, *Lonchocarpus urucu*, *Derris elliptica*, *Derris involuta*, *Mundulea sericea*, *Piscidia piscipula*, *Millettia*, and *Tephrosia* spp. (Nellis 1994; Barton and Meth-Cohn 1999; Fang and Casida 1999). The rotenone molecule shows the quality trait of being rapidly biodegraded, and the risk of hazardous residues associated with it is low.

Nicotine is another important stimulant alkaloid found among Solanaceae. It has been widely used as an insecticide in the past (Ujváry 1999; Rodgman and Perfetti 2009). Nicotine is found in the leaves of *Nicotiana tabacum*, *N. rustica*, *Duboisia hopwoodii*, and *Asclepias syriaca*, in the range of 2–14% (Metcalf 2007). The active compound was isolated in 1828 by Posselt and Reimann. It is effective against various aphid species including other soft-bodied insects (Matsumura 1975) and acts as an anti-herbivore, disrupting the feeding and consequently killing the insects. Nicotine is an extremely nerve toxic alkaloid that competes with acetylcholine neurotransmitters causing uncontrolled nerve fringe in both insects and mammals (Weinzierl and Henn 1994).

Pyrethrin is derived from pulverized dried flowers of *Chrysanthemum cinerariifolium* and acts on the nervous systems of insects. The Chinese culture has been documented to be the first to use crushed *Chrysanthemum* plants as an insecticide, as early as 1000 BC. To extract pyrethrin, the dried flowers are grinded to produce a fine powder. The extract with organic solvents contains six types of pyrethrins: pyrethrin I, pyrethrin II, cinerin I, cinerin II, jasmolin I, and jasmolin II (Metcalf 2000). It is known to affect the closure of voltage-gated sodium ion channels in the nerve cells of insects, resulting in repeated and extended nerve firings that cause hyperexcitations. This condition leads to the loss of motor coordination and paralysis, followed by the insect death. Pyrethrin is also a potent insect repellent when applied at low doses. The active ingredient is easily photodegraded under the sunlight (Stenersen 2004). After the discovery of synthetic forms of pyrethrins, the use of natural pyrethrin sharply declined. Increase in pyrethroid use resulted in several environmental hazards such as the decline in pollinators (Gemmill-Herren 2016).

The compound sabadilla is an alkaloid derived from a plant known as tropical lily, also known as cevadilla or Indian caustic barley (*Schoenocaulon officinale*) in South and Central America. The extract of dried seed contains cevadine and veratridine alkaloids which are the most active compounds that kill insects (Hayes 1982). The insecticide can be prepared by heating or through an alkaline treatment, in order to activate the active compounds, followed by seed grinding to prepare the dust (Allen et al. 1944). Sabadilla acts similar to pyrethrins and affects the voltage-dependent sodium channels at axons in the nervous systems (Bloomquist 1996). Historically, sabadilla has been used against insect pests of crops, mammals, and human beings.

Ryanodine is another natural plant compound and a slow-acting stomach poison. It proceeds from the wood of Salicaceae plants (Roskov et al. 2014). This naturally occurring insect killer was isolated from the stem wood of *Ryania speciosa.* The alkaloid ryanodine is the major toxic component, corresponding approximately to

0.2% of the total dry wood weight. Much detail on mode of action is not available. However, after ingestion, the insects stop feeding followed by a slow paralysis. The compound is known to interact with open-form ryanodine receptors, a group of calcium channels found in the muscular system of the body (Gaetano and Andrew 2015). The purified form of ryanodine is 700 times toxic as compared to the crude wood dust form (Weinzierl and Henn 1994).

8.3 Botanicals Tested Against Termites

8.3.1 Extracts

Researchers recognized many plants with anti-termitic activities (Sakasegawa et al. 2003; Park and Shin 2005; Cheng et al. 2007; Ding and Hu 2010). Duke et al. (2010) suggested the term "greener termiticide" for the first time and elaborated the importance of botanicals in termite management. Plants are naturally supplied with certain chemicals like terpenoids, flavonoids, saponins, etc. or mixtures of chemicals that repel and kill termites or interfere with their gut flora (Boué and Raina 2003; Park and Shin 2005; Verma et al. 2009). Recently, few plant species such as *Pseudotsuga menziesii* (Mirb.), *Lysiloma seemanii* Britton & Rose, *Tabebuia guayacan* (Seem.), *Diospyros sylvatica* Roxb. (Ganapaty et al. 2004), *Curcuma aromatica* Salisb. and *Euphorbia kansui* Gan-Sui (Shi et al. 2008), *Eucalyptus globules* L., lemon grass, *Eucalyptus citriodora* (Hook.), cedar wood, clove bud and vetiver grass (Zhu et al. 2001a), *Taiwania cryptomerioides* Hay. (Chang et al. 2001), *Dodonaea viscosa* (L.) Jacq. (purple hop bush, a termite-resistant shrub), *Ocimum basilicum* L., *Cymbopogon winterianus* Jowitt, *Cinnamomum camphora* (L.) Nees and Eberm., *Rosmarinus officinalis* L. (Sbeghen et al. 2002), and *Coleus amboinicus* (Lour.) (Singh et al. 2004) have been explored for their antifeedant and termiticidal activities. Meepagala et al. (2006) isolated the active compounds "Vulgarone B" (from *Artemisia douglasiana*, Asteraceae), "cnicin" (from *Centaurea maculosa*, Asteraceae), and "apiol" (from *Ligusticum hultenii*, Apiaceae) that significantly caused mortality to subterranean termites in laboratory bioassays. Vulgarone B, cnicin, and apiol also possess other biological activities such as phytotoxic and antifungal properties, suggesting the ecological importance of secondary metabolites in these plants (Meepagala et al. 2003). These compounds reflect the importance of plant secondary metabolites in natural defense mechanisms.

8.3.1.1 Leaf Extracts

Plant leaves are one of the major flavonoid reservoirs with number of attributes. Alshehry et al. (2014) found hexane leaf extract of plants, namely, *Rhazya stricta* Decne, *Lantana camara* L., *Ruta chalepensis* L., and *Heliotropium bacciferum*

Forssk., as promising against the subterranean termite *Psammotermes hybostoma* (Desneux). Yuan and Hu (2012) also reported the repellent, antifeedant, and toxic activities of *Lantana camara* leaf extract against *Reticulitermes flavipes*. Hexane and methanol leaf extracts of *Juniperus* sp. also showed effective performance against termites (Adams et al. 1988). Addisu et al. (2014) used water extracts of *Azadirachta indica* and *Jatropha curcas* leaves against *Macrotermes* spp. with satisfactory results. Similarly, Grace and Yates (1992) discovered Margosan-O (a neem insecticide formulation) with 0.3% azadirachtin and 14% neem oil, highly toxic against the subterranean termite. The leaf extract of *Flourensia cernua* with hexane, diethyl ether, and ethanol showed a high degree of termite toxicity (Tellez et al. 2001). Sharma et al. (1999) studied the toxic effects of six plants, viz., *Acorus calamus*, *L. camara*, *Parthenium hausteneum*, *Pongamia glabra*, *J. curcas*, and *Tagetes erecta*, for their toxic action against *Odontotermes obesus* and reported *A. calamus* rhizomes and aerial parts of *T. erecta* as the most toxic. Thambidurai (2002) achieved the successful prevention of termites for up to 50 days with the use of fermented leaf extracts of *Musa paradisiaca*. Fokialakis et al. (2006) reported *Echinops* sp. as most effective against termites, out of 220 crude extracts tested. A 5% chloroform extract of *L. camara* var. *aculeata* was effective against termites (Verma and Verma 2006), whereas the leaf extract of *Detarium microcarpum* with methanol appeared as an effective antifeedant. Adedeji et al. (2017) treated *Triplochiton scleroxylon* and *Vitex doniana* woods with stem bark and leaf extract compounds of henna (*Lawsonia inermis* Linn.) and reported a reduction in the attack of termites. Shiberu et al. (2013) observed 100% termite mortality with leaf extracts of *N. tabacum* and *Phytolacca dodecandra* after 24 h. Elango et al. (2012) further extended the list of anti-termitic plants with *Andrographis lineata* Wallich, *Aristolochia bracteolata* Lam. (Aristolochiaceae), *Datura metel* L. (Solanaceae), and *Eclipta prostrata* L. (Asteraceae).

8.3.1.2 Root Extracts

The toxic and repellent effect of *Zingiber officinale*, *Allium sativum*, *Dennettia tripetala*, and *Capsicum annuum*, as mixed and individual extracts, against *Macrotermes bellicocus* was studied under laboratory and field conditions. All extracts appeared promising, but the mixture of *Z. officinale* with *A. sativum* was the most toxic (Cynthia et al. 2016). Ganapaty et al. (2004) extracted quinines using chloroform from roots of *Diospyros sylvatica* which exhibited high toxicity against *Odontotermes obesus*. *Jatropha curcas* is another plant with insecticidal, molluscicidal, and fungicidal properties (Nwosu and Okafor 1995; Liu et al. 1997; Solsoloy and Solsoloy 1997). The root, stem, and bark solvent extracts of *Jatropha* exhibited potential termiticidal effects (Verma et al. 2013).

8.3.1.3 Fruit and Seed Extracts

The potentiality of plant seeds and fruit peels as termiticides has been investigated across the globe. The seeds of Indian neem tree, *Azadirachta indica*, are a major source of botanical insecticides. They contain many azadirachtin analogs, dominantly "azadirachtin A," with the remaining analogs sharing a little efficacy (Duke et al. 2010). Hexane extract of *Xylopia aethiopica* fruits and aqueous methanol extract of its seeds were studied for their antifeedant activity against workers of *Reticulitermes speratus*. The crude extract exhibited strong antifeedant activity at 1% concentration. Further isolation of hexane extracts resulted in the discovery of six ent-kaurane diterpenes of which ent-kaur-16-en-19-oic acid exhibited the strongest antifeedant activity against termites (Lajide et al. 1995). Addisu et al. (2014) found that seed extracts in water of *Maesa lanceolata*, *Chenopodium ambrosioides*, and *Vernonia hymenolepis* were effective against *Macrotermes* sp. Escoubas et al. (1995) prepared n-hexane and methanolic seed extracts of *Aframomum melegueta* and isolated various compounds. Among them, [6]-gingerol and [6]-shogoal showed the strongest termite antifeedant activity. Verma et al. (2011, 2013) observed 100% termite mortality with water extracts of nonedible oil seed cake of *J. curcas*. A 100% mortality of termites was also recorded with seed extract of Birbira (*Militia ferruginea*) after 24 h (Shiberu et al. 2013).

8.3.1.4 Wood Extracts

There are few woody plants showing resistance against termite attacks, due to the presence of some active components. These have been isolated and studied for their efficacy against various insect pests. The sapwood extracts of sugar pine, *Pinus lambertiana* Dougl., and related compounds were found having a feeding-deterrent effect against the western drywood termite, *Incisitermes minor* (Hagen) (Scheffrahn and Rust 1983). Similarly, the heartwood extract of *Taxodium distichum* (L.) Rich exhibited feeding deterrence against *Coptotermes formosanus* Shiraki (Scheffrahn et al. 1988). Besides promising termiticidal activity of leaf extracts, the fresh heartwood sawdusts of 12 *Juniperus* species exhibited termiticidal activities (Adams et al. 1988). *Juniperus procera* contains cedrol and cedrene, potentially toxic against termites (Kinyanjui et al. 2000). The heartwoods of *Erisma* sp., *Tabebuia* sp., and *Chamaecyparis thyoides* exhibit an apparent natural resistance to *Reticulitermes flavipes* (Arango et al. 2006). Teak wood and heartwood of *Caesalpinia echinata* Lam. are naturally resistant to termite attacks (Roszaini et al. 2006; Silva et al. 2007). The wood extract of *Catalpa bignonioides* contains four toxic compounds, catalponol, epicatalponol, catalponone, and catapalactone, of which the latter two were the most effective against *R. flavipes* (McDaniel 1992).

The heartwood of *Lonchocarpus castilloi* Standley is a source of some flavonoids such as castillen D and castillen E and showed dose-dependent feeding deterrence against *Cryptotermes brevis* (Walker) (Reyes-Chilpa et al. 1995). Toxifolin and quercetin isolated from Japanese larch wood might be useful as termite control agents (Ohmura et al. 2000). Also, the water extracts of Japanese larch wood, containing flavonoids in large quantities, proved to be an excellent feeding deterrent against termites (Chen et al. 2004). Two compounds, cedrol and a-cadinol, isolated from heartwood of *Taiwania cryptomerioides*, showed a high anti-termitic potential (Chang et al. 2001). Boue and Raina (2003) used five plant flavonoids, genistein, biochanin A, apigenin, quercetin, and glyceollin, for their impact on life attributes of termite *C. formosanus*. Apigenin proved to be the most fatal among all, while biochanin A was also noticed to reduce the fecundity of subterranean termites. The black heartwood of *Camellia japonica* holds some sesquiterpenes, showing termiticidal activity against *C. formosanus* (Arihara et al. 2004). Heartwood of white cypress pine *Callitris glaucophylla* exhibited the potential termite repellency (Watanabe et al. 2005). Finally, the wood vinegar from mixed wooden chips of *C. japonica*, *Pseudotsuga menziesii*, *Quercus serrata*, and *Pinus densiflora* showed high toxicity against *R. speratus* (Yatagai et al. 2002).

8.3.2 Plant Resins

Dipterocarp (Dipterocarpaceae) timber plants are resin exudation trees. Almost all Southeast Asian dipterocarp timbers exude resins (Schulte and Schone 1996). These timber trees exhibit natural resistance against insect pests such as termites and cause substantial mortality to insects feeding on them (Moi 1980). Wood resins of the dipterocarp timber trees contain a variety of terpenoids (Diaz et al. 1966; Bisset et al. 1971). The dipterocarps with sesquiterpenes exhibit greater degree of defense against insects (Messer et al. 1990). The crude resin of *Dipterocarpus kerrii*, containing four sesquiterpenoids closely related to α-gurjunene, is responsible for termiticidal activity against *Zootermopsis angusticollis* (Richardson et al. 1989). Richardson et al. (1991) isolated two uncharacterized sesquiterpenes (1 and 20), from *D. kerrii* resin. Sesquiterpene 20 was more effective against *Neotermes delbergiae* as compared to sesquiterpene 1. Chemical components demonstrating insecticidal properties were isolated and identified from crude resins of many *Dipterocarpus* species. The compounds, viz., alloaromadendrene, humulene, and caryophyllene, are most effective against Southeast Asian termites *Neotermes* sp. (Messer et al. 1990). Sen-Sarma (1963) and Sen-Sarma and Chatterjee (1968) found *Shorea robusta* highly resistant to *Heterotermes indicola* and *Microcerotermes beesoni*. Particle boards made from the wood of *Shorea* spp. exhibited natural protection from *Cryptotermes cynocephalus* (Moi 1980).

8.3.3 Plant Essential Oils

Plant-derived essential oils are mixtures of natural organic compounds responsible for plants' defense against general herbivores (Lima et al. 2013). Also, the essential oils derived from plants may be considered as the most efficient alternative in controlling insect pests like termites (Alavijeh et al. 2014). The active compounds of essential oils are highly volatile and of short persistence in the environment because of their low molecular weights (Isman 2006). Hence, these oil-based compounds appear environmentally safe and as a feasible alternative to the hazardous insecticides used in pest control (Lima et al. 2010). Various oils from plants such as clove, peppermint, etc. are used by pest control operators against cockroaches, ants, and termites (Isman et al. 2011). Bultman et al. (1979) tested 42 tropical African woods and suggested that insecticidal and termiticidal activities of essential oils may be due to the volatile compounds they contain (Bultman et al. 1979). Pandey et al. (2012) reported a strong termiticidal activity of essential oils extracted from lemongrass (*Cymbopogon citratus*), eucalyptus (*Eucalyptus globulus*), clove (*Syzygium aromaticum*), oregano (*Origanum vulgare*), rosemary (*Rosmarinus officinalis*), cinnamon (*Cinnamomum verum*), and thyme (*Thymus vulgaris*) when tested against *Odontotermes assamensis*. Candlenut (*Aleurites moluccana*) oil, obtained by mechanical pressing of the nut from the kukui plant, was used to treat southern yellow pine (*Pinus* sp.) wood against *Coptotermes formosanus* attacks, with an excellent feeding-deterrent effect (Nakayama and Osbrink 2010). Himmi et al. (2013) applied defatted neem oil (DNO) formulation against *C. gestroi* and reported it as superior to the azadirachtin fraction (91% purity). Lima et al. (2013) tested toxicity of essential oils from *Corymbia citriodora*, *Croton sonderianus*, *Cymbopogon martini*, *Lippia alba*, *L. gracilis*, *L. sidoides*, and *Pogostemon cablin* against *Nasutitermes corniger*. All essential oils were noticed relatively toxic against workers rather than soldiers. The oils from *P. cablin* and *L. sidoides* were up to 25 times more toxic than that from *C. sonderianus*.

Scheffrahn et al. (1988) found ferruginol, manool, and nezukol, isolated from bald cypress *Taxodium distichum* L. wood, as antifeedant against *C. formosanus*. The oil extracted from marigold (*T. erecta*) and sweet orange *Citrus sinensis* proved to be an excellent repellent against *O. obesus*. The marigold oil exhibited maximum repellency, followed by sweet orange oil (Verma et al. 2016). Nagnan and Clement (1990) also studied the commercially available geranyllinalool for toxicity against termites. Carter et al. (1978) and Ikeda et al. (1978) isolated the essential oil components, viz., chamaecynone, an acetylenic terpenoid from false cypress *Chamaecyparis pisifera*; 7-methyljuglone, a naphthoquinone from American persimmon *Diospyros virginiana*; and torreyal from Japanese kaya, *Torreya nucifera*, and successfully tested their toxicity against termites. Adams (1991) discovered the repellent activity of essential oils of cedar wood, *Litsea cubeba*, and *Cinnamomum* sp. against termites, later reconfirmed by Lin and Yin (1995). The terpenoids of essential oils, viz., cedrol, citral, citronellal, eugenol, ferruginol, geraniol, limonene, manool, nezukol, and piperitone, are the repelling agents against subterranean termites (Sharma et al. 1994; Cornelius et al. 1997). Zhu et al. (2001b) found nootkatone (sesquiterpene) and other terpenoids in vetiver oil as highly efficient repellents and toxicants against subterranean termites.

Elango et al. (2012) investigated efficacy of solvent extracts of eight medicinal plants including *T. erecta* against the subterranean termite *C. formosanus*. All the crude extracts showed anti-termite activity in a dose-dependent manner and exhibited significant activity after 24 and 48 h of exposure. Data suggested they were a novel, safer, and renewable source of natural wood preservatives and termiticides. A field study was conducted by Aschalew et al. (2008) to evaluate the efficacy of 11 pesticidal plants against *Microtermes adschaggae* on hot pepper crops. They recorded lower percentages of damaged plants, higher stand counts at harvest, and higher amounts of dry pod yields with *T. minuta*. Similar results were obtained by Ahmed and Abraham (2014) on hot pepper. Aschalew et al. (2005) also reported that *Croton macrostachyus* and *T. minuta* have repellent properties against termites. Raina et al. (2007) used citrus peel and orange oil extract to control *C. formosanus*. They reported that d-limonene (92% constituent of orange oil extract) was responsible for termite mortality. Acda (2009) reported antifeedant, anti-tunneling, repellent, and termiticidal activities of *J. curcas* oil against *C. vastator*. Ede and Demissie (2013) obtained 100% termite mortality with *Jatropha* seed oil in 48 h against *Odontotermes obesus*. Singh and Kumar (2008) tested the leaf, root, and bark of *J. curcas* and oils of *T. erecta* and *Citrus sinensis* against *O. obesus* and *Microcerotermes beesoni* that showed significant losses in termite body weights.

8.4 Plant Extracts and Termite Gut Microbes

Apart from affecting the life and development of termites, some plant compounds are able to disrupt the activity of their gut microbes. However, studies of this kind are relatively rare. Soil treated with seeds of *Withania somnifera*, *Croton tiglium*, and *Hygrophila auriculata* disrupted the bacterial activities in the gut of *Microtermes obesi*. Seed extracts of *W. somnifera* and *H. auriculata* were noticed as highly toxic in a 6-day period. The areas of tunneling and the number of bacterial colonies were also reduced at 100% concentration of *W. somnifera* and *H. auriculata* (Ahmed et al. 2006). Three species of flagellated protists (*Spirotrichonympha leidyi*, *Holomastigotoides hartmanni*, and *Pseudotrichonympha grassii*) inhabit the hindgut of Formosan subterranean termites (Ohkuma et al. 2000). Doolittle et al. (2007) investigated the ability of three natural products (neem extract, capsaicin, and gleditschia) to reduce their numbers together with other spirochaetes. They noticed that neem extract significantly reduced the population of *P. grassi* and spirochaetes, with most potent effects at 1 ppm concentration, causing 100% termite mortality.

8.5 Conclusion

The list of plants with pesticidal traits appears long enough, but active or principle compound responsible for their effects is just being uncovered. Most studies are confined to grinding plant parts and using them as hexane, ethanol, or water extracts. More detailed plant analyses about compounds characterizing oils extracted from wood, seeds, and bark and crude chemicals extracted from leaves, flowers, fruits, seeds, barks, stem, and root are needed. Finally, sustainable exploitation of these natural products requires the production of more experimental data on their selectivity and/or target ranges, as most bioassays involving botanical insecticides have been only conducted thus far against a low number of termite species.

References

Acda, M. N. (2009). Toxicity, tunneling and feeding behavior of the termite, *Coptotermes vastator*, in sand treated with oil of the physic nut, *Jatropha curcas*. *Journal of Insect Science, 9*, 1–8.

Adams, R. P. (1991). Cedar wood oil-analysis and properties. *Modern Method of Plant Analysis, 12*, 159–173.

Adams, R. P., McDaniel, C. A., & Carter, F. L. (1988). Termiticidal activities in the heartwood, bark/sapwood and leaves of *Juniperus* species from the United States. *Biochemical Systematics and Ecology, 16*, 453–456.

Addisu, S., Mohamed, D., & Waktole, S. (2014). Efficacy of botanical extracts against termites, *Macrotermes* spp., (Isoptera: Termitidae) under laboratory conditions. *International Journal of Agricultural Research, 9*, 60–73.

Adedeji, G. A., Ogunsanwo, O. Y., & Elufioye TO. (2017). Quantifications of phytochemicals and biocide actions of *Lawsonia inermis* linn. extracts against wood termites and fungi. *International Biodeterioration and Biodegradation, 116*, 155–162.

Ahmed, I., & Abraham, T. (2014). Studies on the effects of some potential botanicals against termite damage on hot pepper (Marakofana) at Mendi, West Wellega, Ethiopia. *International Journal of Innovative and Applied Research, 2*, 29–34.

Ahmed, S., Riaz, M. A., & Shahid, M. (2006). Response of *Microtermes obesi* (Isoptera: Termitidae) and its gut bacteria towards some plant extracts. *Journal of Food, Agriculture and Environment, 4*, 317–320.

Alavijeh, E. S., Habibpour, B., Moharramipour, S., & Rasekh, A. (2014). Bioactivity of *Eucalyptus camaldulensis* essential oil against *Microcerotermes diversus* (Isoptera: Termitidae). *Journal of Crop Protection, 3*, 1–11.

Allen, T. C., Dicke, R. J., & Harris, H. H. (1944). Sabadilla, *Schoenocaulon* spp., with reference to its toxicity to houseflies. *Journal of Economic Entomology, 37*, 400–408.

Alshehry, A. Z., Zaitoun, A. A., & Abo-Hassan, R. A. (2014). Insecticidal activities of some plant extracts against subterranean termites, *Psammotermes hybostoma* (Desneux) (Isoptera: Rhinotermitidae). *International Journal of Agricultural Science, 4*, 257–260.

Ambrose, A. M., & Harvey, B. H. (1936). Toxicological study of *Derris*. *Industrial and Engineering Chemistry, 28*, 815–821.

Arango, R. A., Green, F., III, Hintz, K., Lebow, P. K., & Miller, R. B. (2006). Natural durability of tropical and native wood against termite damage by *Reticulitermes flavipes* (Kollar). *International Biodeterioration and Biodegradation, 57*, 146–150.

Arihara, S., Umeyama, A., Bando, S., Imoto, S., Ono, M., & Yoshikawa, K. (2004). Three new sesquiterpenes from the black heartwood of *Cryptomeria japonica*. *Chemical & Pharmaceutical Bulletin, 52*, 463–465.

Aschalew, S., Diriba, G., & Demisse, A. (2005). *Termite in Mana-Sibu district of Oromiya, Special Research Report*. Addis Ababa: Ethiopian Agricultural Research Organization.

Aschalew, S., Ahimad, I., & Tadele, T. (2008). Management of termite (*Microtermes adschaggae*) on hot pepper using powdered leaves and seeds of some plant species at Bako, Western Ethiopia. *East African Jorunal of Sciences, 2*, 41–44.

Barton, D. H. R., & Meth-Cohn, O. (1999). *Comprehensive natural products chemistry* (Vol. 1). Oxford: Pergamon.

Birkinshaw, C. R., & Colquhoun, I. C. (1998). Pollination of *Ravenala madagascariensis* and *Parkia madagascariensis* by *Eulemur macaco* in Madagascar. *Folia Primatologica, 69*, 252.

Bisset, N. G., Chavanel, V., Lantz, J. P., & Wolff, R. E. (1971). *Constituants sesquiterpeniques* et triterpeniques des rdsines du gem'e shorea. *Phytochemistry, 10*, 2451–2463.

Bloomquist, J. R. (1996). Ion channels as targets for insecticides. *Annual Review of Entomology, 41*, 163–190.

Boué, S. M., & Raina, A. K. (2003). Effects of plant flavonoids on fecundity, survival, and feeding of the Formosan subterranean termite. *Journal of Chemical Ecology, 29*, 2575–2584.

Breuer, M., & Schmidt, G. H. (1995). Influence of a short period treatment with *Melia azedarach* extract on food intake and growth of the larvae of *Spodoptera frugiperda* (Lepidoptera; Noctuidae). *Journal of Plant Diseases and Protection, 102*, 633–654.

Bultman, J. D., Beal, R. H., & Ampong, F. F. K. (1979). Natural resistance of some tropical African woods to *Coptotermes formosanus* Shiraki. *Forest Products Journal, 29*, 46–51.

Carter, F. L., Garlo, A. M., & Stanely, J. B. (1978). Termiticidal components of wood extracts: 7-methyljuglone from *Diospyros virginiana*. *Journal of Agricultural and Food Chemistry, 26*, 869–873.

Chang, S. T., Cheng, S. S., & Wang, S. Y. (2001). Antitermitic activity of essential oils and components from Taiwania (*Taiwania cryptomerioides*). *Journal of Chemical Ecology, 27*, 717–724.

Chen, K., Ohmura, W., Doi, S., & Aoyama, M. (2004). Termite feeding deterrent from Japanese larchwood. *Bioresource Technology, 95*, 129–134.

Cheng, S. S., Chang, H. T., CL, W., & Chang, S. T. (2007). Anti-termitic activities of essential oils from coniferous trees against *Coptotermes formosanus*. *Bioresource Technology, 98*, 456–459.

Cornelius, M. L., Grace, J. K., & Yates, J. R. (1997). Toxicity of monoterpenoids and other natural products to the Formosan subterranean termite (Isoptera: Rhinotermitidae). *Journal of Economic Entomology, 87*, 705–708.

Cynthia, O. C., Precious, O., & Lynda, O. S. (2016). The toxicity and repellency of some plant extracts applied as individual and mixed extracts against termites (*Macrotermes bellicosus*). *Journal of Entomology and Zoology Studies, 4*, 406–418.

Diaz, M. A., Ourisson, G., & Bisset, N. G. (1966). Etudes chimio-taxonomiques dans la famille des dipterocarpacaes–I. *Phytochemistry, 5*, 855–863.

Ding, W., & Hu, X. P. (2010). Antitermitic effect of the *Lantana camara* plant on subterranean termites (Isoptera: Rhinotermitidae). *Insect Science, 17*, 427–433.

Doolittle, M., Raina, A., Lax, A., & Boopathy, R. (2007). Effect of natural products on gut microbes in Formosan subterranean termite, *Coptotermes formosanus*. *International Biodeterioration and Biodegradation, 59*, 69–71.

Duke, S. O., Cantrell, C. L., Meepagala, K. M., Wedge, D. E., Tabanca, N., & Schrader, K. K. (2010). Natural toxins for use in pest management. *Toxins (Basel), 2*, 1943–1962.

Ede, A. G., & Demissie, A. G. (2013). Comparative bio-activity guided characterization of biocide from *Jatropha curcas* and *Ricinus communis* L seeds oil. *Journal of Pharmacognosy and Phytochemistry, 2*, 176–181.

Elango, G., Rahuman, A. A., Kamaraj, C., Bagavan, A., Zahir, A. A., Santhoshkumar, T., Marimuthu, S., Velayutham, K., Jayaseelan, C., Vishnu Kirthi, A., & Rajakumar, G. (2012). Efficacy of medicinal plant extracts against Formosan subterranean termite, *Coptotermes formosanus*. *Industrial Crops and Products, 36*, 524–530.

Escoubas, P., Lajide, L., & Mizutani, J. (1995). Termite antifeedant activity in *Aframomum melegueta*. *Phytochemistry, 40*, 1097–1099.

Fang, N., & Casida, J. (1999). Cubé resin insecticide: Identification and biological activity of 29 rotenoid constituents. *Journal of Agricultural and Food Chemistry, 47*, 2130–2136.

Fimrite, P. (2007). Lake poisoning seems to have worked to kill invasive pike. San Francisco Chronicle. Published on October 02, 2007.

Fokialakis, N., Osbrink, W. L. A., Mamonov, L. K., Gemejieva, N. G., Mims, A. B., Skaltsounis, A. L., Lax, A. R., & Cantrell, C. L. (2006). Antifeedant and toxicity effects of thiophenes from four *Echinops* species against the Formosan subterranean termite, *Coptotermes formosanus*. *Pest Management Science, 62*, 832–838.

Gaetano, S., & Andrew, M. (2015). Essential roles of intracellular calcium release channels in muscle, brain, metabolism, and aging. *Current Molecular Pharmacology, 8*, 206–222.

Ganapaty, S., Thomas, P. S., Fotso, S., & Laatsch, H. (2004). Antitermitic quinines from *Diospyros sylvatica*. *Phytochemistry, 65*, 1265–1271.

Gemmill-Herren, B. (2016). *Pollination services to agriculture: Sustaining and enhancing a key ecosystem service. Food and Agriculture organization of the United Nations*. New York: Routledge. 292pp.

Grace, J. K., & Yates, J. R. (1992). Behavioural effects of a neem insecticide on *Coptotermes formosanus* (Isoptera: Rhinotermitidae). *Tropical Pest Management, 38*, 176–180.

Hayes, W. J. (1982). Pesticides derived from plants and other organisms. In W. J. Hayes (Ed.), *Pesticides studied in man* (pp. 75–111). Baltimore: Williams and Wilkins.

Hayes, W. J. (1991). *Handbook on pesticides* (Vol. 1). Jamestown Road: Academic Press. 1174 pp.

Himmi, S. K., Tarmadi, D., Ismayati, M., & Yusuf, S. (2013). Bioefficacy performance of neem-based formulation on wood protection and soil barrier against subterranean termite, *Coptotermes gestroi* Wasmann (Isoptera: Rhinotermitidae). *Procedia Environmental Sciences, 17*, 135–141.

Hiremath, I. G., Young Joon, A., Kim-Soon, I., & Kim, S. I. (1997). Insecticidal activity of Indian plant extracts against *Nilaparvata lugens* (Homoptera: Delphacidae). *Applied Entomology and Zoology, 32*, 159–166.

Ikeda, T., Takahashi, M., & Nishimoto, K. (1978). Antitermitic components of Kaya wood, *Torreya nucifera* Sieb. Et Zucc. *MokuzaiGakkaishi, 24*, 262–266.

Isman, M. B. (2006). Botanical insecticides, deterrents, and repellents in modern agriculture and an increasingly regulated world. *Annual Review of Entomology, 51*, 45–66.

Isman, M. B., Miresmailli, S., & Machial, C. (2011). Commercial opportunities for pesticides based on plant essential oils in agriculture, industry and consumer products. *Phytochemistry Reviews, 10*, 197–204.

Khan, Z. R., & Saxena, R. C. (1986). Effect of steam distillate extracts of resistant and susceptible rice cultivars on behavior of *Sogatella furcifera* (Homoptera: Delphacidae). *Journal of Economic Entomology, 79*, 928–935.

Kinyanjui, T., Gitu, P. M., & Kamau, G. N. (2000). Potential antitermite compounds from *Juniperus procera* extracts. *Chemosphere, 41*, 1071–1074.

Lajide, L., Escoubas, P., & Mizutani, J. (1995). Termite antifeedant activity in *Xylopia aethiopica*. *Phytochemistry, 40*, 1105–1112.

Lima, R. K., Cardoso, M. G., Moraes, J. C., Andrade, M. A., Melo, B. A., & Rodrigues, V. G. (2010). Caracterização química e atividade inseticida do óleo essencial de *Ageratum conyzoides* L. sobre a lagarta-do-cartucho *Spodoptera frugiperda* (Smith, 1797) (Lepidoptera: Noctuidae). *Bioscience, 26*, 1–5.

Lima, J. K. A., Albuquerque, E. L. D., Santos, A. C. C., Oliveira, A. P., Araújo, A. P. A., Blank, A. F., Arrigoni-Blank, M. F., Alves, P. B., & Santos, D. A. (2013). Biotoxicity of some plant essential oils against the termite *Nasutitermes corniger* (Isoptera: Termitidae). *Industrial Crops and Products, 47*, 246–251.

Lin, T. S., & Yin, H. W. (1995). The effects of *Cinnamomum* spp. oils on the control of the termite *Coptotermes formosanus* Shiraki. *Taiwan Forestry Research Institute New Series, 10*, 459–464.

Liu, S. Y., Sporer, F., Wink, M., Jouurdane, J., Henning, R., Li, Y. L., & Ruppel, A. (1997). Anthraquinones in *Rheum palmantum* and *Rumex dentatus* (Polygonaceae) and phorbol esters in *Jatropha curcas* (Euphorbiaceae) with molluscicidal activity against the *Schistosome* vector snails *Oncomelania*, *Biomphalaria* and *Bulinus*. *Tropical Medicine & International Health, 2*, 179–188.

Matsumura, F. (1975). *Toxicology of insecticides Fumio Matsumura*. Lower John Street: Plenium Press. 493pp.

McDaniel, C. A. (1992). Major antitermitic components of the heartwood of southern catalpa. *Journal of Chemical Ecology, 18*, 359–369.

Meepagala, K. M., Kuhajek, J. M., Sturtz, G., Wedge, D. E., & Vulgarone, B. (2003). The antifungal constituent in the steam-distilled fraction of *Artemisia douglasiana*. *Journal of Chemical Ecology, 29*, 1771–1780.

Meepagala, K. M., Osbrink, W., Sturtz, G., & Lax, A. (2006). Plant-derived natural products exhibiting activity against Formosan subterranean termites (*Coptotermes formosanus*). *Pest Management Science, 62*, 565–570.

Messer, A., McCormick, K., Sunjaya Hagedorn, H. H., Tumbel, F., & Meinwald, J. (1990). Defensive role of tropical tree resins: Antitermitic sesquiterpenes from Southeast Asian dipterocarpaceae. *Journal of Chemical Ecology, 16*, 3333–3352.

Metcalf RL (2000) Insect control: Ullmann's Encyclopedia of Industrial Chemistry. doi:https://doi.org/10.1002/14356007.a14_263.

Metcalf, R. L. (2007). *Insect control: Ullmann's Encyclopedia of Industrial Chemistry* (7th ed.). New York: Wiley. 9 pp.

Moi, L. C. (1980). A new laboratory method for testing the resistance of particle boards to the drywood termite *Cryptotermes cynocephalus*. *Malaysian Forest, 43*, 350–355.

Muthukrishnan, J., & Pushpalatha, E. (2001). Effects of plant extracts on fecundity and fertility of mosquitoes. *Journal of Applied Entomology, 125*, 31–35.

Nagnan, P., & Clement, J. L. (1990). Terpenes from the maritime pine *Pinus pinaster*: Toxins for subterranean termites of the genus *Reticulitermes* (Isoptera: Rhinotermitidae). *Biochemical Systematics and Ecology, 18*, 13–16.

Nakayama, F. S., & Osbrink, W. L. (2010). Evaluation of kukui oil (*Aleurites moluccana*) for controlling termites. *Industrial Crops and Products, 31*, 312–315.

Nellis, D. N. (1994). *Seashore plants of South Florida and the Caribbean* (p. 160). Sarasota: Pineapple Press.

Nwosu, M. O., & Okafor, J. I. (1995). Preliminary studies of the antifungal activities of some medicinal plants against *Basidiobolus* and some other pathogenic fungi. *Mycoses, 38*, 191–195.

Ohkuma, M., Ohtoko, K., Iida, T., Tokura, M., Moriya, S., Usami, R., Horikoshi, K., & Kudo, T. (2000). Phylogenetic identification of hypermastigotes, *Pseudotrichonympha*, *Spirotrichonympha* and *Holomastigotoides*, and parabasalian symbionts in the hindgut of termites. *The Journal of Eukaryotic Microbiology, 47*, 249–259.

Ohmura, W., Doi, S., Aoyama, M., & Ohara, S. (2000). Antifeedant activity of flavonoids and related compounds against the subterranean termite *Coptotermes formosanus* Shiraki. *Journal of Wood Science, 46*, 149–153.

Pandey, A., Chattopadhyay, P., Banerjee, S., Pakshirajan, K., & Singh, L. (2012). Antitermitic activity of plant essential oils and their major constituents against termite *Odontotermes assamensis* Holmgren (Isoptera: Termitidae) of North East India. *International Biodeterioration and Biodegradation, 75*, 63–67.

Park, I. K., & Shin, S. C. (2005). Fumigant activity of plant essential oils and components from garlic (*Allium sativum*) and clove bud (*Eugenia caryophyllata*) oils against the Japanese termite (*Reticulitermes speratus* Kolbe). *Journal of Agricultural and Food Chemistry, 53*, 4388–4392.

Raina, A., Bland, J., Doolittle, M., Lax, A., Boopathy, R., & Folkins, M. (2007). Effect of orange oil extract on the Formosan subterranean termite (Isoptera: Rhinotermitidae). *Journal of Economic Entomology, 100*, 880–885.

Reyes-chilpa, R., Viveros-rodriguez, L. N., Gomez-garibay, Z. F., & Alavez-solano, D. (1995). Antitermitic activity of lonchocarpus castilloi flavonoids and heartwood extracts. *Journal of Chemical Ecology, 21*, 455–463.

Richardson, D. P., Messer, A. C., Greenberg, S., Hagedorn, H. H., & Meinwald, J. (1989). Defensive sesquiterpenoids from a dipterocarp (Dipterocarpus kerrii). *Journal of Chemical Ecology, 15*, 731–747.

Richardson, D. P., Messer, A. C., Newton, B. A., & Lindeman, N. I. (1991). Identification and preparation of antiinsectan dienols from *Dipterocarpus kerrii* tree resins. *Journal of Chemical Ecology, 17*, 663–685.

Rodgman, A., & Perfetti, T. A. (2009). *The chemical components of tobacco and tobacco smoke* (p. 1840). Boca Raton: CRC Press.

Roskov, Y., Kunze, T., Orrell, T., Abucay, L., Paglinawan, L., Culham, A., Bailly, N., Kirk, P., Bourgoin, T., Baillargeon, G., Decock, W., De Wever, A., & Didžiulis, V. (2014) .Species 2000 and ITIS catalogue of life: 2014 annual checklist. Species 2000: Reading, UK. Retrieved 20 March 2017.

Roszaini, K., Salmiah, U., & MohdDahlan, J. (2006). Natural resistance of timbers to attack: Laboratory evaluation of the resistance of Malaysian wood against *Coptotermes curvignathus* (Holmgren). *Journal of the Institute of Wood Science, 17*, 178–182.

Sakasegawa, M., Hori, K., & K Yatagi, K. (2003). Composition and anti-termite activities of essential oils and *Melaleuca* species. *Journal of Wood Science, 49*, 181–187.

Sbeghen, A. C., Dalfovov, V., Serafini, L. A., & De-Barros, N. M. (2002). Repellence and toxicity of basil, citronella, ho-sho and rosemary oils for the control of the termite, *Cryptotermes brevis* (Isoptera: Kalotermitidae). *Sociobiology, 40*, 585–594.

Scheffrahn, R. H., & Rust, M. K. (1983). Drywood termite feeding deterrents in sugar pine and antitermitic activity of related compounds. *Journal of Chemical Ecology, 9*, 39–55.

Scheffrahn, R. H., Hsu, R. C., NY, S., Huffman, J. B., Midland, S. L., & Sims, J. J. (1988). Allelochemical resistance of bald cypress, *Taxodium distichum*, heartwood to the subterranean termite, *Coptotermes formosanus*. *Journal of Chemical Ecology, 14*, 765–776.

Schulte, A., & Schone, D. (1996). *Dipterocarp forest ecosystems: towards sustainable management*. Ltd, Ferrar Road: World Scientific Publishing co. Pvt. 949 pp.

Sen-Sarma, P. K. (1963). Studies on the natural resistance of timbers to termites. I. Observations on the longevity of the test termite *Heterotermes indicola* Wasm in the saw dust from forty common Indian timbers. *Indian Forest Bulletin (N.S.) Entomology, 220*, 1–3.

Sen-Sarma, P. K., & Chatterjee, P. N. (1968). Studies on the natural resistance of timbers to termite attack. V. Laboratory evaluation of the resistance of Indian wood to *Microcerotermes beesoni* Snyder (Termitidae, Amitermitinae). *Indian Forest, 94*, 694–704.

Sharma, R. N., Tungikar, V. B., Pawar, P. V., & Vartak, P. H. (1994). Vapour toxicity and repellency of some oils and terpenoids to the termite, *Odontoterme brunneus*. *Insect Science Application, 15*, 495–498.

Sharma, R. N., Tare, V., & Pawan, P. (1999). Toxic action of some plant extracts against selected insect pest and vectors. *Pestology, 23*, 30–37.

Shi, J., Li, Z., Izumi, M., Baba, N., & Nakajima, S. (2008). Termiticidal activity of diterpenes from the roots of *Euphorbia kansui*. *Zeitschrift für Naturforschung. Section C, 63*, 51–58.

Shiberu, T., Ashagre, H., & Negeri, M. (2013). Laboratory evaluation of different botanicals for the control of termite, *Microterms* spp (Isoptera: Termitidae). *Open Access Scientific Reports, 2*, 1–3.

Silva, A. C., Monteiro, M. B. B., Brazolin, S., Lopez, G. A. C., Richter, A., & Braga, M. R. (2007). Biodeterioration of brazilwood *Caesalpinia echinata* lam. (Leguminosae: Caesalpinioideae) by rot fungi and termites. *International Biodeterioration and Biodegradation, 60*, 285–292.

Singh, N., & Kumar, S. (2008). Anti termite activity of *Jatropha curcas* Linn. Biochemical. *Journal of Applied Sciences and Environmental Management, 12*, 67–69.

Singh, G., Singh, O. P., Prasad, Y. R., De-Lampasona, M. P., & Catalan, C. (2004). Chemical and insecticidal investigations in leaf oil of *Coleus amboinicus* Lour. *Flavour and Fragrance Journal, 17*, 440–442.

Solsoloy, A. D., & Solsoloy, T. S. (1997). Pesticidal efficacy of the formulated physic nut, *Jatropha curcas* oil on pests of selected field crops. *The Philippine Journal of Science, 124*, 59–74.

Stenersen, J. (2004). *Chemical pesticides: Mode of action and toxicology* (p. 265). Boca Raton: CRC Press/LLC.

Tellez, M., Estell, R., Fredrickson, E., Powell, J., Wedge, D., Schrader, K., & Kobaisy, M. (2001). Extracts of *Flourensia cernua* (l): volatile constituents and antifungal, antialgal, and antitermite bioactivities. *Journal of Chemical Ecology, 27*, 11.

Thambidurai, S. (2002). Termite control using natural products. *Indigenous Agriculture News, 1* 9.

Ujváry, I. (1999). Nicotine and other insecticidal alkaloids. In I. Yamamoto & J. Casida (Eds.), *Nicotinoid insecticides and the nicotinic acetylcholine receptor* (pp. 29–69). Tokyo: Springer-Verlag.

Verma, R. K., & Verma, S. K. (2006). Phytochemical and termiticidal study of *Lantana camara* Var. *aculeata* leaves. *Fitotrapia, 77*, 466–468.

Verma, M., Sharma, S., & Parshad, R. (2009). Biological alternatives for termite control. *International Biodeterioration and Biodegradation, 63*, 959–972.

Verma, M., Pradhan, S., Sharma, S., Naik, S. N., & Prasad, R. (2011). Efficacy of karanjin and phorbol ester fraction against termites (*Odontotermes obesus*). *International Biodeterioration and Biodegradation, 65*, 877–882.

Verma, S., Verma, M., Sharma, S., & Malik, A. (2013). Determination of phytocomponents by GC-MS analysis of *Jatropha curcas* root and its termiticidal activity. *Journal of Ecology and Environmental Sciences, 39*, 159–169.

Verma, S., Sharma, S., & Malik, A. (2016). Termiticidal and repellency efficacy of botanicals against *Odontotermes obesus*. *International Journal of Research in Biosciences, 5*, 52–59.

Watanabe, Y., Mihara, R., Mitsunaga, T., & Yoshimura, T. (2005). Termite repellent sesquiterpenoids from *Callitris glaucophylla* heartwood. *Journal of Wood Science, 51*, 514–519.

Weinzierl, R., & Henn, T. (1994). Botanical insecticides and insecticidal soaps. In A. R. Leslie (Ed.), *Handbook of integrated pest management for turf and ornamentals* (p. 627). Washington, DC: US Environmental Protection agency.

Wheeler, D. A., & Isman, M. (2001). Antifeedant and toxic activity off *Trichilia Americana* extract against the larvae of *Spodoptera litura*. *Entomologia Experimentalis et Applicata, 98*, 9–16.

Yatagai, M., Nishimoto, M., Ohira, K. H. T., & Shibata, A. (2002). Termiticidal activity of wood vinegar, its components and their homologues. *Journal of Wood Science, 48*, 338–342.

Yuan, Z., & Hu, X. P. (2012). Repellent, antifeedant and toxic activities of *Lantana camara* leaf extract against *Reticulitermes flavipes* (Isoptera: Rhinotermitidae). *Journal of Economic Entomology, 105*, 2115–2121.

Zhao, B., Grant, G. G., Langevin, D., & MacDonald, L. (1998). Deterring and inhibiting effects of quinolizidine alkaloids on the spruce budworm (Lepidoptera: Tortricidae) oviposition. *Environmental Entomology, 27*, 984–992.

Zhu, B. C. R., Henderson, G., Chen, F., Fei, H., & Laine, R. A. (2001a). Evaluation of vetiver oil and seven insect-active essential oils against the Formosan subterranean termite. *Journal of Chemical Ecology, 27*, 1617–1625.

Zhu, B. C. R., Henderson, G., Chen, F., Maistrello, L., & Laine, R. A. (2001b). Nootkatone is a repellent for Formosan subterranean termite (*Coptotermes formosanus*). *Journal of Chemical Ecology, 27*, 523–531.

Chapter 9
Microbes in Termite Management: Potential Role and Strategies

Priyanka Verma, Ajar Nath Yadav, Vinod Kumar, Md. Aslam Khan, and Anil Kumar Saxena

Contents

P. Verma
Department of Microbiology, Akal College of Basic Sciences, Eternal University, Sirmour 173101, India

A.N. Yadav (✉) • V. Kumar
Department of Biotechnology, Akal College of Agriculture, Eternal University, Sirmour 173101, India
e-mail: ajarbiotech@gmail.com

M.A. Khan
Department of Biology, Faculty of Science, Jazan University, Jazan, Saudi Arabia

A.K. Saxena
ICAR-National Bureau of Agriculturally Important Microorganisms, Kusmaur, Mau Nath Bhanjan 275103, India

M.A. Khan, W. Ahmad (eds.), *Termites and Sustainable Management*, Sustainability in Plant and Crop Protection,
https://doi.org/10.1007/978-3-319-68726-1_9

Abstract Several control methods like physical, chemical, and biological are adopted to control termites in various localities. Biological control methods are eco-friendly and target-specific; hence they could represent a suitable alternative to chemical control methods. Microbial biological control is based on the use, and proper adjustment, of natural enemies via microbial organisms, such as bacteria, fungi, virus, and nematodes with the aim of suppression and management of insect populations. A broad range of species, from different groups of microbial organisms, have strong association with termites, and some have been recorded as parasites, including species currently used as commercial biological control agents.

Keywords Biological control • Termite • Management • Microorganisms

9.1 Introduction

During the last three decades, research on alternative measures in termite control has increased. Various control methods like physical, chemical, and biological are effective in management. Although chemical insecticides are effectively used against termites, they are environmentally hazardous. Furthermore, the cryptic life style of termites makes their direct application difficult. Biological control agents appear as effective, eco-friendly, economically viable, and socially acceptable methods of management. Biological control is generally perceived as both providing more permanent insect control and as having less potential damage to the environment or nontarget organisms. Therefore, biocontrol should be considered as a long-range research goal rather than an immediate solution.

Recent prominences on biological processes to improve agricultural productivity are essential for sustainability in agriculture system. The shortcomings associated with conventional chemical control methods have prompted policy makers and scientists to evaluate the potential for natural enemies to suppress termite populations. Brazil has a history of success with biological control projects involving the use of insect pathogenic fungi and viruses (Campanhola et al. 1995). The use of biological control agents to hunt or to infect termites within their hidden galleries is appealing. Reviews by Grace (1997), Culliney and Grace (2000), and Verma et al. (2009) on biological control strategies for the suppression of termite or on biological alternatives for their control provide complete knowledge for nonchemical approaches.

The use of naturally occurring pathogens offers unique advantages over chemically based termiticides. The study of pathogens for termite control started as early as 1965 (Yendol and Paschke 1965). Since then there has been renewed interest in using pathogenic organisms, such as bacteria, viruses, nematodes, and most fungi for controlling termites in recent years. Several lineages of pathogenic fungi, such as *Beauveria*, *Metarhizium*, *Aspergillus*, and *Entomophthora*; bacteria, such as *Bacillus*, *Serratia*, *Pseudomonas*, *Photorhabdus*, and *Xenorhabdus*; and nematodes, such as genera *Heterorhabditis*, *Steinernema*, and *Neosteinernema*, have been reported as termite biocontrol agents (Kanzaki et al. 2010; Muralidhara et al. 2013).

Development of recombinant DNA techniques has made it possible to significantly improve the insecticidal efficacy of some microorganisms (Inceoglu et al. 2006; Wang and St. Leger 2007a, b, c). These new types of biological insecticides offer a range of environment eco-friendly options for cost-effective control of insect pests (Federici et al. 2008).

Phylogenetic sequence of termites is depicted in Fig. 9.1 with the Termitidae (the largest termite family) as responsible to damage most of crops (Fig. 9.2). The present chapter reviews the biological methods for termite management, focusing on its microbial share. The relationship between termites and microbes and their possible contribution toward termiticidal activity is reviewed and discussed.

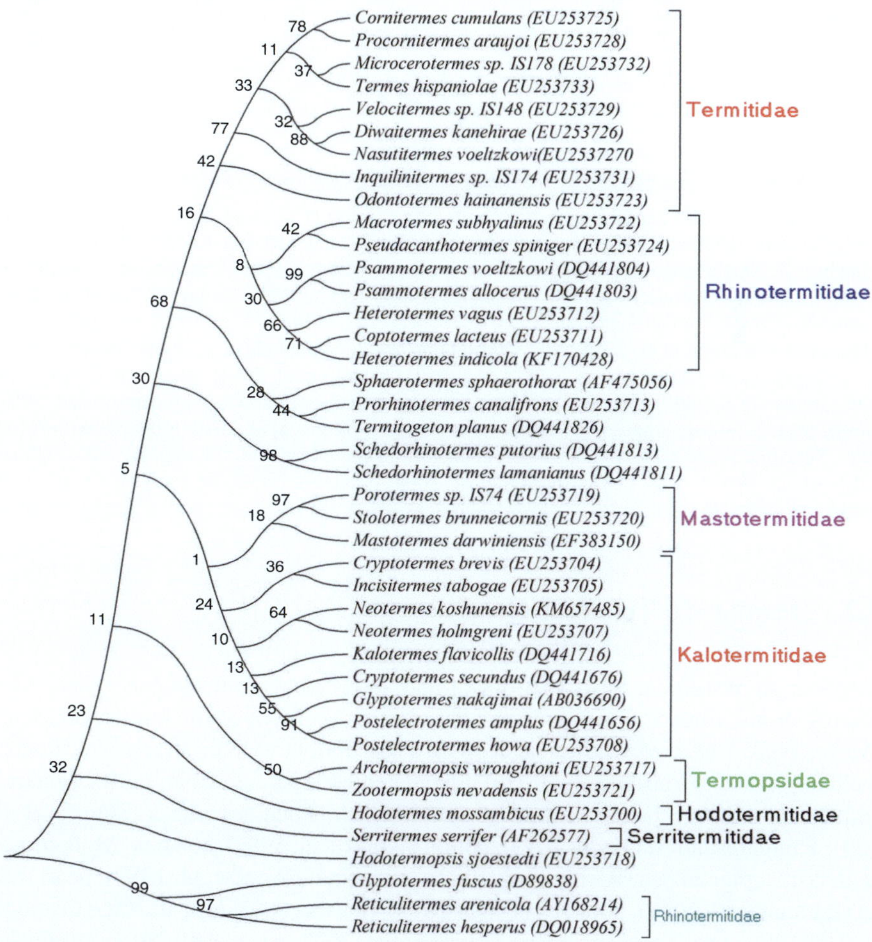

Fig. 9.1 Phylogenetic characterizations of termites, using 12S rRNA gene sequences obtained from NCBI GenBank databases. The sequence alignment was performed using the CLUSTAL W program, and tree was constructed using maximum likelihood method with algorithm using MEGA6 software

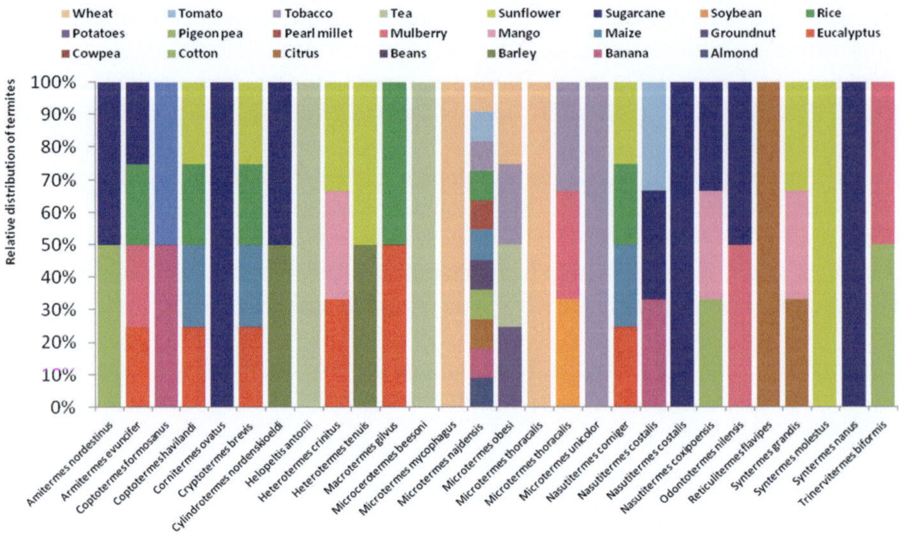

Fig. 9.2 Relative distributions of termites affecting different tropical crops (See for almond: Faragalla and Al Qhtani 2013; banana: Lai et al. 1983; Faragalla and Al Qhtani 2013; citrus: Stansly et al. 2001; Faragalla and Al Qhtani 2013; cotton: Wood et al. 1987; eucalyptus: Constantino 2002; Groundnut: Wood et al. 1987; Faragalla and Al Qhtani 2013; maize: Wood et al. 1987; Constantino 2002; Faragalla and Al Qhtani 2013; pearl millet: Wood et al. 1987; Faragalla and Al Qhtani 2013; Rathour et al. 2014; rice: Constantino 2002; Agunbiade et al. 2009; Maayiem et al. 2012; Togola et al. 2012b; Tomar 2013; Acda 2013; Oyetunji et al. 2014; sugarcane: Constantino 2002; Ahmed et al. 2007; Alam et al. 2012; sunflower: Aslam et al. 2000; Ashfaq and Aslam 2001; Sileshi et al. 2009; tea: Ahmed 2012; Singha et al. 2011; Singha et al. 2010; tobacco: Wood et al. 1987; Shah and Shah 2013; tomato: Wood et al. 1987; Pearce et al. 1995; Faragalla and Al Qhtani 2013; wheat: Wood et al. 1987; Sharma et al. 2004; Ahmed et al. 2004; Pardeshi et al. 2010; Rathour et al. 2014)

9.2 Damage in Tropical Crops

Termites are economic pests in tropical and subtropical environments where they destroy crops, forests along with wood and wooden products of human buildings (Meyer et al. 1999). Most tropical crops are susceptible to their attacks worldwide, including maize (Constantino 2002; Faragalla and Al Qhtani 2013), barley (Kharub and Chander 2012), beans (Sileshi et al. 2009), chickpea and citrus (Stansly et al. 2001; Faragalla and Al Qhtani 2013), cotton (Wood et al. 1987), cowpea (Mohammed et al. 2014), pigeon pea (Reddy et al. 1992), groundnut (Wood et al. 1987), pearl millet (Rathour et al. 2014), potatoes (Tomar 2013), rice (Agunbiade et al. 2009; Oyetunji et al. 2014), soybean and sugarcane (Ahmed et al. 2007; Alam et al. 2012), sunflower

(Ashfaq and Aslam 2001; Sileshi et al. 2009), tomato (Pearce et al. 1995), wheat (Ahmed et al. 2004; Pardeshi et al. 2010; Rathour et al. 2014), tea (Singha et al. 2011), tobacco (Shah and Shah 2013), eucalyptus (Faragalla and Al Qhtani 2013), mango (Tomar 2013), mulberry (Ahmed and Qasim 2011), and almond (Faragalla and Al Qhtani 2013) (Table 9.1).

Table 9.1 Crops attacked by termites and their regional distribution

Location	Banana	Barley	Beans	Chickpea	Citrus	Coconut	Cotton	Cowpea	Eucalyptus	Groundnut	Maize	Mango	Millet	Pigeon pea	Potatoes	Rice	Sesame	Sorghum	Soybean	Sugarcane	Sunflower	Tea	Tomato	Wheat
Afghanistan					■																			
Africa						■	■		■									■						
Algeria					■																			
America					■																			
Argentina																				■				
Australia					■				■	■					■					■				
Bangladesh										■										■		■		
Benin																■								
Brazil									■		■					■				■				
China							■			■			■	■										
China						■			■											■		■		
Colombia																				■				
Ethiopia	■	■			■					■	■	■					■							■
Ghana								■		■	■					■								
Hawaii	■											■												
India		■		■	■		■			■	■	■	■	■	■	■		■		■	■	■		■
Iran					■																			
Iraq					■																			
Israel					■																			
Kenya																			■	■		■		
Malawi	■						■			■	■							■						
Mexico																				■				
Nigeria								■					■	■						■				
Pakistan							■					■						■		■	■			■
Philippines																								
Saudi Arabia	■	■			■				■	■	■	■	■					■					■	
Sri Lanka																						■		
Sudan			■				■						■				■						■	
Tanzania			■			■	■												■			■		■

(continued)

Table 9.1 (continued)

Location	Banana	Barley	Beans	Chickpea	Citrus	Coconut	Cotton	Cowpea	Eucalyptus	Groundnut	Maize	Mango	Millet	Pigeon pea	Potatoes	Rice	Sesame	Sorghum	Soybean	Sugarcane	Sunflower	Tea	Tomato	Wheat
Uganda										■	■			■				■						
Uganda							■		■						■					■				
Yemen							■			■			■				■						■	■
Zambia			■				■	■						■					■		■			
Zimbabwe											■													

Sources:
Afghanistan: Stansly et al. (2001); Algeria: Stansly et al. (2001); America: Stansly et al. (2001); Africa: Zida et al. (2011); Tomar (2013); Rouland-Lefevre (2011); Argentina: Constantino (2002); Australia: Werner et al. (2008); Alam et al. (2012); Tomar (2013); Biswas (2014); Bangladesh: Alam et al. (2012); Biswas (2014); Benin: Togola et al. (2012a), b; Brazil: Rouland-Lefevre (2011); Constantino (2002); China: Muraleedharan (1992); Gui-Xiang et al. (1994); Rao et al. (2002); Zeng (2004); Tang et al. (2006); Tomar (2013); Maayiem et al. (2012); Colombia: Tomar (2013); Ethiopia: Cowie et al. (1990); Wood (1991); Maayiem et al. (2012); Ghana: Maayiem et al. (2012); Hawaii: Lai et al. (1983); India: Reddy et al. (1992); Basappa (2004); Kharub and Chander (2012); Tomar (2013); Pandey et al. (2013); Rathour et al. (2014); Iran: Faragalla and Al Qhtani (2013); Iraq: Faragalla and Al Qhtani (2013); Israel: Faragalla and Al Qhtani (2013); Kenya: Adoyo et al. (1997); Malawi: Munthali et al. (1999); Mexico: Collins (1984); Nigeria: Nwilene et al. (2008); Agunbiade et al. (2009); Pakistan: Wood (1991); Ahmed et al. (2007); Javaid and Afzal (2001); Aslam et al. (2000); Ahmed et al. (2004); Philippines: Orikiriza et al. (2012); Acda (2013); Saudi Arabia: Badawi et al. (1986); Faragalla and Al Qhtani (2013); Sri Lanka: Hemachandra et al. (2014); Sudan: Wood et al. (1987); Pearce et al. (1995); Tanzania: Bigger (1966); Mwalongo et al. (1999); Wood et al. (1987); Uganda: Wood et al. (1987); Nahdy et al. (1994); Nyeko and Nakabonge (2008); Orikiriza et al. (2012); Yemen: Wood et al. (1987); Zambia: Sileshi et al. (2008), (2009); Zimbabwe: Thierfelder et al. (2013)

9.3 Management

Different control methods (chemical, biological, and physical) have been adopted for termite management (Fig. 9.3). Majority of control practices are ineffective and ecologically unsustainable and, above all, do not address the root cause of termite infestation, merely providing a temporary relief to the problem (Mugerwa 2015). Most termite management practices are focused on total elimination rather than regulating their populations. Various natural enemies have shown the potential for use in biological control of termites and serve as an alternative to broad-spectrum chemical insecticides.

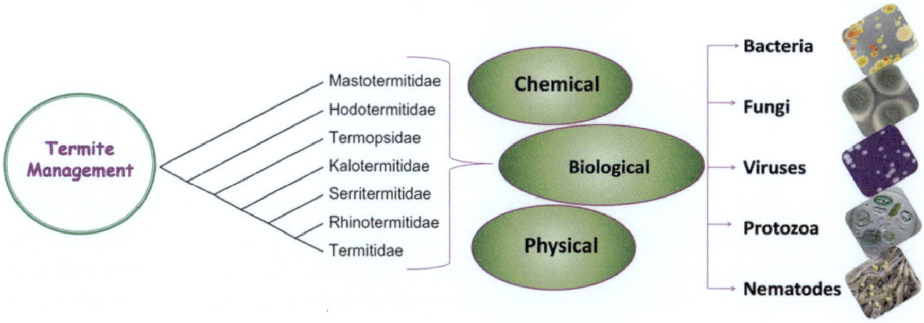

Fig. 9.3 Diagrammatic representation of different control measures of termites

9.3.1 Biological Control

With the growing realization of hazards and side effects associated with the extensive and indiscriminate use of synthetic chemical insecticides, entomologists have adopted a new concept of pest control, termed as integrated pest management (IPM). This term refers to a system that utilizes all suitable techniques and methods, in an as compatible manner as possible, in order to maintain the pest population at levels below a threshold causing significant economic losses (Mahtur and Kishor 1987). In this context, the role of biocontrol agents, viz., predators, parasitoids, and microbes, needs no emphasis due to their specificity, effectiveness, and safety to nontargeted organisms, besides other components in relation to man and biosphere. Microbial products have a long history of safety in use, and most of the microbial agents are compatible with other methods of pest control. In recent years, entomologists are leaning their attention on the exploration of microbial agents for pest suppression. Facultative pathogens of some insect species are commonly used as a biopesticide or microbial control agent. Interestingly, some of them have been widely tested and proved very effective against pernicious insect pests of agricultural crops. In certain developing and developed countries, a number of microbial biopesticides have been registered for field application on various vegetables, fruits, and other crops of agricultural, horticultural, and forest importance.

Microbial control includes all aspects of utilization of microorganisms or their by-products for pest control. Microbial biocontrol agents are relatively host specific and do not upset other biotic systems. They are safe to humans, vertebrates, and beneficial organisms and do not cause environmental pollution. The microorganisms isolated from termites mounds/nests or rhizosphere soil could be screened for their termiticidal activities for subsequent use in biocontrol plans. Due to their compatibility, the synergistic combinations of microbial control agents with other technologies will have excellent potential for use in IPM programs.

9.3.1.1 Bacteria

The large number of microbial species includes members of all major groups such as bacteria, virus, fungi, and nematodes. However, bacterial pathogens have been exploited the most and are recommended as potential biocontrol agent for the control of major insect pests. Bacteria are prokaryotic, unicellular organisms varying in size from less than 1 μm to several μm in length and are characterized by spherical, spiral, or rod-shaped cells. Most of the insect pathogenic bacteria occur under the families Bacillaceae, Pseudomonadaceae, Enterobacteriaceae, and Streptococcaceae (Kalha et al. 2014). Members of Bacillaceae, particularly *Bacillus* spp., have received maximum attention as microbial control agents. *Bacillus thuringiensis* Berliner, which occupied 90% of the world biopesticide market, is pathogenic to more than 525 insect species belonging to various orders (Jayaraj 1986).

Bacillus thuringiensis is a rod-shaped, gram-positive, spore-forming, aerobic soil bacterium common in many ecosystems (Yadav et al. 2015c). It was first discovered in Japan in 1901 by Ishiwata and then reported in 1911 in Germany by Berliner (Baum et al. 1999). It is distributed worldwide in soil, stored products, insects, insect-breeding environments, and the phylloplane (Hofte and Whiteley 1989). Vegetative cells of *B. thuringiensis* are 0.2–5 μm in size with peritrichous flagella. They divide by binary fission and frequently occur in chains. During the sporulation phase of growth, this bacterium accumulates insecticidal crystal proteins (ICPs)/δ-endotoxin. They accumulate as inclusion bodies in the mother cell compartment and are finally released in the environment along with the spore, at the end of sporulation phase. Some ICPs have toxicity comparable to that of widely used organophosphate pesticides. There are around 34 recognized subspecies of *B. thuringiensis* with two distinct groups of toxin proteins: *Cry* (crystal delta-endotoxins) and *Cyt* (cytolysins), both pathogenic to insect pests (Schnepf et al. 1998).The insecticidal proteins of *B. thuringiensis* are highly specific as gut toxins with a superior safety record in regard to their effects on nontarget organisms (Lacey and Goettel 1995; Sarwar 2015). Castilhos-Fortes et al. (2002) evaluated the effects of *B. thuringiensis* subspecies against *Nasutitermes ehrhardti* under laboratory conditions. They reported that *B. thuringiensis* subspecies *kurstaki* registered <72% mortality at the seventh day after the bacterial application. Singha et al. (2010) evaluated *B. thuringiensis* and *B. thuringiensis* subsp. *israelensis* for their pathogenicity against two species of tea termites, viz., *Microtermes obesi* and *Microcerotermes beesoni*. They reported that *B. thuringiensis* strains caused >80% mortality in both the termite species. *Bacillus thuringiensis* subsp. *israelensis*, however, was noticed to be more virulent compared to other *B. thuringiensis*.

Similarly, other species showed similar efficacy level, such as *B. subtilis* that was also reported effective against termite species (Omoya and Kelly 2014). Natsir and Dali (2014) described the pathogenicity of *B. licheniformis* against termites, using feeding (baiting) and contact (spraying), which is further analyzed quantitatively by calculating host mortality during 2 weeks of observation. These authors reported that chitin deacetylase from *B. licheniformis* HSA3-1a induced a satisfactory level of pathogenicity, by inhibiting the growth of termites. The mortality obtained using spraying method reached 100% by the sixth day of observation. By the feeding method, mortality reached 100% by the eleventh day of observation. Data suggest that chitin deacetylase is effective against termite, to replace conventional termiticide.

Osbrink et al. (2001) isolated *Serratia* from dead termites and reported that three *Serratia* isolates induce >85% mortality within 19 days in petri dish tests. Connick et al. (2001) reported that *S. marcescens* strain T8 was highly virulent at the concentration of 3.4×10^{10} CFU/ml against *Coptotermes formosanus*. They reported termite mortality around 24% by 2 days, reaching 99% of termites after 19 days of assay. Omoya and Kelly (2014) also described that *S. marcescens* was much effective against termites.

Many rhizobacteria are known to produce and excrete hydrogen cyanide (HCN) into the rhizosphere (Verma et al. 2013; Verma et al. 2014; Yadav et al. 2016b). Release of HCN by rhizospheric bacteria into the soil can be toxic to subterranean termites. HCN-producing rhizobacteria could be useful for control by introducing them into termite mounds, thereby localizing cyanide production and minimizing potential deleterious effects on other soil fauna.

Nonparasitic rhizobacteria that produce harmful metabolites might also facilitate the biocontrol of termites. Three different species of HCN-producing rhizobacteria, *Rhizobium radiobacter*, *Alcaligenes latus*, and *Aeromonsa caviae*, were tested for their potential to kill *Odontotermes obesus*. The three bacterial species were found to be effective in killing the termites under in vitro conditions (Devi et al. 2007). Ivermectin is a metabolite produced by the bacterium *Streptomyces avermitilis*. Sublethal concentrations of ivermectin decreased the food consumption and tunneling capacity of *C. formosanus* (Mo et al. 2006). Further bacteria tested against termites include *Pseudomonas fluorescens* which blocked the pest respiratory system by producing hydrogen cyanide, resulting in a high termites mortality (Devi and Kothamasi 2009).

9.3.1.2 Fungi

Fungal diseases are known to cause in nature epidemics with high insect mortality levels (Vimaladevi and Prasad 2001). Entomopathogenic or disease-causing fungi have received considerable attentions as they are exceptionally virulent, being lethal parasites of insect pests. Fungi were among the first microorganisms to be used for the biological control of insect pests. They are cosmopolitan organisms and have been isolated from soils and infected insects from around the world. Conidia of entomopathogenic fungi could be spread through the colony by contact and grooming between contaminated and uncontaminated hosts. Entomopathogenic fungi possess added advantages over other microbial biocontrol agents, as they are capable of attacking all developmental stages of their hosts (Ferron 1978; Anand et al. 2009). None of the entomopathogenic fungi currently in use or under consideration are invasively pathogenic to humans (Kubicek and Druzhinina 2007).

Soil is a natural environment for entomopathogenic fungi, as many insects spend at least part of their life in soil. Such natural behavior is related to the insects' biology, due to accumulation in soil or leaf litter for wintering or pupation, a stage conducive to fungal infections and leading to natural regulation of many pests. Asexually produced fungal spores or conidia are generally responsible for infection and are dispersed throughout the environment in which the insect hosts are present.

More than 700 fungal species from around 90 genera are pathogenic to insects (Wraight et al. 2007; Hemasree 2013). However, only a few have been thoroughly investigated for their use against insect pests in agriculture. When a spore adheres to the host cuticle, a germ tube is generated which penetrates through the integument by mechanic and enzymatic processes (e.g., through the action of chitinases, proteases, and lipases). When the hyphae reach the insect hemocoel, they produce blastospores which are the final pathogenic step for completing the host infection (Vincent et al. 2007). The development of fungal infections in terrestrial insects is largely influenced by environmental conditions. High humidity is vital for germination of fungal spores and transmission of the pathogens from one insect to another.

Two fungal pathogens, *Metarhizium anisopliae* and *Beauveria bassiana*, have been extensively evaluated for termite control. *Metarhizium anisopliae* is a biological control agent that requires special application and handling techniques. Rath and Tidbury (1996) found that *Coptotermes acinaciformis* and *Nasutitermes exitiosus* were equally susceptible to direct conidial applications of both Australian and American strains of *M. anisopliae.* Ahmed et al. (2009) described that three strains of *M. anisopliae* were isolated from swarmed termite, *C. heimi* at Gujranwala. The exposure methods of these isolates against termites included soil substrate and filter paper treatments, at different concentrations of conidial suspension (1×10^4, 1×10^6, 1×10^8, 1×10^{10} propagules mL^{-1}). LT_{50} for these strains against *C. heimi* were comparatively higher (65–106 h) in soil than on filter paper (50–83 h). Ravindran et al. (2015) isolated four strains of *M. anisopliae* using insect bait (*Galleria*) method and examined their sporulation characteristics and virulence against *C. formosanus. Metarhizium* sp. (Tk 4) was a high virulent strain identified showing 86.6% mortality rate on 4th day post inoculation (dpi). Wang and Powell (2003) reported six *B. bassiana* isolates obtained from *Reticulitermes flavipes* and *C. formosanus* termites in the USA and China. These isolates, along with *B. bassiana* isolate 26,037 from American Type Culture Collection, proceeding from a Colorado potato beetle, *Leptinotarsa decemlineata* (Say), were compared against *C. formosanus* and *R. flavipes* in the laboratory. Most *B. bassiana* isolates caused termite mortality within 4–8 days after treatment.

Conidiobolus coronatus isolates were found to be pathogenic against *C. formosanus*, *R. flavipes*, and *Nasutitermes exitiosus* (Wells et al. 1995). *Aspergillus* sp. has also been studied for its entomopathogenic activity against termites. In a study conducted by Muralidhara et al. (2013), *Aspergillus* sp. TK inoculation resulted in profuse colonization on the surface of *Microcerotermes beesoni*, with 100% mortality within 5 days, confirming the high potential of entomopathogenic fungus as biocontrol agents.

9.3.1.3 Viruses

A large number of viruses offer potential as microbial control agents of termites. Al Fazairy and Hassan (1993) reported that *Kalotermes flavicollis* died 2–10 dpi after inoculation by Nuclear polyhedrosis virus (NPV, Baculoviride), under laboratory conditions. Further authors suggested that termites control with NPV might be feasible. However, the potential of viruses for termite control has yet to be evaluating under field conditions. Accessibility of the target pest to control is the prime factor affecting the efficacy of viral pathogens. The efficacy, specificity, and production of secondary inocula make Baculoviruses an attractive alternative to broad-spectrum insecticides, and ideal components of IPM systems, due to their lack of effects on beneficial insects including other biological control organisms (Sindhu et al. 2011). Unfortunately, there are other drawbacks to the use of viruses to suppress pest populations: they kill their hosts slowly, when compared to other pathogens; environmental factors such as rainfall and solar radiation may reduce the viral persistence in soil; mass production of viruses is hampered by the need for living hosts or tissue culture; and finally, viral formulations experienced difficulties in competing successfully, on the basis of performance and cost, with other pest control products such as chemical insecticides or even other microbial agents (Fuxa 1990).

9.3.1.4 Nematode

Entomopathogenic nematodes (EPNs) are beneficial nematodes offering excellent potential for control of insects in soil habitats and commercially used to control many pests. EPNs like steinernematids and heterorhabditis are obligate insect parasites associated with symbiotic bacteria of the genera *Xenorhabdus* sp. and *Photorhabdus* sp., respectively (Akhurst and Boemare 1990). These bacteria are motile, gram-negative, facultative, non-spore-forming anaerobic rod members of the family *Enterobacteriaceae*. Together, nematodes and their symbiotic partners form an insecticidal complex that is effective against a wide range of hosts (Kaya and Gaugler 1993). Most *Photorhabdus* spp. are luminescent and catalase-positive, whereas *Xenorhabdus* spp. have no luminescence and are catalase-negative. Poinar and Thomas (1966) demonstrated the location of bacteria in the infective-stage juveniles, using light and electron microscopy. In *Steinernema*, the bacterial symbiont cells are harbored in a specialized structure known as "bacterial receptacle." In

Heterorhabditis, the symbionts are distributed along a broad stretch of the anterior portion of the nematode intestine.

In laboratory tests, *S. carpocapsae* alone infected more than 250 species of insects from over 75 families in 11 orders (Poinar 1975). The broad host range and high virulence of entomopathogenic nematodes make them suitable for use as augmentative-release biocontrol agents (Hui and Webster 2000). The nematodes are compatible with many pesticides, can be mass produced and formulated, and are exempt from registration in many countries. Lacey and Georgis (2012) highlight EPN development for control of insect pests, above and below ground, including those from foliar, soil surface, cryptic, and subterranean habitats. Shahina and Tabassum (2010) reported higher mortalities in the subterranean termite *Macrotermes* caused by *S. pakistanense* in a filter paper and sand assay. Yu et al. (2010) compared virulence of three novel strains of *S. riobrave* (3-8b, 7-12, and TP) against subterranean termites *Heterotermes aureus*, *R. flavipes*, and *C. formosanus* workers. *Heterotermes aureus* was noticed as most susceptible to all the *S. riobrave* strains, and termites in all nematode treatments died after 4 days. Differential susceptibility of two termite species, *Macrotermes bellicosus* and *Trinervitermes occidentalis*, against EPNs isolates *H. indica* Ayogbe1, *H. sonorensis* Azohoue2, *H. sonorensis* Ze3, and *Steinernema* sp. Bembereke, from Benin (West Africa), was studied by Zadji et al. (2014). They reported that all tested EPN isolates can be recycled in both *M. bellicosus* and *T. occidentalis*, and the soldiers of both termites studied were noticed as more susceptible than workers. Current use of *Steinernema* and *Heterorhabditis* nematodes as biological control organisms has been summarized by Shapiro-Ilan and Gaugler (2010).

9.4 Biological Control Strategies

There are three main factors affecting microbial biological agent efficacy, reviewed as follows.

9.4.1 Toxin Production

Most of the insecticidal activity of *B. thuringiensis* is associated with the proteinaceous toxins located in the parasporal inclusion bodies, also known as parasporal crystals. Collectively, the toxins found in the parasporal crystals are referred to as δ-endotoxins. The *Cry*1 proteins which are found in the crystals are biologically inactive. Following ingestion and solubilization in the alkaline midgut, cleavage by gut proteases produces a 60–65 kDa activated protein that recognizes specific binding sites at the brush border membrane surface of the epithelial columnar cells, lining the host gut lumen (Lacey and Goettel 1995; Sarwar 2015).

Exposure of laboratory colonies of the subterranean species, *R. flavipes* and *R. hesperus*, to a mixture of soluble endotoxin, spores, and inclusion bodies of *B.*

thuringiensis resulted in 95% mortality after 6 days (Blaske and Hertel 2001; Lax and Osbrink 2003).

Pearce (1997) described that also the spores of entomopathogenic fungi may contain toxins, which would kill the termite host when ingested. Insecticidal cyclic depsipeptides were found to be produced by entomopathogenic fungi, including the destruxins from *M. anisopliae* var. *major*. It has been suggested that depsipeptides are localized on the surface of spores of *Beauveria* sp., whereas *Metarhizium* spp. destruxins are generally associated with in vivo or in vitro mycelial growth (Jegorov et al. 1989).

9.4.2 Siderophore Production

Siderophores are iron-chelating extracellular metabolites produced by different groups of bacteria and fungi (Verma et al. 2015a; Yadav et al. 2015a; Verma et al. 2016b; Suman et al. 2016). Extracellular siderophores of the brown-rot wood decay fungus *Gloeophyllum arabeum* were found to inhibit feeding of *C. formosanus* termite (Grace et al. 1992). Natural products, such as ant semiochemicals and fungal metabolites, or their synthetic analogues, might be valuable in termite control programs as repellents or insecticides in wood treatments or soil applications (Grace et al. 1992). However, the development of more stable formulations, such as microencapsulation, would be necessary to ensure their long-term, residual action.

9.4.3 Hydrolytic Enzymes Production

A large number of microorganisms including bacteria, fungi, and actinomycetes are capable of producing hydrolytic enzymes such as proteases, lipase, and chitinase from various types of natural resources (Pandey et al. 2013; Verma et al. 2015b; Yadav et al. 2015b; Yadav et al. 2017). Lysenko and Kucera (1971) described that *Serratia marcescens* produced extracellular proteases that could be underpin pathogenicity of these bacteria in termites. Osbrink et al. (2001) reported that 15 bacteria and 1 fungus were associated from dead termites as possible biological control agents against subterranean termites, *C. formosanus*. Bacterial isolates from dead termites were primarily *S. marcescens* that caused septicemia in *C. formosanus* and was found to contain proteolytic enzymes. Lipases (triacylglycerol acyl hydrolases) are one of the most important class of hydrolytic enzymes that catalyze both the hydrolysis and the synthesis of ester formed from glycerol and long-chain fatty acids. Lipases are ubiquitous enzymes produced by all animals, plants, and microorganisms (Yadav 2015; Yadav et al. 2016a). Lysis by hydrolytic enzymes excreted by microorganisms is a well-known feature of mycoparasitism.

Vaidya et al. (2003) isolated mutants of *Alcaligenes xylosoxidans* producing hyperchitinases and developed a rapid technique for screening chitinolytic bacteria using chitin-binding dye calcofluor, white M2R, in chitin agar. Microorganisms possessing high chitinolytic activity gave clear zone under ultraviolet light after 24–48 h of incubation. The mutant *A. xylosoxidans* EMS 33 was found to produce three to four times more chitinase then wild type. Chitinase production has also been reported in *S. marcescens*, *Pseudomonas* sp., *Bacillus* strains, *Paenibacillus* sp., and *Pseudomonas maltophilia* (Suyal et al. 2015; Verma et al. 2016a; Yadav et al. 2016a, b). Dua (2014) reported ten bacterial strains having termite killing ability along with two control strains for biocontrol against termites. Different bacterial strains showed >80% killing of termites at 5 days of incubation. Four bacterial strains (KBM79, KPM35, PPM147 and PBM195) caused 100% killing at 10 days of observation. The cell-free culture filtrate of these cultures showed that the antagonistic substance was extracellular. Bacterial strains of *B. subtilis* KBM79 and *Pseudomonas synxantha* KPM35 possessed proteolytic, lipolytic, and chitinolytic enzyme activities and caused 100% killing of termites at 10 dpi.

9.5 Conclusion

Termite management is a challenging task for researchers. While various approaches have been tried to manage termite populations, the relevant insecticides are associated with certain risks to the environment and human health. Biological control agents should be seen as one tool among others in an integrated approach to managing termite problems. Microbial control has been a component of IPM strategies in developing countries, enjoying particular success in Asia and South America. An increased understanding of the molecular basis of the various microbial pathogenic mechanisms on termites will lead not only to a rational management but also to the development of new biological control strategies. It is therefore worth to mention that the use of entomopathogens can provide a successful and environment friendly avenue for controlling termite pests.

Acknowledgments The authors duly acknowledge the Department of Biotechnology, Akal College of Agriculture, Eternal University and Vice Chancellor, Eternal University for providing the motivation and research infrastructure.

References

Acda, M. N. (2013). Geographical distribution of subterranean termites (Isoptera) in economically important regions of Luzon, Philippines. Philipp. *Agricultural Sciences, 96*, 205–209.

Adoyo, F., Mukalama, J. B., & Enyola, M. (1997). Using Tithonia concoctions for termite control in Busia District, Kenya. ILEIA. *Newsletter, 13*, 24–25.

Agunbiade, T., Nwilene, F., Onasanya, A., Semon, M., Togola, A., Tamo, M., & Falola, O. (2009). Resistance status of upland NERICA rice varieties to termite damage in Northcentral Nigeria. *Journal of Applied Sciences, 9*, 3864–3869.

Ahmed, M. (2012). Ecofriendly pest management of tea in Bangladesh. *Two and a Bud, 59*, 11–16.

Ahmed, S., & Qasim, M. (2011). Foraging and chemical control of subterranean termites in a farm building at Faisalabad, Pakistan. *Pakistan Journal of Life and Social Sciences, 9*(1), 58–62.

Ahmed, S., Akbar, W., & Riaz, M. A. (2004). Effect of crop rotation and intercropping on subterranean termites in wheat at Faisalabad. *Pakistan Entomological, 26*, 25–30.

Ahmed S., Khan R. R., Riaz M. A. (2007). Some studies on the field performance of plant extracts against termites (*Odontotermes guptai* and *Microtermes obesi*) in sugarcane at Faisalabad. *International Journal of Agriculture and Biology* 9:398–400.

Ahmed, S., Ashraf, M. R., Hussain, A., & Riaz, M. A. (2009). Pathogenicity of isolates of *Metarhizium anisopliae* from Gujranwala (Pakistan) against *Coptotermes heimi* (Wasmann) (Isoptera: Rhinotermitidae). *International Journal of Agriculture and Biology, 11*, 707–711.

Akhurst, R. J., & Boemare, N. (1990). Biology and taxonomy of *Xenorhabdus*. *In*: R. Gaugler & H. K. Kaya (Eds.), *Entomopathogenic nematodes in biological control,* Boca Raton: CRC Press, pp. 75–87.

Al Fazairy, A. A., & Hassan, F. A. (1993). Histopathology of termite *Kalotermes Flavicollis* Fabr. Infected with a nuclear Polyhedrosis virus. *International Journal of Tropical Insect Science,* 14:127–134.

Alam, M., Alam, M., Abdullah, M., Begum, M., & Ahmed, T. (2012). Effects of insecticides on sugarcane termites in Modhupur Tract. *Bangladesh Journal of Agricultural Research, 37*, 295–299.

Anand, R., Prasad, B., & Tiwary, B. N. (2009). Relative susceptibility of *Spodoptera litura* pupae to selected entomopathogenic fungi. *BioControl, 54*, 85–92.

Ashfaq, M., & Aslam, M. (2001). Response of different insect pests to some sunflower (*Helianthus annuus* L., Compositae) genotypes and their correlation with yield component under field conditions. *Journal of Biological Sciences, 1*, 835–839.

Aslam, M., Suleman, N., Riaz, A., Rehman, A., & Zia, Q. (2000). Insect pests found on *Helianthus annuus* L. (Compositae) in the Potohar region of Pakistan. *Pakistan Journal of Biological Sciences, 3*, 963–964.

Badawi, A., Kady, H. A., & Faragalla, A. (1986). Termites (Isoptera) of Saudi Arabia, their hosts and geographical distribution1. *Journal of Applied Entomology, 101*, 413–420.

Basappa, H. (2004). Integrated pest management in sunflower: An Indian scenario. *In*: *Proceedings of the 16th international sunflower conference,* North Dakota: Fargo, pp. 853–859.

Baum, J. A., Johnson, T. B., & Carlton, B. C. (1999). *Bacillus thuringiensis*: Natural and recombinant bioinsecticide products. *In*: J. J. Menn & F. R. Hall (Eds.), *Biopesticides: Use and delivery*, Totowa: Humana Press, pp. 189–210.

Bigger, M. (1966). The biology and control of termites damaging field crops in Tanganyika. *Bulletin of Entomological Research, 56*, 417–444.

Biswas, G. (2014). Insect pests of groundnut (*Arachis hypogaea* L.), nature of damage and succession with the crop stages. *Bangladesh Journal of Agricultural Research, 39*, 273–282.

Blaske, V. U., & Hertel, H. (2001). Repellent and toxic effects of plant extracts on subterranean termites (Isoptera: Rhinotermitidae). *Journal of Economic Entomology, 94*, 1200–1208.

Campanhola, C., de Moraes, G. J., & de Sa, L. A. N. (1995). Review of IPM in South America. *In*: A. N. Mengech, K. N. Saxena, & H. N. B. Gopalan (Eds.), *Integrated pest management in the tropics*, Chichester: Wiley, pp. 121–152.

Castilhos-Fortes, R., Matsumura, A. T. S., Diehl, E., & Fiuza, L. M. (2002). Susceptibility of *Nasutitermes ehrhardti* (Isoptera: Termitidae) to *Bacillus thuringiensis* subspecies. *Brazilian Journal of Microbiology, 33*, 219–222.

Collins, N. (1984). Termite damage and crop loss studies in Nigeria–assessment of damage to upland sugarcane. *International Journal of Pest Management, 30*, 26–28.

Connick, W. J., Jr., Osbrink, W. L. A., Wright, M. S., Williams, K. S., Daigle, D. J., Boykin, D. L., & Lax, A. R. (2001). Increased mortality of *Coptotermes formosanus* (Isoptera: Rhinotermitidae) exposed to eicosanoid biosynthesis inhibitors and *Serratia marcescens* (Eubacteriales: Enterobacteriaceae). *Environmental Entomology, 30*, 449–455.

Constantino, R. (2002). The pest termites of South America: Taxonomy, distribution and status. *Journal of Applied Entomology, 126*, 355–365.
Cowie, R., Wood, T., Barnett, E., Sands, W., & Black, H. (1990). A checklist of the termites of Ethiopia with a review of their biology, distribution and pest status. *African Journal of Ecology, 28*, 21–33.
Culliney TW, Grace JK (2000) Prospects for biological control of subterranean termite (Isoptera: Rhinotermitidae), with special reference to *Coptotermes formosanus*. *Bulletin of Entomological Research, 90*, 9–21.
Devi, K. K., & Kothamasi, D. (2009). *Pseudomonas fluorescens* CHA0 can kill subterranean termite *Odontotermes obesus* by inhibiting cytochrome c oxidase of the termite respiratory chain. *FEMS Microbiology Letters, 300*, 195–200.
Devi, K. K., Seth, N., Kothamasi, S., & Kothamasi, D. (2007). Hydrogen cyanide-producing rhizobacteria kill subterranean termite *Odontotermes obesus* (rambur) by cyanide poisoning under in vitro conditions. *Current Microbiology, 54*, 74–78.
Dua, S. (2014). Biological control of subterranean termites (Isoptera: Termitidae) with soil bacteria. Ph.D. thesis, Chaudhary Charan Singh Haryana Agricultural University, pp. 225
Faragalla, A. R. A., & Al Qhtani, M. H. (2013). The urban termite fauna (Isoptera) of Jeddah city, Western Saudi Arabia. *Life Science Journal, 10*, 1695–1701.
Federici, B. A., Bonning, B. C., & St. Leger, R. J. (2008). Improvement of insect pathogens as insecticides through genetic engineering. *In*: C. Hill & R. Sleator (Eds.), *Pathobiotechnology*, Austin: Landes Bioscience, pp. 15–40.
Ferron, P. (1978). Biological control of insect pests by entomopathogenic fungi. *Annual Review of Entomology, 23*, 409–442.
Fuxa, J. R. (1990). New directions for insect control with baculoviruses. *In*: R. R. Baker & P. E. Dunn (Eds.), *New directions in biological control: Alternatives for suppressing agricultural pests and diseases*. Alan R. Liss, Inc., New York, pp. 97–113.
Grace, J. K. (1997). Biological control strategies for suppression of termites. *Journal of Agricultural Entomology, 14*, 281–289.
Grace, J., Goodell, B., Jones, W., Chandhoke, V., & Jellison, J. (1992). *Inhibition of termite feeding by fungal siderophores*. The International Research Group on Wood Preservation, Document No: IRGAVP/1558-92 Biological Problems (Fauna), pp. 1–4.
Gui-Xiang L., Zi-Rong D., & Biao, Y. (1994). Introduction to termite research in China. *Journal of Applied Entomology, 117*, 360–369.
Hemachandra, I. I., Edirisinghe, J. P., Karunaratne, W. I. P., Gunatilleke, C. S., & Fernando, R. S. (2014). Diversity and distribution of termite assemblages in montane forests in the Knuckles Region, Sri Lanka. *International Journal of Tropical Insect Science, 34*, 41–52.
Hemasree, E. (2013). A critical review on the natural occurrence of entomopathogenic fungi in agricultural ecosystem. *International Journal of Applied Biology and Pharmaceutical Technology, 4*, 372–375.
Hofte, H., & Whiteley, H. R. (1989). Insecticidal crystal proteins of *Bacillus thuringiensis*. *Microbiological Reviews, 53*, 242–255.
Hui, E., & Webster, J. (2000). Influence of insect larvae and seedling roots on the host-finding ability of *Steinernema feltiae* (Nematoda: Steinernematidae). *Journal of Invertebrate Pathology, 75*, 152–162.
Inceoglu, A. B., Kamita, S. G., & Hammock, B. D. (2006). Genetically modified baculoviruses: A historical overview and future outlook. *Advances in Virus Research, 68*, 323–360.
Javaid, A., & Afzal, M. (2001). Incidence of termite attack on trees in University of the Punjab, Quaid-e-Azam Campus, Lahore, Pakistan. *Pakistan Journal of Zoology, 33*, 80–81.
Jayaraj, S. (1986). Role of insect pathogens in plant protection. *Proceedings of the Indian National Science Academy, 1*, 91–107.
Jegorov, A., Kadlec, J., Novak, J., Matha, V., Sedmera, P., Triska, J., & Zahradnickova, H. (1989). Are the depsipeptides of *Beauveria brongniartii* involved in the entomopathogenic process?.

In *Proceedings of the international conference on biopesticides, theory and practice*. Ceske Budejovice, pp. 71–81.

Kalha, C. S., Singh, R. P., Kang, S. S., Hunjan, M. S., Gupta, V., & Sharma, R. (2014). Entomopathogenic viruses and bacteria for insect-pest control. *In*: D. P. Abrol (Ed.), *Integrated pest management: Current concepts and ecological perspectives,* Academic Press, San Diego, pp. 225–244.

Kanzaki, N., Giblin-Davis, R. M., Herre, E. A., Scheffrahn, R. H., & Center, B. J. (2010). *Pseudaphelenchus vindai* n. sp.(Tylenchomorpha: Aphelenchoididae) associated with termites (Termitidae) in Barro Colorado Island, Panama. *Nematology, 12*, 905–914.

Kaya, H. K., & Gaugler, R. (1993). Entomopathogenic nematodes. *Annual Review of Entomology, 38*, 181–206.

Kharub, A., & Chander, S. (2012). Production technology for malt barley. *In*: Verma, R. P. S., Kharub, A. S., Kumar, D., Sarkar, B., Selvakumar, R., Malik, R., & Sharma, I. (Eds.), *Compendium on malting quality improvement in Barley, Sorghum & Corn,* Directorate of Wheat Research, Karnal, pp. 56–62.

Kubicek, C. P., & Druzhinina, I. S. (2007). *Environmental and microbial relationships, The Mycota IV*), Springer, Berlin, pp. 159–187.

Lacey, L. A., & Georgis, R. (2012). Entomopathogenic nematodes for control of insect pests above and below ground with comments on commercial production. *Journal of Nematology, 44*, 218–225.

Lacey, L., & Goettel, M. (1995). Current developments in microbial control of insect pests and prospects for the early 21st century. *Entomophaga, 40*, 3–27.

Lai, P., Tamashiro, M., Yates, J., Su, N., Fujii, J., & Ebesu, R. (1983). Living plants in Hawaii attacked by *Coptotermes formosanus*. *Proceedings of the Hawaiian Entomological Society, 24*, 283–286.

Lax, A. R., & Osbrink, W. L. (2003). United States Department of Agriculture–agriculture research service research on targeted management of the Formosan subterranean termite *Coptotermes formosanus* Shiraki (Isoptera: Rhinotermitidae). *Pest Management Science, 59*, 788–800.

Lysenko, O., & Kucera, M. (1971). Microorganisms as sources of new insecticidal chemicals: toxins. *In*: H. D. Burges & N. W. Hussey (Eds.), *Microbial control of insects and mites,* Academic Press, New York, pp. 205.

Maayiem, D., Bernard, B. N., & Irunuoh, A. O. (2012). Indigenous knowledge of termite control: A case study of five farming communities in Gushegu District of Northern Ghana. *Journal of Entomology and Nematology, 4*, 58–64.

Mahtur, Y. K., & Kishor, P. (1987). Recent concept of integrated management of key pests of agriculture crops. *In*: Y. K. Mathur, A. K. Bhattacharya, N. D. Pandey, K. D. Upadhyaya, & J. P. Srivastava (Eds.), *Recent advances in entomology,* Gopal Prakashan, Kanpur, pp. I–X.

Meyer, V., Braack, L., Biggs, H., & Ebersohn, C. (1999). Distribution and density of termite mounds in the northern Kruger National Park, with specific reference to those constructed by *Macrotermes Holmgren* (Isoptera: Termitidae). *African Entomology: Journal of the Entomological Society of Southern Africa, 7*, 123–130.

Mo, J., Wang, Z., Song, X., Guo, J., Cao, X., & Cheng, J. (2006). Effects of sublethal concentrations of ivermectin on behaviors of *Coptotermes formosanus* (Isoptera: Rhinotermitidae). *Sociobiology, 47*, 687–696.

Mohammed, M., Abiodun, A. F., & Jibia, A. B. (2014). Denudation effect of termitaria and characterization of associated termite species in Lafia Nasarawa State, Nigeria. *European Scientific Journal, 10*, 185–195.

Mugerwa, S. (2015). Infestation of African savanna ecosystems by subterranean termites. *Ecological Complexity, 21*, 70–77.

Munthali, D., Logan, J., Wood, T., & Nyirenda, G. (1999). Termite distribution and damage to crops on smallholder farms in southern Malawi. *International Journal of Tropical Insect Science, 19*, 43–49.

Muraleedharan, N. (1992). Pest control in Asia. *In*: K. C. Willson & N. C. Michael (Eds.), *Tea: Cultivation to consumption,* Springer, Dordrecht, pp. 375–412.
Muralidhara, S., Prasad, B., & Singh, R. (2013). Colonization and antagonistic activity of entomopathogenic *Aspergillus* sp. against tea termite (*Microcerotermes beesoni* Snyder). *Current Science, 105*, 1216–1619.
Mwalongo, G. C., Mkayula, L. L., Mubofu, E. B., & Mwingira, B. A. (1999). Preventing termite attack. Environmentally friendly chemical combinations of cashew nut shell liquid, sulfited wattle tannin and copper (II) chloride. *Green Chemistry, 1*, 13–16.
Nahdy, M. S., Musaana, M., Ugen, M., & Areeke, E. (1994). Survey of pigeonpea production and postproduction systems in three districts of Uganda. *In*: S. N. Silim, S. Tuwafe, & S. Laxman (Eds.), *Pigeonpea improvement in Eastern and Southern Africa,* International Crops Research Institute for the Semi-Arid Tropics, pp. 67–73.
Natsir, H., & Dali, S. (2014). Production and application of chitin deacetylase from *Bacillus licheniformis* HSA3-1a as Biotermicide. Marina. *Chimica Acta, 15*, 1–12.
Nwilene, F., Agunbiade, T., Togola, M., Youm, O., Ajayi, O., Oikeh, S., Ofodile, S., & Falola, O. (2008). Efficacy of traditional practices and botanicals for the control of termites on rice at Ikenne, southwest Nigeria. *International Journal of Tropical Insect Science, 28*, 37–44.
Nyeko, P., & Nakabonge, G. (2008). Occurrence of pests and diseases in tree nurseries and plantations in Uganda. Study Commissioned by the Sawlog Production Grant Scheme, pp. 1–42.
Omoya, F., & Kelly, B. (2014). Variability of the potency of some selected entomopathogenic bacteria (*Bacillus* spp. and *Serratia* spp.) on termites, *Macrotermes bellicosus* (Isoptera: Termitidae) after exposure to magnetic fields. *International Journal of Tropical Insect Science, 34*, 98–105.
Orikiriza, L., Nyeko, P., & Sekamatte, B. (2012). Farmers knowledge, perceptions and control of pestiferous termites in Nakasongola district, Uganda. *Uganda Journal of Agricultural Science, 13*, 71–83.
Osbrink, W. L., Williams, K. S., Connick, W. J., Wright, M. S., & Lax, A. R. (2001). Virulence of bacteria associated with the *Formosan subterranean* termite (Isoptera: Rhinotermitidae) in New Orleans, LA. *Environmental Entomology, 30*, 443–448.
Oyetunji, O., Peluola, C., Nwilene, F., & Togola, A. (2014). Root and Stem Damage Caused by Termite-fungi Interaction on Rice. *Journal of Applied Sciences,* 14:1851–1857.
Pandey, S., Singh, S., Yadav, A. N., Nain, L., & Saxena, A. K. (2013). Phylogenetic diversity and characterization of novel and efficient cellulase producing bacterial isolates from various extreme environments. *Bioscience, Biotechnology, and Biochemistry, 77*, 1474–1480.
Pardeshi, M., Kumar, D., & Bhattacharyya, A. (2010). Termite (insecta: isoptera) fauna of some agricultural crops of vadodara, gujarat (India). *Records of the Zoological Survey of India, 110*, 47–59.
Pearce, M. J. (1997). Termites: Biology and pest management. *CAB International, 36*, 296–304.
Pearce, M. J., Logan, J., & Tiben, A. (1995). Termites (Isoptera) from the Darfur region of the Sudan with comments on their pest status. *Journal of Arid Environments, 30*, 197–206.
Poinar, G. O., Jr. (1975). Entomogenous nematodes: A manual and host list of insect-nematode associations. EJ Brill, Leiden, p. 317.
Poinar, G. O., & Thomas, G. M. (1966). Significance of *Achromobacter nematophilus* Poinar and Thomas (Achromobacteraceae: Eubacteriales) in the development of the nematode, DD-136 (Neoaplectana sp. Steinernematidae). *Parasitology, 56*, 385–390.
Rao, G., Saxena, K., Shiying, Y., Wen, P., & Tian, W. G. (2002). Insect pest problems of pigeonpea in Guangxi and Hainan provinces. Int Chickpea Pigeonpea. *Newsletter, 9*, 48–49.
Rath, A., & Tidbury, C. (1996). Susceptibility of *Coptotermes acinaciformis* (Isoptera: Rhinotermitidae) and *Nasutitermes exitiosus* (Isoptera: Termitidae) to two commercial isolates of *Metarhizium anisopliae*. *Sociobiology, 28*, 67–72.
Rathour, K. S., Ganguly, S., Das, T., Singh, P., Kumar, A., & Somvanshi, V. S. (2014). Biological management of subterranean termites (*Odontotermes obesus*) infesting wheat and pearl millet crops by entomopathogenic nematodes. *Indian Journal of Nematology, 44*, 97–100.

Ravindran, K., Qiu, D., & Sivaramakrishnan, S. (2015). Sporulation characteristics and virulence of *Metarhizium anisopliae* against subterranean termites (*Coptotermes formosanus*). *International Journal of Microbiological Research, 6*, 01–04.

Reddy, M. V., Yule, D., Reddy, V. R., & George, P. (1992). Attack on pigeonpea (*Cajanus cajan* (L.) Millsp.) by *Odontotermes obesus* (Rambur) and *Microtermes obesi* Holmgren (Isoptera: Microtermitinae). *International Journal of Pest Management, 38*, 239–240.

Rouland-Lefèvre, C. (2011). Termites as pests of agriculture. *In*: D. E. Bignell, Y. Roisin, & N. Lo (Eds.), *Biology of termites: A modern synthesis*. Springer Netherlands, Dordrecht, pp. 499–517.

Sarwar, M. (2015). Microbial insecticides-an ecofriendly effective line of attack for insect pests management. *International Journal of Engineering and Advanced Research Technology, 1*, 4–9.

Schnepf, H. E., Cricmore, N., Vanrie, J., Lereclus, D., Baum, J., Feitelson, J., Zfi, d. D. R., & Dean, D. H. (1998). *Bacillus thuringiensis* and its pesticidal crystal proteins. *Microbiology and Molecular Biology Reviews, 62*, 775–806.

Shah, R., & Shah, I. (2013). Report of termites infestation in tobacco (*Nicotiana tabacum*) from Khyber Pakhtoonkhwa, Pakistan. *Journal of the Entomological Research, 37*, 243–247.

Shahina, F., & Tabassum, K. A. (2010). Virulence of *Steinernema pakistanense* against different insect species in laboratory condition. *Pakistan Journal of Nematology, 28*, 279–284.

Shapiro-Ilan, D. I., & Gaugler, R. (2010). Nematodes: Rhabditida: Steinernematidae & Heterorhabditidae. In A. Shelton (Ed.), *Biological control: A guide to natural enemies in North America*. New York: Cornell University. Available from: http://www.biocontrol.entomology.cornell.edu/pathogens/nematodes.php

Sharma, A., Babu, K., Nagarajan, S., Singh, S., & Kumar, M. (2004). Distribution and status of termite damage to wheat crop in India. *Indian Journal of Entomology, 66*, 235–237.

Sileshi, G. W., Kuntashula, E., Matakala, P., & Nkunika, P. O. (2008). Farmers' perceptions of tree mortality, pests and pest management practices in agroforestry in Malawi, Mozambique and Zambia. *Agroforestry Systems, 72*, 87–101.

Sileshi, G. W., Nyeko, P., Nkunika, P. O., Sekematte, B. M., Akinnifesi, F. K., & Ajayi, O. C. (2009). Integrating ethno-ecological and scientific knowledge of termites for sustainable termite management and human welfare in Africa. *Ecology and Society, 14*, 48.

Sindhu, S. S., Rakshiya, Y. S., & Verma, M. K. (2011). Biological Control of Termites by Antagonistic Soil Microorganisms. *In*: A. Singh, N. Parmar, & R. C. Kuhad (Eds.), *Bioaugmentation, biostimulation and biocontrol*. Springer, Berlin/Heidelberg, pp. 261–309.

Singha, D., Singha, B., & Dutta, B. (2010). In vitro pathogenicity of *Bacillus thuringiensis* against tea termites. *Journal of Biological Control, 24*, 279–281.

Singha, D., Singha, B., & Dutta, B. K. (2011). Potential of *Metarhizium anisopliae* and *Beauveria bassiana* in the control of tea termite *Microtermes obesi* Holmgren *in vitro* and under field conditions. *Journal of Pest Science, 84*, 69–75.

Stansly, P. A., Nan-Yao, S., & Conner, J. M. (2001). Management of subterranean termites, *Reticulitermes* spp. (Isoptera: Rhinotermitidae) in a citrus orchard with hexaflumuron bait. *Crop Protection, 20*, 199–206.

Suman, A., Yadav, A. N., & Verma, P. (2016). Endophytic Microbes in Crops: Diversity and Beneficial Impact for Sustainable Agriculture. *In*: D. P. Singh, H. B. Singh, & R. Prabha (Eds.), *Microbial inoculants in sustainable agricultural productivity,* Vol. 1.: Research Perspectives. Springer, India, pp. 117–143.

Suyal, D. C., Yadav, A., Shouche, Y., & Goel, R. (2015). Bacterial diversity and community structure of Western Indian Himalayan red kidney bean (*Phaseolus vulgaris*) rhizosphere as revealed by 16S rRNA gene sequences. *Biologia, 70*, 305–313.

Tang, B., Tang, M., Chen, C., Qiu, P., Liu, Q., Wang, M., & Li, C. (2006). Characteristics of soil fauna community in the Dongjiao coconut plantation ecosystem in Hainan, China. *Acta Ecologica Sinica, 26*, 26–32.

Thierfelder, C., Cheesman, S., & Rusinamhodzi, L. (2013). Benefits and challenges of crop rotations in maize-based conservation agriculture (CA) cropping systems of southern Africa. *International Journal of Agricultural Sustainability, 11*, 108–124.

Togola, A., Kotoklo, E., Nwilene, F., Amevoin, K., Glitho, I., Oyetunji, O., & Kiepe, P. (2012a). Specific diversity and damages of termites on upland rice in Benin. *Journal of Entomology, 9*, 352–360.

Togola, A., Nwilene, F., Kotoklo, E., Amevoin, K., Glitho, I., Oyetunji, O., & Niang, A. (2012b). Effect of upland rice varieties and cultural practices on termite populations and damage in the field. *Journal of Applied Sciences, 12*, 675–680.

Tomar, S. P. S. (2013). Characteristics of agro-ecological knowledge of farmers on termites and their devastation in semi-irrigated farming system of central India. *Insect Environment, 19*, 142–152.

Vaidya, R., Macmil, S., Vyas, P., & Chhatpar, H. (2003). The novel method for isolating chitinolytic bacteria and its application in screening for hyperchitinase producing mutant of *Alcaligenes xylosoxidans*. *Letters in Applied Microbiology, 36*, 129–134.

Verma, M., Sharma, S., & Prasad, R. (2009). Biological alternatives for termite control: A review. *International Biodeterioration and Biodegradation, 63*, 959–972.

Verma, P., Yadav, A. N., Kazy, S. K., Saxena, A. K., & Suman, A. (2013). Elucidating the diversity and plant growth promoting attributes of wheat (*Triticum aestivum*) associated acidotolerant bacteria from southern hills zone of India. *National Journal of Life Sciences, 10*, 219–226.

Verma, P., Yadav, A. N., Kazy, S. K., Saxena, A. K., & Suman, A. (2014). Evaluating the diversity and phylogeny of plant growth promoting bacteria associated with wheat (*Triticum aestivum*) growing in central zone of India. *International Journal of Current Microbiology and Applied Sciences, 3*, 432–447.

Verma, P., Yadav, A. N., Khannam, K. S., Panjiar, N., Kumar, S., Saxena, A. K., & Suman, A. (2015a). Assessment of genetic diversity and plant growth promoting attributes of psychrotolerant bacteria allied with wheat (*Triticum aestivum*) from the northern hills zone of India. *Annales de Microbiologie, 65*, 1885–1899.

Verma, P., Yadav, A. N., Shukla, L., Saxena, A. K., & Suman, A. (2015b). Hydrolytic enzymes production by thermotolerant *Bacillus altitudinis* IARI-MB-9 and *Gulbenkiania mobilis* IARI-MB-18 isolated from Manikaran hot springs. *International Journal of Advanced Research, 3*, 1241–1250.

Verma, P., Yadav, A. N., Khannam, K. S., Kumar, S., Saxena, A. K., & Suman, A. (2016a). Molecular diversity and multifarious plant growth promoting attributes of *Bacilli* associated with wheat (*Triticum aestivum* L.) rhizosphere from six diverse agro-ecological zones of India. *Journal of Basic Microbiology, 56*, 44–58.

Verma, P., Yadav, A. N., Khannam, K. S., Mishra, S., Kumar, S., Saxena, A. K., & Suman, A. (2016b). Appraisal of diversity and functional attributes of thermotolerant wheat associated bacteria from the peninsular zone of India. *Saudi Journal of Biological Sciences*. https://doi.org/10.1016/j.sjbs.2016.01.042.

Vimaladevi, P. S., & Prasad, Y. G. (2001). *Nomuraea rileyi*: A potential mycoinsecticide. *In*: R. K. Upadyay, K. G. Mukherji, & B. P. Chamola (Eds.), *Biocontrol potential and its exploitation in sustainable agriculture insect Pests,* vol 2. Kluwer Academic Plenum. New York, pp. 23–38.

Vincent, C., Goettel, M. S., & Lazarovits, G. (2007). *Biological control, a global perspective*. Oxfordshire: CABI.

Wang, C., & Powell, J. E. (2003). Isolation and evaluation of *Beauveria bassiana* for control of *Coptotermes formosanus* and *Reticulitermes flavipes* (Isoptera: Rhinotermitidae). *Sociobiology, 41*, 369–382.

Wang, C. S., & St. Leger, R. J. (2007a). A scorpion neurotoxin increases the potency of a fungal insecticide. *Nature Biotechnology, 25*, 1455–1456.

Wang, C. S., & St. Leger, R. J. (2007b). The *Metarhizium anisopliae* perilipin homolog MPL1 regulates lipid metabolism, appressorial turgor pressure, and virulence. *The Journal of Biological Chemistry, 282*, 21110–21115.

Wang, C. S., & St. Leger, R. J. (2007c). The MAD1 adhesin of *Metarhizium anisopliae* links adhesion with blastospore production and virulence to insects, and the MAD2 adhesin enables attachment to plants. *Eukaryotic Cell, 6*, 808–816.

Wells, J., Fuxa, J., & Henderson, G. (1995). Virulence of four fungal pathogens to *Coptotermes formosanus* (Isoptera: Rhinotermitidae). *Journal of Entomological Science, 30*, 208–215.

Werner, P. A., Prior, L. D., & Forner, J. (2008). Growth and survival of termite-piped *Eucalyptus tetrodonta* and *E. miniata* in northern Australia: Implications for harvest of trees for didgeridoos. *Forest Ecology and Management, 256*, 328–334.

Wood, T. G. (1991). Termites in Ethiopia: The environmental impact of their damage and resultant control measures. *Ambio, 20*, 136–138.

Wood, T., Bednarzik, M., & Aden, H. (1987). Damage to crops by *Microtermes najdensis* (Isoptera, Macrotermitinae) in irrigated semi-desert areas of the Red Sea coast 1. The Tihama region of the Yemen Arab Republic. *International Journal of Pest Management, 33*, 142–150.

Wraight, S. P., Inglis, G. D., & Goettel, M. S. (2007). Fungi. *In*: L. A. Lacey & H. K. Kaya (Eds.), *Field manual of techniques in invertebrate pathology*, 2nd ed. Springer, Dordrecht, pp. 223–248.

Yadav, A. N. (2015). Bacterial diversity of cold deserts and mining of genes for low temperature tolerance. PhD thesis, Birla Institute of Technology, Mesra, Ranchi/Indian Agricultural Research Institute, New Delhi, pp. 234. doi:10.13140/RG.2.1.2948.1283/2.

Yadav, A. N., Sachan, S. G., Verma, P., & Saxena, A. K. (2015a). Prospecting cold deserts of north western Himalayas for microbial diversity and plant growth promoting attributes. *Journal of Bioscience and Bioengineering, 119*, 683–693.

Yadav, A. N., Sachan, S. G., Verma, P., Tyagi, S. P., Kaushik, R., & Saxena, A. K. (2015b). Culturable diversity and functional annotation of psychrotrophic bacteria from cold desert of Leh Ladakh (India). *World Journal of Microbiology and Biotechnology, 31*, 95–108.

Yadav, A. N., Verma, P., Kumar, M., Pal, K. K., Dey, R., Gupta, A., Padaria, J. C., Gujar, G. T., Kumar, S., Suman, A., Prasanna, R., & Saxena, A. K. (2015c). Diversity and phylogenetic profiling of niche-specific Bacilli from extreme environments of India. *Annals of Microbiology, 65*, 611–629.

Yadav, A. N., Sachan, S. G., Verma, P., Kaushik, R., & Saxena, A. K. (2016a). Cold active hydrolytic enzymes production by psychrotrophic Bacilli isolated from three sub-glacial lakes of NW Indian Himalayas. *Journal of Basic Microbiology, 56*, 294–307.

Yadav, A. N., Sachan, S. G., Verma, P., & Saxena, A. K. (2016b). Bioprospecting of plant growth promoting psychrotrophic Bacilli from cold desert of north western Indian Himalayas. *Indian Journal of Experimental Biology, 54*, 142–150.

Yadav, A. N., Verma, P., Kumar, V., Sachan, S. G., & Saxena, A. K. (2017). Extreme cold environments: A suitable niche for selection of novel psychrotrophic microbes for biotechnological applications. *Advances in Biotechnology and Microbiology, 2*, 1–4.

Yendol, W. G., & Paschke, J. D. (1965). Pathology of an entomophthora infection in the eastern subterranean termite, *Reticulitermes fl avipes* (Kollar). *Journal of Invertebrate Pathology, 7*, 414–422.

Yu, H., Gouge, D. H., & Shapiro-Ilan, D. I. (2010). A novel strain of *Steinernema riobrave* (Rhabditida: Steinernematidae) possesses superior virulence to subterranean termites (Isoptera: Rhinotermitidae). *Journal of Nematology, 42*, 91–95.

Zadji, L., Baimey, H., Afouda, L., Moens, M., & Decraemer, W. (2014). Characterization of biocontrol traits of heterorhabditid entomopathogenic nematode isolates from South Benin targeting the termite pest *Macrotermes bellicosus*. *BioControl, 59*, 333–344.

Zeng, T. (2004). Control of insect pests in sugarcane: Ipm approaches in China. *Sugar Tech, 6*, 273–279.

Zida, Z., Ouedraogo, E., Mando, A., & Stroosnijder, L. (2011). Termite and earthworm abundance and taxonomic richness under long-term conservation soil management in Saria, Burkina Faso, West Africa. *Applied Soil Ecology, 51*, 122–129.

Chapter 10
Sustainable Termite Management Using Physical Barriers

Menandro N. Acda

Contents

Abstract Termites are highly socialized insects that caused serious damage to wood products and timber structures worldwide. Termite activity results in billions of dollars spent on control and replacement of damaged wooden members. Traditional termite control method involves injecting hundreds of liters of synthetic insecticides to the soil or use of termite bait products containing insect growth regulators. The former poses risk to both man and environment and the latter is relatively expensive. Slow-acting, non-repellent termiticides have also been developed recently for colony management. However, due to inherent problems and difficulties associated with these methods, their general use is considered non-sustainable.

M.N. Acda (✉)
Department of Forest Products and Paper Science, University of the Philippines Los Banos, College, Laguna 4031, Philippines
e-mail: mnacda@yahoo.com

M.A. Khan, W. Ahmad (eds.), *Termites and Sustainable Management*, Sustainability in Plant and Crop Protection,
https://doi.org/10.1007/978-3-319-68726-1_10

In recent years, the use of physical barriers for sustainable termite management gained popularity due to inherent risks with conventional termite control treatments and erratic performance of bait products in tropical climates. Physical barriers using sand and lahar aggregates are alternative, nonchemical control method that can be used to prevent tunneling and penetration of subterranean termites into wood structures. The installations of these barriers are relatively simple, requiring no expensive equipment with barrier remaining effective for an indefinite period of time against various species of temperate and tropical termites. Although these methods offer a sustainable and environmentally friendly alternative, limited commercial applications of these techniques have been developed to date.

Keywords Termites • Lahar • Physical barrier • Sustainable management

10.1 Introduction

Termites are widely distributed and cause significant damage in tropical and subtropical regions of the world. They are most abundant throughout the so-called termite belt, which is 40° north and south of thc cquator. This rcgion includcs South and Southcast Asia, northern Australia, most of Africa, and South America and the southern states of the USA. Their activities result in extensive damage to wood products and reduction of service life of timber structures. Building materials such as lumber, plywood, wood-based composites, paper, and textiles containing cotton are susceptible to their activity. In some cases, agricultural crops, seedlings, and living trees are attacked by termites (Edwards and Mill 1986; Logan et al. 1990). The damage in terms of worldwide annual control and repair cost is estimated to be about USD 40 billion (Rust and Su 2012).

There are over 3000 reported living and fossil termite species worldwide (Krishna et al. 2013). However, only 10% of the reported species are destructive and considered as pests (Edwards and Mill 1986). Most termites are beneficial because they contribute to the decomposition of organic materials lying on the ground, such as tree branches and twigs, grasses, leaf litter, etc. In addition, termites contribute to soil ecology by mixing soil nutrients through their burrowing and foraging (Holt and Lepage 2000). However, they are regarded as structural pest because of their natural appetite for wood.

Subterranean termites have large colonies that live in the soil and require constant source of moisture for survival. Drywood termites consist of small colonies that live inside wood, require little moisture, and never enter the ground. Several species belonging to the family Rhinotermitidae, Termitidae, and Kalotermitidae are considered serious structural pests. Genera belonging to *Coptotermes*, *Reticulitermes*, *Odontotermes*, *Nasutitermes*, *Macrotermes*, *Microcerotermes*, and *Cryptotermes* are regarded to have significant economic importance. Two species of subterranean termites belonging to the genus *Coptotermes*, viz., *C. formosanus* Shiraki and *C. gestroi* Wasmann, are most destructive and widely distributed. Majority of the drywood termites considered pests include *Cryptotermes brevis* Walker, *Cr. dudleyi* Banks, and *Incisitermes minor* Hagen.

10.2 Termite Control Methods

10.2.1 Chemical Barrier

The traditional control method to prevent subterranean termite infestation involves injection of hundreds of liters of liquid termiticide to the soil beneath structures before or after construction. The objective is to create a chemical barrier between the soil and the structure to be protected that is toxic or repellent to foraging termites in the ground. The toxicity or repellence of the chemical barrier prevents tunneling and penetration of termites into the structure, thus preventing infestation. Active ingredients of currently available termiticides are either contact or systemic poisons. Typical liquid termiticides contain organophosphates (e.g., chlorpyrifos) and synthetic pyrethroids (e.g., permethrin, cypermethrin, bifenthrin, etc.). However, the use of persistent synthetic insecticides associated with chemical barrier treatments poses risks to both health and environment.

A recent development in termite control is the use of slow-acting, non-repellent termiticides for colony management. Non-repellent termiticides (e.g., imidacloprid, fipronil, chlorfenapyr, and chlorantraniliprole) are metabolic inhibitors that affect nerve impulses and the normal functioning of the insect's nervous system. Due to the delayed toxicity of these chemicals, termites tunnel through treated soil contaminating their bodies and ingesting chemically laden soil. Social grooming and trophallaxis feeding facilitate transfer of chemical to unexposed members of the colony leading to convulsion, hyperactivity, or inability to move muscles and eventually death of the insect. However, due to its delayed toxicity, effective control could take several weeks to months after treatment.

10.2.2 Termite Baits

Termite baits contain insect growth regulators (e.g., chitin synthesis inhibitors, juvenile hormone analogues, etc.) impregnated into wood or cellulose-based material. The baits are placed in underground stations along the perimeter of the structure or above ground along natural pathways (mud tubes) of termites. The workers foraging randomly in the soil find the treated materials, feed on them, and carry it back to the nest. Social grooming and trophallaxis facilitate the transfer of toxicant to other members of the colony. By this process of food transfer from exposed colony member to another, termites in the colony eventually receive a lethal dose to cause death or colony suppression (Su 1994). Since the toxicants used for termite baiting (e.g., hexaflumuron, noviflumuron, chlorfluazuron, diflubenzuron, etc.) have very low mammalian toxicity and used in small amount, it is considered an environmentally friendly control method. However, the cost of treatment is relatively expensive. Consequently, only a small number of families and property owners were able to afford such treatments. In addition, poor or inconsistent performance of termite baits containing chitin synthesis inhibitors was reported in tropical countries, due primarily to the presence of other termite species belonging to the Termitidae, among others (Bajo and Acda 2016).

10.2.3 Biological Control

Concerns over effects of persistent insecticides on the environment and problems associated with current termite control treatments prompted demand for a safe and affordable alternative. Nonchemical, biological, and physical control methods were studied to response to the challenge to offer an alternative termite control method. The use of pathogens as biological control agents has been considered as an alternative technology for termite control (Grace 1997). Various virulent entomopathogenic organisms such as predatory nematodes (*Steinernema* sp.), fungi (*Metarhizium anisopliae*, *Beauveria bassiana*), and bacteria (*Bacillus thuringiensis*, etc.) were investigated against various species of termite worldwide (Connick et al. 2001; Osbrink et al. 2001; Chouvenc et al. 2008; Ibrahim and Abd El-Latif 2008; Husseneder et al. 2010; Shahina et al. 2011). However, laboratory results using inundative treatment were inconsistent, and field trials have been generally unsuccessful (Lai 1977; Mauldin and Beal 1989; Chouvenc et al. 2011). Factor that could have contributed to the poor performance of biological agents is the difficulty of introducing a pathogen or inoculating enough individuals to trigger an outbreak of disease or infection within the colony (Chouvenc et al. 2008).

10.2.4 Botanical Insecticides

Plant extracts have been studied as potential sources of botanical insecticides to control termites (Verma et al. 2009). Plant families belonging to Meliaceae, Rutaceae, and Annonaceae, among others, have been investigated for their termiticidal properties. Botanical insecticides are generally regarded as an alternative to chemical insecticides and considered safe, with little or no threat to man and environment (Isman 2006). Recent studies showed that plant derivatives such as pyrethrins, terpenoids, azadirachtin, saturated and unsaturated fatty acids, and flavanoids have excellent termiticidal activity (Grace and Yates 1992; Sharma et al. 1994; Cornelius et al. 1997; Ohmura et al. 1999, 2000; Maistrello et al. 2001; Zhu et al. 2001a, b; Chang et al. 2001; Doolittle et al. 2007; Acda 2014a, b). Botanical insecticides are reportedly toxic and repellent and have anti-feeding effects on termites. However, effective concentrations of plant extracts to cause mortality in termites are generally high, compared with synthetic insecticides. Apparently, isolation and use of the pure active component may offer a more effective termiticidal formulation.

10.2.5 Physical Barriers

Termite control using physical barrier uses inert particles to prevent tunneling and entry of termites into wood structures. To prevent termite penetration, the barrier size must be large enough to prevent them from moving with their mandibles but small enough so that spaces between particles are too small for termites to pass through. The effective particle size is dependent on the mandible and head capsule dimensions of the target termite species (Table 10.1). The barrier must be laid under slabs and foundation walls prior to the pouring of concrete during construction. Laboratory and field studies using particles of sand (Ebeling and Pence 1957; Tamashiro et al. 1987; Myles 1997), glass shards (Pallaske and Igarashi 1991), granite (Smith and Rust 1990; French 1991; French and Ahmed 1993), crushed basalt (Tamashiro et al. 1987, 1991), quartz and coral sand (Su et al. 1991), crushed cement-stabilized sludge (Yanase et al. 2000), lahar aggregates (Acda and Ong

Table 10.1 Effective size and materials reported for particle barriers against various species of subterranean termites

Species	Material	Effective particle size (mm)	References
Reticulitermes hesperus	Sand (silica)	1.2–2.7	Ebeling and Pence (1957)
	Granite	1.7–2.4	French et al. (2003)
	Granite	0.84–2.36	Smith and Rust (1990)
Reticulitermes flavipes	Sand (silica)	2.0–2.8	Su and Scheffrahn (1992)
	Sand (limestone)	1.4–2.8	Myles and Grace (1991)
	Sand (beach)	1.18–2.26	Myles (1997)
	Coral (crushed)	1.0–2.36	Su et al. (1991)
Coptotermes formosanus	Sand (silica)	1.7–2.4	Tamashiro et al. (1987)
	Sand (silica)	2.0–2.8	Su and Scheffrahn (1992)
	Coral (crushed)	1.7–2.36	Su et al. (1991)
	Polynite	1.7–2.0	Yanase et al. (2000)
C. lacteus	Granite	1.7–2.4	French et al. (2003)
C. acinaciformis	Granite	1.7–2.4	French et al. (2003)
	Glass (sintered)	1.7–2.4	Ahmed and French (2011)
C. gestroi (*C. vastator*)	Lahar aggregates	1.18–2.36	Acda and Ong (2005a)
Nasutitermes luzonicus	Lahar aggregates	1.18–2.36	Acda and Ong (2005b)
Microcerotermes losbanosensis	Lahar aggregates	1.18–2.36	Acda and Ong (2005b)
Macrotermes gilvus	Lahar aggregates	1.70–2.26	Acda and Ong (2005b)
M. beesoni	Marble chips	1.18–2.36	Singh and Rawat (1999)
Heterotermes indicola	Glass beads	0.5–3.0	Pallaske and Igarashi (1991)

2005a, b), etc. screened to specific particle sizes have proven to be effective in preventing termite penetration. However, the range of effective particle size differs from one termite species to another (Su and Scheffrahn 1992).

The success of laboratory and field trials of particle barriers resulted in commercial applications. For example, crushed basalt (Basaltic Termite Barrier®, Ameron HC&D, Honolulu) and granite aggregate (Granitgard®, Granitgard Pty. Ltd., Victoria) are already available as alternative method of termite control in the USA and Australia. However, installation issues including unstable or not compacted soil, irregular surfaces at the edges of the barrier, protection from contamination, or mixing with adjacent soil were reported (Grace et al. 1996). Other types of nontoxic physical barriers were also investigated. Concrete slabs (Lenz et al. 1997), solid sheet material (e.g., high-grade stainless steel, marine-grade aluminum), plastic sheets impregnated with insecticide (Su et al. 1994), kaolin-based particle film (Wiltz et al. 2010), and woven stainless steel mesh were reported to be effective in preventing termite penetration as those of particle barrier (Grace et al. 1996). A commercial stainless steel mesh barrier (Termi-Mesh®, Termi-Mesh Australia Pty. Ltd.) was also developed in Australia. However, the role of physical barriers in the future of subterranean termite control may depend on continuous ban on persistent organic insecticides and willingness of property owners to absorb the higher cost of treatment.

10.3 Sustainable Termite Management

Termite management as outlined above represents various termite control methods, but each has its own inherent shortcomings. It is in this context that the concept of integrated pest management (IPM) for termite control came about. As discussed by Su and Scheffrahn (1998), IPM originated in the 1990s as a philosophy to address agricultural crop problems such as pest resistance, pest outbreaks, environmental pollution, etc. It is essentially a knowledge-based decision-making process requiring an understanding of the pest biology, in order to take action or intervention aimed at reducing the economic impact of the pest (Forschler 2011). The type of intervention is dictated by available technologies and by its economics and capability to reduce pest population. The definition of IPM for termite control, however, is dependent on the point of view and priorities of various stakeholders (Su and Scheffrahn 1998). Home and property owners look at IPM as eliminating termite infestation at the shortest time and most possible cost-effective way. Researchers perceived IPM in terms of effective and safe control methods with limited risk to the environment (Robinson 1996). Chemical companies look at IPM as the use of its own product plus the use of all other control measures, to achieve acceptable performance and remedy shortcoming of its own product (Ballard 1997). For pest control operators, IPM is the use of their preferred method (i.e., chemical barrier, spot treatments, or baiting) plus all other cultural methods (e.g., removal of wood debris, drainage, leaks and excessive moisture problems, mounds, etc.) to solve client's termite problems. These concepts and programs are often used for marketing purposes (Robinson 1996).

In contrast to IPM, sustainable termite management is relatively a new concept. It originated from the concept of sustainable development used in ecology and environmental science. For termite control, sustainable termite management may generally be described as an effective control measure that is safe to man, with no ecological damage or loss of ecosystem benefits derived from termite activity, conservation of nontarget organisms, and the use of products and technology that do not contribute to the depletion of natural resources. No, or limited, threat to man and environments for the protection of wood products, timber structures, and other nontarget organisms is essential in any sustainable termite management program. The description is somewhat contradictory since the aim of an effective termite control measure is to eliminate or kill termite colonies. However, killing termites would deprive the environment of an efficient decomposer of organic materials lying on the ground. Apparently, a working compromise between these two objectives must be reached, to arrive at an acceptable definition. An acceptable solution is the modification of the original definition of sustainable termite management into eliminating or suppressing destructive termite colonies near wood structures or areas of economic activities, such as farms, around utility poles, golf courses, etc. In view of the above, termite control method such as chemical barrier using toxic or repellent termiticides would not meet sustainable termite management criteria. The use of termite baits and slow-acting termiticides may pass the requirements (Su 1994; Su et al. 1995, 2001; Evans 2010). However, the use of physical barriers such as sand and lahar aggregates would truly fit the requisites of sustainable termite management. Both sand and lahar aggregates have been shown to be effective in preventing tunneling and penetrating of various species of subterranean termites into structures (Tamashiro et al. 1987, Myles and Grace 1991; Myles 1997; Acda and Ong 2005a, b). These materials are also safe, widely available, and cost-effective barriers against subterranean termites. Assuming no break in the barrier is made during service due to remodeling or landscaping activity, the protection offered by physical barriers could last indefinitely.

10.3.1 Sand Barrier

Sand as barriers to prevent tunneling of termites was discovered by Ebeling and Pence (1957), and Tamashiro et al. (1987), and later confirmed by others (Smith and Rust 1990; Su et al. 1991; Su and Scheffrahn 1992; Lewis et al. 1996). Commercial sand barrier is available in Hawaii (Honolulu Construction & Draying Co., Ltd., Honolulu) and Australia. However, despite studies finding that sand barrier excludes termites from wood structures, its use by the pest management industry has been mostly overlooked for a number of reasons (Yates et al. 2002). These include consumer unawareness of the product, a slightly higher initial cost compared with chemical barrier treatments, resistance on the part of the pest control industry to accept and implement this nonchemical technology, and the absence of a performance warranty from the licensed manufacturer (Yates et al. 2002). In addition, architects and building contractors have little understanding of installation requirements for this barrier. Furthermore, termites can build over physical barriers, and regular inspections of the building are necessary. However, mud tubes over the barrier reveal evidence of their presence facilitating control during regular inspections. Recent study involving engineering analysis of sand aggregate particles indicated that angularity, fineness modulus, and weighted particle size were variables related to the success of particle barriers against subterranean termites (Keefer et al. 2013).

10.3.2 Lahar Barrier

Another material that can be used as physical barrier to prevent entry of subterranean termites into wood structures is lahar. Lahar is a saturated mixture of ash, solid rock particles, and other volcanic debris washed down by rainwater from the slope of recently erupted volcano. Once dried, lahar could be described as a sandy aggregate (Fig. 10.1) consisting mainly of feldspar, hornblende, quartz, mica, and magnetite (Cabillon et al. 1997). The northern part of the island of Luzon in the

Fig. 10.1 Lahar (10×) consists of sandy aggregates prescreened to 1.18–2.36 mm used as physical barrier to prevent tunneling and penetration of Philippine subterranean termites

Philippines has huge volume of lahar deposits in several provinces made during the eruption of Mt. Pinatubo in 1991. It is estimated that about 11 billion cubic meters of ash and volcanic debris ejected during the eruption could potentially be carried by monsoon rains downslope as lahar to clog streams and rivers in low-lying areas (Newhall and Punongbayan 1996). Twenty-five years after the eruption and after continuous quarrying, lahar still clogs major river systems in the northern provinces of the Philippines. Similar lahar deposits were made during volcanic eruptions in Nevado del Ruiz and Nevado del Huila volcanoes in Colombia, Mount St. Helens and Redoubt Volcano in the USA, Mount Ontake volcano in Japan, and other eruptions worldwide (Pierson et al. 1994).

Laboratory and field trials showed that the effective particle size of 1.18–2.36 mm would prevent tunneling and penetration of *C. gestroi*, *N. luzonicus*, and *Microcerotermes losbanosensis* (Acda and Ong 2005a, b). However, a slightly larger lahar particle size range of 1.7–2.36 mm would be required for *Macrotermes gilvus*, due to the large head capsule and mandibles of this species. The natural sharp edges of lahar particles also proved to be detrimental to the termites as they cut their appendages (i.e., antennae and legs) and die when they tunnel and burrow through the barrier.

A small wooden house was built in 1997 where a protective barrier consisting of prescreened lahar particles was installed beneath floor and concrete foundation walls (Fig. 10.2, Acda 2013). Regular inspections made over a 7-year period showed no signs of subterranean termite penetration inside and outside of the structure (Fig. 10.3). The study showed that lahar barrier could be used to protect wooden structures from entry of subterranean termites and offer a nonchemical alternative to

Fig. 10.2 Installation of lahar barrier underneath floor and foundation walls prior to the pouring of concrete

Fig. 10.3 A small wooden house protected underneath by lahar barrier remained free of subterranean termites 7 years after construction

commercially available termiticides. Promoting the use of lahar barrier against termites offers several economic and environmental benefits compared to commercially available termite control products.

The utilization of lahar from affected river systems could help improve the profile of streambeds and river channels, thus benefiting flood control, irrigation, and water quality. There would also be potential for a commercial bagged product that could be sold at garden shops and hardware stores for small-scale applications such as protection of utility poles, wooden posts and signs, under stacked firewood, etc. The technique is beneficial to property owners not only in the Philippines but also in other neighboring countries like Malaysia, Indonesia, Thailand, China, India, Australia, etc. with structural problems involving similar species of termites. Lahar barrier offers an alternative and affordable method of protecting properties against ongoing threat of subterranean termites. It is environment friendly and could greatly reduce the load of toxic chemicals in the urban environment. However, despite the availability of excellent performance in laboratory and field trials, the use of lahar has not seen commercial application. The exact reason for this is unclear. Various pest control operators hinted that the use of liquid termiticides or baits has better economic gains compared with physical barriers.

10.4 Conclusion

Termites are serious structural pests of wood products and timber structures worldwide. Financial damage to structures and agricultural crops results in billions of dollars spent in termite control and replacement of wooden members. Traditional control method injecting hundreds of liters of synthetic insecticides to the soil poses risk to both man and environment. Recent development in colony management resulted in the use of termite baiting technology and slow-acting, non-repellent termiticides. However, due to inherent problems and difficulties associated with these methods, the general use for sustainable management remains in doubt. A truly sustainable termite control method using physical barriers gained popularity and acceptance due to the ban on persistent inorganic termiticides. Sand and lahar aggregates are examples of alternative, nonchemical control method that can be used to prevent tunneling and penetration of subterranean termites into wood structures. Although they offer a sustainable and environmentally friendly alternative, limited commercial applications of these techniques have been developed to date.

References

Acda, M. N. (2013). Evaluation of lahar barrier to protect wood structures from Philippine subterranean termites. *The Philippine Journal of Science, 142*, 21–25.

Acda, M. N. (2014a). Repellent effects of Annona crude seed extract on the Asian subterranean termite *Coptotermes gestroi* Wasmann (Isoptera: Rhinotermitidae). *Sociobiology, 61*, 332–337.

Acda, M. N. (2014b). Chemical composition of ethanolic seed extracts of *Annona squamosa* and *A. muricata* (Annonaceae) using GC-MS analysis. *Philippine Agricultural Sciences, 97*, 422–426.

Acda, M. N., & Ong, H. B. (2005a). Use of volcanic debris as physical barrier to prevent tunneling of the Philippine milk termite (Isoptera: Rhinotermitidae). *Sociobiology, 46*, 117–129.

Acda, M. N., & Ong, H. B. (2005b). Penetration of lahar aggregates by Philippine subterranean termites (Isoptera: Termitidae). *Sociobiology, 47*, 189–200.

Ahmed, B. M. S., & French, J. R. J. (2011). Termite foraging behavior and biological response against sintered glass screenings as a potential graded physical barrier. *African Journal of Entomology and Zoology Studies, 5*, 327–336.

Ballard, J. (1997). Termite baiting 101. *Pest Control, 65*, 30–40.

Cabillon, S. L., Kimura, K., Bernardo, S. T., & Visaya, A. (1997). Lahar gems, glasses and glass-ceramics. In *Proceedings on the utilization of Pinatubo Volcanic ejecta and other related R&D studies* (p. 93). ITDI-PCIERD-DOST, Manila.

Chang, S. T., Cheng, S. S., & Wang, S. Y. (2001). Antitermitic activity of essential oils and components from taiwania (*Taiwania cryptomerioides*). *Journal of Chemical Ecology, 27*, 717–724.

Chouvenc, T., NY, S., & Elliott, M. L. (2008). Interaction between the subterranean termite *Reticulitermes flavipes* (Isoptera: Rhinotermitidae) and the entomopathogenic fungus *Metarhizium anisopliae* in foraging arenas. *Journal of Economic Entomology, 101*, 885–893.

Chouvenc, T., Su, N. Y., & Grace, J. K. (2011). Fifty years of attempted biological control of termites – Analysis of a failure. *Biological Control, 59*, 69–82.

Connick, W. J., Jr., Osbrink, W. L. A., Wright, M. S., Williams, K. S., Daigle, D. J., Boykin, D. L., & Lax, A. R. (2001). Increased mortality of *Coptotermes formosanus* (Isoptera: Rhinotermitidae) exposed to eicosanoid biosynthesis inhibitors and *Serratia marcescens* (Eubacteriales: Enterobacteriaceae). *Environmental Entomology, 30*, 449–455.

Cornelius, M., Grace, J. K., & Yates, J. R. (1997). Toxicity of monoterpenoids and other natural products to the Formosan subterranean termites. *Journal of Economic Entomology, 90*, 320–325.

Doolittle, M., Raina, A., Lax, A., & Boopathy, R. (2007). Effect of natural products on gut microbes in Formosan subterranean termite, *Coptotermes formosanus*. *International Biodeterioration & Biodegradation, 59*, 69–71.

Ebeling, W., & Pence, R. J. (1957). Relation of particle size of the penetration of subterranean termites through barriers of sand or cinders. *Journal of Economic Entomology, 50*, 690–692.

Edwards, R., & Mill, A. E. (1986). *Termites in buildings. Their biology and control* (p. 261). East Grinstead: Rentokil Ltd.

Evans, T. A. (2010). Rapid elimination of field colonies of subterranean termites (Isoptera: Rhinotermitidae) using bistrifluron solid bait pellets. *Journal of Economic Entomology, 103*, 423–432.

Forschler, B. (2011). Sustainable termite management using an integrated pest management approach. In P. Dhang (Ed.), *Urban pest management: An environmental perspective* (pp. 133–144). Wallingford: CABI.

French, J. R. J. (1991). *Physical barriers and bait toxicants: The Romeo and Juliet of future termite control*. International Research Group on Wood Preservation IRG/WP/1503

French, J. R. J., & Ahmed, B. (1993). *Termite physical barrier: Is retrofitting with granitgard and option?* International Research Group on Wood Preservation IRG/WP/93-40011

French, J. R. J., Ahmed, B., & Trajstman, A. (2003). Laboratory and field evaluation of Granite aggregate as a physical barrier against subterranean termites of the genus *Coptotermes* spp. (Isoptera: Rhinotermitidae). *Sociobiology, 42*, 129–149.

Grace, J. K. (1997). Biological control strategies for suppression of termites. *Journal of Agricultural Entomology, 14*, 281–289.

Grace, J. K., & Yates, J. R. (1992). Behavioural effects of a neem insecticide on *Coptotermes formosanus* (Isoptera: Rhinotermitidae). *Tropical Pest Management, 38*, 176–180.

Grace, K. J., Yates, J. R., Tome, C. H. M., & Oshiro, R. J. (1996). Termite resistance construction: Use of stainless steel mesh to exclude *Coptotermes formosanus* (Isoptera: Rhinotermitidae). *Sociobiology, 28*, 365–372.

Holt, J. A., & Lepage, M. (2000). Termites and soil properties. In Y. Abe, D. E. Bignell, & T. Higashi (Eds.), *Termites: Evolution, sociality, symbioses, ecology* (pp. 389–407). Dordrecht: Kluwer Academic Publisher.

Husseneder, C., Ho, H. Y., & Blackwell, M. (2010). Comparison of the bacterial symbionts composition of the Formosan subterranean termite from its native and introduced range. *The Open Microbiology Journal, 4*, 53–66.

Ibrahim, S. A. M., & Abd El-Latif, N. A. (2008). A laboratory study to control subterranean termites. *Psammotermes hybostoma* (Desn.) (Isoptera: Rhinotermitidae) using entomopathogenic nematodes. *Egypt Journal of Biological Pest Control, 18*, 99–103.

Isman, M. B. (2006). Botanical insecticides, deterrents and repellents in modern agriculture and increasingly regulated world. *Annual Review of Entomology, 51*, 45–66.

Keefer, T. C., Zollinger, D. G., & Gold, R. E. (2013). Evaluation of aggregate particles as a physical barrier to prevent subterranean termite incursion into structures. *South West Entomology, 38*, 447–464.

Krishna, K., Grimaldi, D. A., Krishna, V., & Engel, M. S. (2013). Treatise on the Isoptera of the World. *Bulletin of the American Museum of Natural History*, no. 377. American Museum of Natural History, New York

Lai, P. Y. (1977). Biology and ecology of the formosan subterranean termite, *Coptotermes formosanus*, and its susceptibility to the entomogenous fungi, *Beauveria bassiana* and *Metarhizium anisopliae*. PhD dissertation, University of Hawaii, Honolulu

Lenz, M., Schafer, B. L., Runko, S., & Glossop, L. (1997). The concrete slab as part of a termite barrier system: Response of Australian subterranean termites to cracks of different width in concrete. *Sociobiology, 30*, 103–118.

Lewis, V. R., Haverty, M. I., Carver, D. S., & Fouche, C. (1996). Field comparison of sand or insecticide barriers for control of Reticulitermes sp. (Isoptera: Rhinotermitidae) infestation in homes in northern California. *Sociobiology, 28*, 327–335.

Logan, J. W. M., Cowie, R. H., & Wood, T. G. (1990). Termite (Isoptera) control in agriculture and forestry by non-chemical methods: A review. *Bulletin of Entomological Research, 80*, 309–330.

Maistrello, L., Henderson, G., & Laine, R. A. (2001). Efficacy of vetiver oil and nootkatone as soil barriers against Formosan subterranean termite (Isoptera: Rhinotermitidae). *Journal of Economic Entomology, 94*, 1532–1537.

Mauldin, J. K., & Beal, R. H. (1989). Entomogenous nematodes for control of subterranean termites, *Reticulitermes* sp. (Isoptera: Rhinotermitidae). *Journal of Economic Entomology, 82*, 1638–1642.

Myles, M. G. (1997). Comparison of the penetrability of smooth and crushed sand by subterranean termites (Isoptera: Rhinotermitidae). *Sociobiology, 30*, 295–302.

Myles, T. G., & Grace, J. K. (1991). Behavioral ecology of the Eastern subterranean termite in Ontario as basis for control. In *Proceedings of technology transfer conference* (Vol. 2, pp. 547–554). Toronto: Ontario Ministry of the Environment.

Newhall, C. G., & Punongbayan, R. S. (1996). *Fire and mud: Eruptions and lahars of Mt. Pinatubo, Philippines*. Seattle/London: Philippine Institute of Volcanology and Seismology Quezon City University of Washington Press.

Ohmura, W., Doi, S., Aoyama, M., & Ohara, S. (1999). Components of steamed and non-steamed Japanese larch (*Larix leptolepis* (Sieb. et Zucc.) Gord.) heartwood affecting the feeding behavior of the subterranean termites, *Coptotermes formosanus* Shiraki (Isoptera: Rhinotermitidae). *Holzforschung, 53*, 569–574.

Ohmura, W., Doi, S., Aoyama, M., & Ohara, S. (2000). Antifeedant activity of flavonoids and related compounds against the subterranean termite *Coptotermes formosanus* Shiraki. *Journal of Wood Science, 46*, 149–153.

Osbrink, W. L. A., Williams, K. S., Connick, J. W. J., Wright, M. S., & Lax, A. R. (2001). Virulence of bacteria associated with the Formosan subterranean termite (Isoptera: Rhinotermitidae) in New Orleans, LA. *Environmental Entomology, 30*, 443–448.

Pallaske, M., & Igarashi, A. (1991). Glass splinters as physical barriers: Optimized material properties in use with and without insecticidal pretreatment minimizes environmental contaminations. International Research Group on Wood Preservation IRG/WP 1476.

Pierson, T. C., Wood, N. J., & Driedger, C. L. (1994). Reducing risk from lahar hazards: Concepts, case studies, and roles for scientists. *Journal of Applied Volcanology, 3*, 16.

Robinson, W. H. (1996). Integrated pest management in the urban entomology. *American Entomologist, 42*, 76–78.

Rojo, M. J. A., & Acda, M. N. (2016). Interspecific agonistic behavior of Macrotermes gilvus (Isoptera: Termitidae): Implication on termite baiting in the Philippines. *Journal of Insect Behavior, 29*, 273–282.

Rust, M. K., & Su, N. Y. (2012). Managing social insects of urban importance. *Annual Review of Entomology, 57*, 355–375.

Shahina, F., Tabassum, K. A., Salma, J., & Mahreen, G. (2011). Biopesticidal effect of *Photorhabdus luminescens* against *Galleria mellonella* larvae and subterranean termite (Termitidae: Macrotermis). *Pakistan Journal of Nematology, 29*, 35–43.

Sharma, R. N., Tungikar, V. B., Pawar, P. V., & Vartak, P. H. (1994). Vapor toxicity and repellency of some oils and terpenoids to the termite *Odontotermes brunneus*. *Insect Science and Its Application, 15*, 495–498.

Singh, Y., & Rawat, B. S. (1999). Studies on inert barrier to prevent entry of subterranean termites into building. *Hexapoda, 11*, 91–95.

Smith, J. L., & Rust, M. K. (1990). Tunneling response and mortality of the western subterranean termite (Isoptera: Rhinotermitidae) to soil treated with termiticides. *Journal of Economic Entomology, 83*, 1395–1401.

Su, N. Y. (1994). Field evaluation of hexaflumuron bait from population suppression of subterranean termites (Isoptera: Rhinotermitidae). *Journal of Economic Entomology, 87*, 389–397.

Su, N. Y., & Scheffrahn, R. H. (1992). Penetration of sized-particle barriers by field populations of subterranean termites (Isoptera: Rhinotermitidae). *Journal of Economic Entomology, 85*, 2275–2278.

Su, N. Y., & Scheffrahn, R. H. (1998). A review of subterranean termite control practices and prospects for integrated pest management programmes. *Integrated Pest Management Review, 3*, 1–13.

Su, N. Y., Scheffrahn, R. H., & Ban, P. (1991). Uniform size particle barriers: A physical exclusion device against subterranean termites (Isoptera: Rhinotermitidae). *Journal of Economic Entomology, 83*, 912–916.

Su, N. Y., Ban, P., & Scheffrahn, R. H. (1994). Polyethylene barrier impregnated with lambdacyhalothrin for exclusion of subterranean termites (Isoptera: Rhinotermitidae) from structures. *Journal of Economic Entomology, 97*, 570–574.

Su, N. Y., Thoms, E. M., Ban, P. M., & Scheffrahn, R. H. (1995). Monitoring/baiting station to detect and eliminate foraging populations of subterranean termites (Isoptera: Rhinotermitidae) near structures. *Journal of Economic Entomology, 88*, 932–936.

Su, N. Y., Ban, P. M., & Scheffrahn, R. H. (2001). Control of subterranean termites (Isoptera: Rhinotermitidae) using commercial prototype above ground stations and hexaflumuron baits. *Sociobiology, 37*, 111–120.

Tamashiro, M., Yales, J. R., & Ebesu, R. H. (1987). The Formosan subterranean termite in Hawaii: Problems and control. In M. Tamashiro & S. N-Y (Eds.), *Biology and control of the Formosan subterranean termite* (pp. 15–22). Honolulu: College of Tropical Agriculture and Human Resources, University of Hawaii.

Tamashiro, M., Yates, J. R., Yamamoto, R. T., & Ebesu, R. H. (1991). Tunneling behavior of the Formosan subterranean termite and basalt barriers. *Sociobiology, 19*, 163–170.

Verma, M., Sharma, S., & Prasad, R. (2009). Biological alternative for termite control: A review. *International Biodeterioration and Biodegradation, 63*, 959–972.

Wiltz, B. A., Woodson, W. D., & Puterka, G. J. (2010). Potential of kaolin-based particle film barriers for Formosan subterranean termite (Isoptera: Rhinotermitidae) control. *Sociobiology, 55*, 405–413.

Yanase, Y., Shibata, M., Yoshihisa, S., Okumura, K., Iwamoto, T., Nogiwa, T, & Yoshimura, I. Y. (2000). *Feasibility of termite control using crushed cement stabilized sludge (polynite) as a physical barrier and acoustic emission (AE) monitoring*. International Research Group on Wood Preservation IRG/WP/00-10381.

Yates, J. R., Grace, J. K., & Reinhardt, J. N. (2002). Creating installation guidelines for a particle barrier for formosan subterranean termites (lsoptera: Rhinotermitidae). *Sociobiology, 40*, 207–210.

Zhu, B. C. R., Henderson, G., Chen, F., Fei, H., & Laine, R. A. (2001a). Evaluation of vetiver oil and seven insect-active essential oils against the Formosan subterranean termite. *Journal of Chemical Ecology, 27*, 1617–1625.

Zhu, B. C. R., Henderson, G., Chen, F., Maistrello, L., & Laine, R. A. (2001b). Nootkatone is a repellent for Formosan subterranean termite (*Coptotermes formosanus*). *Journal of Chemical Ecology, 27*, 523–531.

Chapter 11
Synthetic Insecticides: The Backbone of Termite Management

Bishwajeet Paul, Sharda Singh, K. Shankarganesh, and Md. Aslam Khan

Contents

Abstract Termite management has been a challenge since time immemorial. In good old days, plant products have been used with limited success. With the discovery of organochlorine pesticides, the use of chemicals gained an impetus in management. Due to longer persistence and health hazards, the use of organochlorine pesticides was banned in most countries. However, for several decades, chlordane, heptachlor, lindane, etc. were mainly used for termite management. Subsequently organophosphates and pyrethroids replaced organochlorines in this scenario.

B. Paul (✉) • S. Singh
Division of Entomology, ICAR-Indian Agricultural Research Institute, New Delhi 110012, India
e-mail: bishwajeet_paul2011@yahoo.com

K. Shankarganesh
ICAR-Central Institute for Cotton Research Regional Station, Coimbatore, Tamil Nadu, India

M.A. Khan
Department of Biology, Faculty of Science, Jazan University, Jazan, Saudi Arabia

M.A. Khan, W. Ahmad (eds.), *Termites and Sustainable Management*, Sustainability in Plant and Crop Protection,
https://doi.org/10.1007/978-3-319-68726-1_11

Organophosphates and pyrethroids repel termites, but due to their highly toxic nature, ready availability, and relatively inexpensiveness, they are still being used in developing countries for management.

Termite management practices changed dramatically with the advent of newer molecules, viz., neonicotinoid (imidacloprid), phenylpyrazole (fipronil), pyrrole (chlorfenapyr), oxadiazine (indoxacarb), anthranilic diamide (chlorantraniliprole), etc. Majority of these compounds are of the slow-acting and non-repellent type. The termites fail to detect these insecticides and continue to forage in the treated soil for longer period and carry lethal amount of the toxicant to be later transferred to their nestmates. This behavior and the properties of the new molecules prompted research on bait technologies. Termite baiting has developed tremendously and has been commercialized over last three decades. Various bait matrices and bait stations have been developed successfully. The activity of termiticides varies widely depending on the soil characteristics and properties, as soil plays an important role in the success of management practices. With these new termiticides, fairly long-term barriers could be established around wooden structures and buildings. Optimal termite management still remains a challenge and depends widely on the type of termiticides available, soil type, cropping system, expertise available, type of structure/building, and economics of the procedure.

Keywords Termites • Termiticides • Termite baits

11.1 Introduction

Termites belong to one of the oldest groups of insects comprising of more than 3000 species and are reported to survive in all ecosystems except above snow line and the polar regions. The majority of the termite species are highly beneficial to mankind, and only a handful of species (Edwards and Mill 1986; Logan et al. 1990) are pests. Sustainability of life without termites is unimaginable. They are one of the most efficient organisms in food webs, helping in recycling nutrients in both agricultural and nonagricultural ecosystems. Though termites lack some of the very basic survival requirements, i.e., they are visually impaired, but their olfactory sense is extremely well developed, they are highly sensitive to changes in light, temperature, as well as humidity. With all these lacunae termites pose a formidable threat to agricultural crops. Their ability to withstand adverse environmental conditions is primarily due to their reproduction in large numbers at an astounding rate. Their social structure compliments in their ability to tide over adverse conditions. During last few decades as the demand for food production increased tremendously due to burgeoning population, damage caused to agricultural crops by termites has been realized resulting in a steady increase in the related literature (Vargo and Husseneder 2009). Even though damage control measures have been developed all over the world, including cultural practices, mechanical measures, and biological and chemical measures, management of termites is still in infancy, because very little is known about their biology.

Tremendous amount of data have been generated on termite management with synthetic insecticides. In the USA an estimated 77% of pest control market is represented by termite management using soil insecticides (Anonymous 2002). Various strategies have been developed to use different groups of insecticides, including botanical products (Verma et al. 2009). Investigations were carried out using natural enemies of termites, but their commercial viability is still lacking. Termites are known to thrive in conditions highly conducive for growth of microbes, leading to evaluation of several insect pathogens including bacteria, fungi, and viruses with limited success (Kramm et al. 1982; Rath 2000). In this chapter we shall discuss some of the most landmark studies carried out in termite management, with special reference to synthetic insecticides.

In subterranean termite management, mainly soil applicable formulations were used, and major strategies were followed for (a) prevention by soil treatment and (b) remedial control of active infestations, by applying insecticides directly into the infested soil. Both strategies have their own merits and demerits, the soil characteristics and insecticide properties playing a deciding role regarding the method to be used. For prophylactic treatment the insecticide must have a long residual life in order to repel or kill the termites rapidly. For remedial treatments the insecticide must be odorless/non-repellent and should be relatively slow acting so that the termites are able to carry it to their nests and galleries. The affected termites must be healthy enough so that no panic is created in the nest, whereas the other termites do not exclude the treated area for foraging.

Su et al. (1982) concluded that evaluation of insecticides against termites could not be based on mortality alone. The behavioral responses must be considered indeed, because termites could seal off or avoid treated areas and effectively protect themselves. These authors found that such behavior was due to repellency of insecticide itself or chemical factors associated with dead termites. They categorized tested insecticides into three groups, viz., Type I (including natural and synthetic pyrethroids), Type II (including commonly used insecticides, viz., diazinon, chlordane, and carbaryl), and Type III (including Amdro®). Type I insecticide exhibited repellent action which initially kills some termites, and subsequently the emanating source can be sealed off by the other termites. In Type II group, the termites continued to move around on the treated surface, and only when some individuals have died and decayed, the area was sealed off. This indicates that the insecticides were not as such repellent, the termites realized in a period of time to avoid the treated surface. In Type III group, the termites did not detect any chemical and continued visiting the treated area since the affected termites did not die at one place and the insecticide did not exhibit any repellent action.

The newer insecticides recommended for termite management are of two types, viz., fast-acting repellents, at lethal doses, and slow-acting non-repellents, at lower doses. The organophosphates used in termite management are quick acting with relatively short soil residual period, whereas the pyrethroids used in management have shown repellent properties which make the termites change their foraging area and persist in soil for longer period (Su et al. 1999a). Insecticides, viz., imidacloprid, fipronil, chlorfenapyr, indoxacarb, and chlorantraniliprole, belong to a novel group of insecticides that, when used at recommended concentrations, act as repellents (Osbrink et al. 2001; Ibrahim et al. 2003; Hu 2005; Rust and Saran 2006). These insecticides induce harmful behavioral changes/dysfunctions among termites, and their toxic effects are transmissible from poisoned termites to nonpoisoned individuals in the colony, leading to substantial deaths (Haynes 1988; Hu and Hickman 2006).

Investigations on termite poisoning with insecticides are multidimensional studies. Termite mechanisms for picking up insecticides are walking on treated surfaces, casual contacts, mutual grooming and antennation, trophallaxis (stomodeal or proctodeal), exchange of food and chemicals, coprophagy, necrophoresis, necrophagy, cannibalism, and contacting secondary contaminated surfaces (Smith and Rust 1990; Ibrahim et al. 2003; Kard 2003; Tomalski and Vargo 2004; Hu et al. 2005; Shelton et al. 2006; Song and Hu 2006; Tsunoda 2006; Spomer et al. 2008; Bagnères et al. 2009).

11.2 Termiticides

11.2.1 Repellent

All the organochlorine, organophosphate, carbamate, and pyrethroid insecticides are categorized as repellents. In the past, repellents were mainly used to protect wooden structures and buildings from attacks. One major advantage with a repellent termiticide is that it provides an effective barrier against termites and prevents any damage to structures and buildings. The use of inexpensive pyrethroid termiticides as a barrier was in vogue in the developed world for decades. The use of such barriers has a serious limitation, i.e., perfect barriers cannot be created under fully constructed houses or buildings. The termite workers can locate the gaps, and where the barrier is improper, they gain access to the structures, causing damage by recruiting more workers.

Synthetic pyrethroids, viz., permethrin, cypermethrin, deltamethrin, fenvalerate, cyfluthrin, tralomethrin, lambda-cyhalothrin, tefluthrin, bifenthrin, and flucythrinate, act on the sodium channels located in the insect nervous system. Their contact activity quickly kills termites when applied at the recommended doses or causes a directional change in tunneling, away from the treated area (Su and Scheffrahn 1990; Gahlhoff and Koehler 1999). Furthermore, repellents require rigorous application to all the possible entry points, if the application is meant to create a continuous chemical barrier (a full barrier) around and beneath the structure, and to all interior active infestations, in order to get an immediate control. To ensure that the structure is thoroughly treated, termite professionals need an in-depth understanding of the construction type, methods, and architectural materials and of the building features. Any untreated or poorly treated area or gap can be used by termites to invade and infest.

Efficacy of insecticides depends on toxicity, mode of action, susceptibility of termites to the test compound, soil properties, formulation, and application methods

(Osbrink et al. 2001). These authors further observed that fipronil was relatively slow acting on soldiers but quickly acting on workers. This slow action allows the soldiers to interact more with workers before death. The soldiers can identify the intoxicated individuals and separate or avoid contact with them (Osbrink et al. 2001). Su et al. (1997a) suggested elucidation of termite penetration ability in insecticide-treated soil is essential to evaluate efficacy. If a sufficient amount of insecticide is not acquired by the termites due to repellency, eventually reduced mortalities may be observed, in spite of toxicity (Smith and Rust 1990). Manzoor et al. (2012) observed bifenthrin to be repellent and fipronil repelling termites only above 25 ppm, whereas chlorfenapyr was non-repellent.

11.2.2 Non-repellent

The use of slow-acting insecticides has been reported to be in practice in the first few decades of the twentieth century (Table 11.1). Randall and Doody (1934) reviewed the use of slow-acting arsenic dust in termite control, citing earlier works by Van Zwaluwenberg (1916) and Wolcott (1924) on colonies of the arboreal termite *Nasutitermes costalis* (Holmgren) that were killed by applying powdered arsenic in their runways. Sodium arsenate, DDT, trichlorobenzene, creosote, ethylene dibromide, and pentachlorophenol were used for subterranean termite control in the early twentieth century (USDA 1951). The objectives of using slow-acting non-repellent toxicants for termite control are to impact colony populations, either by suppression or elimination.

In the 1980s imidacloprid (a neonicotinoid) was first studied for management and registered for termite treatment in Japan in 1993 (Potter 1997). With the introduction of new non-repellent termiticides in the 1990s, the application technologies changed drastically. Termites are unable to differentiate between treated and untreated soil when non-repellent insecticides are used. The unique properties of the newer molecules and formulations changed the termite management practices dramatically. The use of neonicotinoids rather than organophosphates is considered the choice for developing a new strategy to manage termites (Rust and Saran 2008; Smith et al. 2008; Ahmed et al. 2014; Ahmed and Saba 2014) as organophosphates are inherently more toxic than neonicotinoids for higher animals. Non-repellent insecticides often maintain the property of non-repellency even at high concentrations (up to 500 ppm in *Reticulitermes hesperus* Banks) (Saran and Rust 2007). Thorne and Breisch (2001) observed that non-repellent insecticide allows termites to enter treated soil, to be killed before the pest is able to cause any damage.

Fipronil (a phenylpyrazole) was first investigated for termites in France in the late 1970s and was registered as a termiticide by BASF in 1999. This termiticide has a relatively low vapor pressure (3.7×10^{-4} mm Hg) and water solubility (1.9–2.4 mg/l at 25 °C) but a soil adsorption coefficient value (K_{oc}) as high as 825 (Gunasekara et al. 2007). Chlorfenapyr (a pyrrole) interferes with an insect's ability to produce energy by disrupting proton shuttles across the mitochondrial inner membrane

Table 11.1 History of termite control

Landmark achievements	References
The first report of the presence of parasitic "head-inhabiting" nematodes in *Reticulitermes lucifugus* (Rossi)	Merrill and Ford (1916)
Slow-acting toxicants such as arsenic dust applied into foraging tubes in an attempt to impact on colony populations	Van Zwaluwenberg (1916) and Wolcott (1924)
Nematodes could not kill *Coptotermes formosanus* Shiraki under soil conditions	Pemberton (1928)
Soil termiticides widely used for control of subterranean termites since the early 1900s	Randall and Doody (1934)
Runway of *Coptotermes formosanus* may extend up to 50 m in length and 0.3–3.0 m in depth	Erhorn (1934)
The presence of the fungus *Conidiobolus* sp. on *Nasutitermes* sp.	Kevorkian (1937)
Two bacterial species, *Bacterium* sp. and *Serratia marcescens* Bizio, killing laboratory colonies of *Zootermopsis angusticollis* Hagen	DeBach and McOmie (1939)
The presence of the fungus *Conidiobolus* sp. on *Coptotermes* sp.	Altson (1947)
Juvenile hormone-regulated soldier formation in termites	LuÈscher (1958)
Serratia marcescens could kill termites with "low vigor." This report marked the debut of termite biological control research	Toumanoff and Toumanoff (1959)
Aspergillus flavus Link as a fungal pathogen of *Reticulitermes* sp.	Beal and Kais (1962)
Serratia sp. and *Aspergillus* sp. used for termite control	Lund (1962)
First field study using *Serratia marcescens* against *Reticulitermes flavipes (Kollar)*	Lund (1965)
Susceptibility of *Reticulitermes flavipes* to a formulation of *Bacillus thuringiensis Berliner*	Smythe and Coppel (1965)
Metarhizium anisopliae (Metsch.) Sorokin and *Beauveria bassiana* (Balsamo) Vuill. as the two most virulent entomopathogenic microorganisms against *Reticulitermes flavipes*	Toumanoff and Rombau (1965)
Isaria sp. (syn. *Paecilomyces* sp.) shown to be pathogenic to *Reticulitermes flavipes*	Smythe and Coppel (1966)
Entomophthora virulenta Hall & Dunn, in association with *B. thuringiensis*, used to control *Coptotermes formosanus* in Hawaii	Page (1966)
The first patented formulations of *Aspergillus flavus* and *Serratia marcescens* against termites	Lund (1966)
Patent of a combination of *Entomophthora virulenta* Hall et Dunn and *Bacillus thuringiensis* as biological control agents against termites	Page (1967)
Proposal to the US Navy to investigate the effect of various pathogens against *Coptotermes formosanus*, including nematodes (*Steinernema* spp.) and fungi (*Metarhizium anisopliae* and *Beauveria bassiana*)	Tamashiro (1968)
Dechlorane (Mirex®), a slow-acting toxicant, proposed to eliminate isolated populations of *Reticulitermes flavipes* in Canada	Esenther and Gray (1968)
Development of biopesticides to control *Anacanthotermes ahngerianus* Jacobs	Stadykov (1970)

(continued)

Table 11.1 (continued)

Landmark achievements	References
Conclusion that field studies with various pathogens demonstrated sufficient pathogenicity to termites	Lund (1971)
Evaluation of various species of fungi against *Reticulitermes lucifugus* Rossi under laboratory conditions	Krejzová (1971)
Studies using Mirex™ bait blocks indicated that the continuous placement of toxic baits suppressed foraging activity of *Reticulitermes* spp.	Beard (1974) and Esenther and Beal (1974)
Field trials using juvenile hormone analogs such as methoprene and hydroprene against *Prorhinotermes simplex* (Hagen)	Lenz (1976)
The use of *Bacillus thuringiensis*, in an effort to develop a biopesticide against two termite species	Khan et al. (1977)
The first use of the mark-recapture method to estimate the cryptic foraging populations of subterranean termites. Importance of defining colony foraging territory using entomopathogenic fungi against *Coptotermes formosanus*	Lai (1977)
Ecdysis inhibition by diflubenzuron (Dimilin®) against *Heterotermes indicola* (Wasmann) and *Reticulitermes flavipes* demonstrated	Doppelreiter and Korioth (1981)
Termites exhibited behavioral mechanisms that reduced the possibility of epizootics	Kramm et al. (1982)
Organophosphates and pyrethroids used in termiticide barrier treatments	Mix (1988)
Surface treatment of borate solution, even at the label rate, did not protect wood beneath the treated surface from field populations of *Coptotermes formosanus*	Grace and Yamamoto (1994)
Using the trap + treat + release approach with sulfluramid, colony populations of *Reticulitermes flavipes* were significantly suppressed in city blocks of Toronto (Canada)	Myles (1996)
The first fipronil product in termite control in the USA (Termidor®) approved by the EPA in 1999	PANNA (2009)

(Silver and Soderlund 2005). Indoxacarb (an oxadiazine proinsecticide) perturbs voltage-gated Na^+ channels in the insect nervous system by binding receptors at a site different from that affected by pyrethroids (Wing et al. 2000; Nauen and Bretschneider 2001). Indoxacarb was initially registered with the Environmental Protection Agency (EPA) in the year 2000 as a “reduced-risk” insecticide for use on vegetable and other crops against Lepidopteran and sucking insect pests (McCann et al. 2001; Tillman et al. 2002). Chlorantraniliprole (anthranilic diamide) targets the ryanodine receptor and causes impaired muscle regulation, paralysis, and eventual death of insects (Cordova et al. 2006). The intoxicated termites become uncoordinated or convulsive, which interferes with the normal colony activities such as foraging, grooming, feeding, and trophallaxis (Glenn and Gold 2003). Chlorantraniliprole binds to soil with an average K_{oc} value of 328, which means it has less potential for leaching (McCall et al. 1979). Ecotoxicological profile of chlorantraniliprole based on its persistence in different soil types makes it an invaluable termiticide (Spomer and Kamble 2011; Wagner et al. 2011; Shelton et al. 2014). Shelton et al. (2014) found that in USDA Forest Service field plot trials, chlorantraniliprole when applied at 0.05% provided 8 years of protection from termites.

11.3 Soil Properties

Soil treatment has been used for termite control since 1920 (Su and Scheffrahn 1998; Peterson 2010). Su et al. (1982) found that soil properties, viz., texture, pH, moisture, temperature, particle size, organic matter content and microorganism diversity and content, and compactness, are important factors that determine the application rate, persistence, and movement of a termiticide in soil. Migration of termiticides is slower in soils with high clay and organic matter, compared to sandy soils. Groundwater contamination by termiticides is restricted in high-clay-content soils. Forschler and Townsend (1996) and Spomer et al. (2009) observed that a higher amount of termiticide is required in soils to attain desired result in general perimeter treatment in soils high in organic matter content. The pH affects the biodiversity of soil flora and their activities leading to degradation of applied termiticides. Gold et al. (1996) reported that in slightly acidic soils, low organic matter content, coupled with low soil temperature and moisture, helps in persistence of termiticides for a relatively longer period.

Non-repellent compounds are preferred for soil treatment because they do not seem to disrupt termite foraging in the treated soil zone and have a delayed mode of action that may contribute to movement of the active ingredient in the colony, through trophallaxis and social grooming (Kard 2003). Today, non-repellent liquid insecticides such as fipronil, imidacloprid, chlorfenapyr, and indoxacarb are gaining popularity, which have a delayed mode of action. The foraging termites fail to detect these insecticides and continue to forage on the treated soils. The termite foragers exposed to these insecticides are not killed immediately; rather they transfer the toxicants in lethal amounts to unexposed nestmates before death. In due course of time, the entire colony is seriously affected (Quarcoo et al. 2010). However, such effects are variable depending on physicochemical properties of soils, inherent toxicity of the insecticides, and termite behavior.

Temperature and other natural conditions in the field affect the uptake and transfer of toxicants. At higher temperatures more intense tunneling, foraging, and feeding activity have been noticed by Spomer et al. (2008), leading to higher uptakes of termiticides. Higher temperatures promote microbial degradation of termiticides leading to shorter residual effects in tropical climates (Reid et al. 2002). Osbrink and Lax (2002) found fipronil-treated sand had significantly greater and faster mortality of the Formosan subterranean termite workers, from susceptible or unsusceptible colonies, than treated soil or clay. This intense substrate effect also was observed by other researchers with other insecticides (Smith and Rust 1990; Forschler and Townsend 1996; Gold et al. 1996; Osbrink and Lax 2002). High-clay-content soils have high colloidal fraction which increases surface area, thus increasing the number of chemical-binding sites that promote hydrogen bonding and hydrophobic binding of hydrophobic insecticides (Saltzman and Yaron 1986). Sand particles retain more of the toxicant on the surface than other substrates (Harris 1972) and may increase pesticide performance. Manzoor and Pervez (2014) found that biflex and fipronil were effective against *Heterotermes indicola* (Wassman) in sandy loam soil compared to sandy clay loam soil. They observed concentration and time both are inversely related to each other, relative to efficacy and biflex was more bioavailable than fipronil. Several workers have reported the use of organic matter to prevent termite damage to crops (Mando et al. 1999; Mando and Stroosnijder 1999; Gould et al. 2001; Bokhtiar and Sakurai 2005).

Saran et al. (2014) reported *Reticulitermes flavipes* (Kollar) workers exposed to sand and soils treated with chlorantraniliprole at 50 ppm exhibited delayed mortality (it took >5 days to observe 90–100% mortality in termite workers). They also observed that exposure to chlorantraniliprole-treated sand (50 ppm) for as little as 1 min stopped feeding and killed 90–100% of workers. Tunneling (~2 h) was observed in different soil types treated with chlorantraniliprole at 50 ppm, even those with high organic matter (6.3%) and clay content by (30%) caused immediate feeding cessation in worker termites and mortality in the next 7–14 days. Worker termites exposed for 1 and 60 min to sand treated with chlorantraniliprole (50 ppm) were able to walk normally for 4 h after exposure. Delayed toxicity, increased aggregation, and grooming were observed in exposed termites leading to horizontal transfer effects within colonies. Yeoh and Lee (2007) demonstrated non-repellency of chlorantraniliprole in tunneling assays. Ramakrishnan et al. (2000) suggested that soil type affected termite worker mortality even after 7 days of continuous exposure to imidacloprid-treated sand or soil.

11.4 Baiting

Research on baiting technology developed during mid-twentieth century in Indonesia (Kalshoven 1955), Canada (Esenther and Gray 1968), the USA (Esenther and Beal 1974, 1978), and Australia (Paton and Miller 1980; Lenz and Evans 2002) especially for protection of timber trees, where barrier treatments were unsuccessful. Before the development of the baiting technology, subterranean termite management was performed by application of slow-acting toxicants such as arsenic dust, directly into the foraging galleries (Randall and Doody 1934). French (1991) reported the use of bait blocks containing dechlorane directly into the active galleries of subterranean termites in Australia. He further reported that termite control professionals would drill a hole in the trunk of the infested tree and connect a conduit box bait container. Dechlorane was the most commonly used bait during these years, but the use of organochlorines was then discontinued by national regulations or as the result of the Stockholm Convention on Persistent Organic Pollutants in 2004.

Baits are an effective method of subterranean termite management, since only a small amount of insecticide is required (Su 1994) and, ideally, is in contact with a relatively small proportion of the foraging population, who then proceeds to transfer the toxicant to other colony members. Baits are typically made of organochlorine/organophosphorus/carbamate insecticides (Rust 1986). Baiting is mainly successful against the lower termite family Rhinotermitidae and is not effective against higher termites (Ngee et al. 2004). Baiting is a long-term approach to termite management because the impact is not visible immediately. The time taken by bait to show desired effect depends upon the type of insecticide used, inherent toxicity of the insecticide, the attractant used, the season when baiting is done, atmospheric temperature, and relative humidity. In this system it is expected that the foraging termites would carry the insecticide and pass it on to unexposed workers in lethal doses. The desired results are hence noticed some days or weeks after treatment. The most important aspect of a baiting program is the proper monitoring and maintenance of the bait stations till the goal is achieved. Baiting research took an impetus with the availability of newer molecules by the end of 1980s. Now the research focuses on the development of baiting active ingredients (Prestwich et al. 1983; Jones 1984; Su et al. 1987) and food matrices (Su et al. 1985; Su and Scheffrahn 1986) that would attract more termites toward the bait stations, eventually leading to commercialization of these technologies (Su et al. 1995, 2001).

In baiting systems, routine monitoring is essential to achieve the desired success. Monitoring devices are placed in the affected areas and left for the termites to visit. Once the presence of termites is established, then the untreated baits are replaced with those containing a small amount of insecticide such as hexaflumuron or noviflumuron. A successful bait trial is assumed when activity is reduced or eliminated at all locations identified as being visited by the targeted termites (Su and Scheffrahn 1996; Thorne and Forschler 2000). Some termite species that are sensitive to disturbance start avoiding the bait stations (Swoboda et al. 2004). Su (2007) evaluated cellulose baits containing 0.5% hexaflumuron in a hermetically sealed closed-cell polyethylene sheet envelope and placed in soil to test their durability and efficacy against field colonies of *Coptotermes formosanus* Shiraki and *R. flavipes*. He suggested that the sealed baits may be placed in soil for months or years without the need of monitoring, would save labor costs by bypassing the monitoring phase, circumvent the station avoidance by some termite species, and enable the use of baiting technologies in large areas such as agricultural fields in which the manual monitoring is impractical.

Ripa et al. (2007) evaluated four different termite control strategies consisting of two soil treatments with cypermethrin and fipronil and two bait treatments with hexaflumuron and sulfluramid for their efficacy and potential for controlling *R. flavipes*, in Quillota and Valparaiso, Chile. They reported that soils treated with fipronil and cypermethrin prevented termite access in 75% of homes. Sekamatte et al. (2003) used baits made of dead animals, meat bones, and sugarcane husks to poison *Macrotermes* mounds. They observed that ants were more attracted to protein-based baits resulting in more number of ant nests near maize plants leading to decrease in damage by termites and increase in maize yields. They suggested use of protein-based baits for integrated management of termites in maize in Uganda.

Su (1994) and Su et al. (1995) showed that colonies of subterranean termites can be eliminated by using hexaflumuron bait matrix. However the same delivery system cannot be used for managing arboreal termites (Su et al. 1989). For managing *Mastotermes* in Australian tropics, higher quantities of termiticides are required (Lenz 2002). Mound excavation is essential for control of *Coptotermes* spp. along the perimeter of baited homes in Malaysia, Thailand, and Singapore (Lee et al. 2007). Su and Scheffrahn (1990) screened and identified several toxicants as candidates to be incorporated into subterranean termite baits under laboratory conditions. Henderson and Forschler (1997) reported that fipronil could be effectively used as baits against the Formosan subterranean termite, at levels 100 times lower than hexaflumuron.

Baiting technology seems to have achieved its aims, i.e., small quantity of toxicant could suppress or eliminate termite colonies. However there was wide variation on the length of time taken to achieve the desired result, i.e., 24–80 weeks approximately (Pawson and Gold 1996; Haagsma and Rust 2005; Austin et al. 2008). Many workers have carried out studies on different bait technologies under laboratory and field conditions with highly encouraging results. However, the matter of concern is that majority of studies do not have untreated controls. These studies only include observations on termite activity either in monitoring or bait stations (Grace et al. 1996; Su et al. 1997b; Tsunoda et al. 1998; Su et al. 2001). The major advantage of the baiting system approach is the capability of reducing populations of subterranean termites, with the possibility of suppressing or eliminating colonies (Lax and Osbrink 2003). Ahmed Shiday and French (2013) showed flufenoxuron as a potential termite bait toxicant, particularly against *Coptotermes* species. There are some drawbacks in baiting systems, viz., initial treatment costs are higher, and it is a highly labor-intensive method and is based mainly on the treatment location area, with regular monitoring and maintenance hassles. However, the cost becomes negligible when weighed against the environmental safety, viz., no soil or ground pollution is reported because very little amount of toxicant is used.

11.5 Capture, Treat, and Release

"Trap and treat" is a method wherein termites are first lured into a trap using a food as bait and then treated with a poison. Dusting with Paris Green was the first documented example (early twentieth century) of trap and treat in tropical Asia, Australia, Hawaii, and California (Froggatt 1905; Fullaway 1920; Keuchenius et al. 1922; Jepson 1930; Kofoid et al. 1934; Cleghorn 1861; Hickin 1971; Roonwal 1979; Watson 1988).

Over the last two decades, many workers investigated transfer of termiticides among nestmates (Thorne and Breisch 2001; Valles and Woodson 2002; Ibrahim et al. 2003). Tomalski and Vargo (2004) proposed a theory stating that pesticides adhering to the integument of exposed termites (donors) are transferred to unexposed nestmates (recipients) through interaction with the donors. Haagsma and

Rust (2007) observed that in termites with sealed mouthparts, transfer of imidacloprid was through body contact, not trophallaxis. However, laboratory studies indicated that toxicant transfer occurs among termites (Hu et al. 2005; Rust and Saran 2006, 2008; Shelton et al. 2006; Song and Hu 2006; Saran and Rust 2007; Bagneres et al. 2009). Transfer of termiticide is a function of simple association among termites, i.e., common occurrence that happens is crowding during termite assays (Peterson et al. 2004). Had it been a simple correlation, then the increase in the number of donors would certainly increase the number of receivers leading to quicker spread of the toxicants. Valles et al. (2000) found a twofold difference in LC_{50} between a pair of colonies of *R. flavipes* to permethrin.

The transfer of fipronil and imidacloprid among workers as well as between workers and soldiers has been studied (Thorne and Breisch 2001; Ibrahim et al. 2003; Saran and Rust 2007). Saran and Rust (2007) observed that body contact including grooming plays a major role in horizontal transmission of a lethal dose of termiticide compared to transmission by trophallaxis. A linear relationship was found between dose uptake and insecticide contact time, in subterranean termites. Ibrahim et al. (2003) observed the transfer of termiticides from soldiers to workers was significantly higher than from workers to soldiers. A study on distance of horizontal transfer in the field showed that the lethal effects in Formosan subterranean termites may be limited (Su 2005).

Rust and Saran (2006) and Saran and Rust (2007) worked with *R. hesperus* and argued that a single donor carries very little amount of termiticide and only the termites directly interacting with it are killed. The receiving individuals in turn do not act as donors. The intoxicated termites are unable to travel long distances limiting the potential of spread of termiticide (Su 2005; Ripa et al. 2007; Quarcoo et al. 2010). Myles (1996) indicated that termites coated in sulfluramid were groomed by other nestmates, passing the toxicant through the colony by trophallaxis. None of the traditional repellent termiticides have been shown to transfer by contact among termites in the laboratory (Shelton et al. 2005). Schoknecht et al. (1994) used microencapsulated permethrin as a bait toxicant, which was transferred among nestmates by trophallaxis. Similarly, Iwata et al. (1989) argued that transmission of microencapsulated fenitrothion among *C. formosanus* individuals was accomplished by grooming. As on date only delayed-action, non-repellent termiticides are known to transfer among termites by bodily contact; however, the speed of lethality varies (Mao et al. 2011). To develop an effective bait technology, various workers have evaluated different ratios and proportions of different termiticides (Hu et al. 2005; Song and Hu 2006; Tsunoda 2006; Rust and Saran 2008; Bagneres et al. 2009).

11.6 Exclusion Barriers

During the 1940s the use of chemical soil barriers was in vogue and considered to be safer and more persistent, replacing the practice of dusting chemicals for termite control. Initially organochlorines (e.g., dieldrin and chlordane) dominated the termite control scenario for more than a decade, followed by organophosphates (e.g., chlorpyrifos) and synthetic pyrethroids (e.g., deltamethrin and bifenthrin). Chemical soil barriers were the most commonly used method of termite management for the next five decades. The decline in the use of chemical soil barriers started in the early 1990s due to environmental concerns (Carson 1962; Cropper et al. 1992).

The application of chemical pesticides against termites is generally aimed at creating a barrier to prevent access to plants. Effective chemicals are those with a degree of persistence or are able to penetrate into the soil profile to provide control. Therefore, insoluble compounds, or those that are not readily adsorbed onto clay particles, would be of no use unless they were thoroughly mixed into the soil. Soil termiticides have been widely used for control of subterranean termites since the early 1900s (Randall and Doody 1934). The major objective of barrier treatments is to exclude soil-inhabiting subterranean termites from structures in ground contact. Termiticides applied to establish soil barriers can be repellent, toxic, or both (Forschler 1994). A wide range of insecticides have been investigated, and their effects varied with environmental factors or soil conditions, which affect the residual activity and longevity of biocides (Smith and Rust 1993; Forschler and Townsend 1996).

Chlorpyrifos is one of the most common insecticides used worldwide against termites since late 1980s, till date (Mix 1988). In 1996, Bayer Corporation (Kansas City, MO) introduced a new termiticide, imidacloprid (Premise® 75), that belongs to chloronicotinyls (Potter 1997). Boucias et al. (1996) reported that imidacloprid-treated termites become sluggish, inhibited or reduced grooming and tunneling activity, eventually followed by death. Premise 75® when applied at the rate of 0.1% in concrete-slab tests in Arizona, Florida, and South Carolina provided 100% control for 5 years (Kard 1998). Kard et al. (1989) observed that 100% prevention of subterranean termite attack resulted when Dursban® TC was applied at the rate of 1% solution in tests carried out in Mississippi for 21 years by the US Department of Agriculture (USDA) Forest Service, based on concrete slabs. Su et al. (1999b) suggested that longevity of termiticides applied underneath and around structures could be underestimated in the USDA Forest Service trials because Dursban TC degraded faster in small plots such as those of the USDA Forest Service field trials compared with that of larger plots (40,000 cm^2). Beal and Carter (1968) observed that 1 day after applying heptachlor (0.47 liter/0.093 m^2) to Florida soils, 95% of active ingredient (a.i.) penetrated down to a depth of 1.9 cm. In another study, Davis and Kamble (1992) found that subslab termiticide concentrations decreased depending on horizontal and vertical distance from the termiticide injection point. Depth of termiticide penetration in soil is a function of soil moisture at application time and presence of rocks, duff, and organic matter (Beal and Carter 1968).

Traditionally, an amount of 0.5–5 kg of a.i. is required to exclude soilborne subterranean termites from a structure (NPCA 1985). A continuous horizontal barrier is created if the a.i. is applied correctly, which prevents entry of termites. However, the success rate depends on various factors, because in a termite colony, the gallery system may extend up to 50–100 m from an infested structure (King and Spink 1969; Su and Scheffrahn 1988; Grace et al. 1989) and soil treatments rarely impact the entire colony population, despite the large quantity of insecticide applied (Su and

Scheffrahn 1998). Soil insecticide treatment is, however, widely used for prevention of structural infestations (Grace et al. 1993; Gahlhoff and Koehler 2001). For several years, non-repellent soil insecticides, e.g., pyrethroids, have become popular alternatives to the use of more repellent materials, as barriers, to termite penetration. Higher mortalities are observed due to lack of repellence and delayed mode of action, the termites moving freely within the treated soil before death (Kard 2001).

Hu et al. (2006) demonstrated that soil-barrier application of fipronil can suppress and eventually eliminate termite colonies. They showed that workers readily tunneled and repeatedly moved in and out of the treated soil barriers, depending on the treatment concentrations. They observed intoxicated behaviors of termites, including erratic walking, body shaking, fluid excretion from the anus and mouth, impaired mobility, and lying on the back, while twitching and shivering. There was no avoiding of dead termites by the live ones, as shown by the increasing number of dead termites at the nest site and the decreasing number of termites in the foraging and treatment areas. Remmen and Su (2005) found that thiamethoxam and fipronil at $\geq$8 ppm and $\geq$1 ppm, respectively, provided an effective barrier against *C. formosanus* and *R. flavipes*. "Exterior-Only" and "Exterior Perimeter plus Localized Interior Treatment" (EP/LIT) strategy was proposed by Potter and Hillery (2002, 2003). EP/LIT is a two-phase strategy: (1) a full volume treatment of the soil outside thc foundation wall, to establish a continuous barrier in soil on the structure's exterior, and (2) targeted applications to all known infested areas inside the structure by foaming, injection, or dust application. This strategy was more economic reducing labor, amount of insecticide, and intrusion into the structures.

Increasing public awareness concerns raised on the environmental fate of the insecticides applied led to development of physical barriers such as stainless steel mesh (Lenz and Runko 1994) and uniform-sized particle barriers (Tamashiro et al. 1987). Su et al. (2004) found that sufficient quantities of λ-cyhalothrin were released from the impregnated polyethylene film into adjacent sand, to prevent termite penetration. The impregnated film has less environmental impact than conventional liquid termiticides, because the a.i. is kept in the polymer.

11.7 Termite Management in Cropped Areas

References to plant protection are found in ancient Indian literature, the Vedas (viz., Rigveda ~3700 BC, Atharvaveda ~2000 BC), Kautilya's Arthashastra (~300 BC), Buddhist literature (~200 BC), Krishi Parashara (~100 BC), Sangam literature of Tamils (200 BC-100 AD), Agni Purana (~400 AD), Brihat Samhita of Varahamihira (~600 AD), Kashyapiyakrisukti (~800–900 AD), Surapala's Vrikshayurveda (~1000 AD, Someshwara Deva's Manasollasa (~ 1100A D), Lokopakara) by Chavundaraya (~1108 AD), Sarangadhara's Upavana Vinoda (~1300 AD), Vishwavallabha of Chakrapani Mishra (~1577 AD), and some documents of the medieval and premodern period. A detailed methodology of seed treatment and systematic strategies of plant protection and grain storage were presented by Surapala

(~1000 AD) in Vrikshayurveda. Therefore, this period may be considered as the starting point of systematic plant protection in Indian agricultural history. Surapala suggested watering the trees with cold water for a week to get rid of insects from branches and roots, smearing the roots with a mixture of white mustard, vaca (*Zingiber zerumbet* Rosc. Ex Smith.), kushta (*Saussurea lappa* C. B. Clarke), and ativisa (*Aconitum heterophyllum* Wall ex Royle). For termite control he suggested application of extract of *aak, Calotropis procera* (Aiton) (8–10 Kg of *aak* is soaked in water for about 24 h and filtered and applied in soil infested with termites). Details of termite control can be found in Agni Purana, an ancient Indian scripture, wherein the remedies mentioned to get rid of termites and fruit cracking consider the application of a paste (containing 200 g of turmeric powder and 1000 ml of mustard oil) on the tree trunks. The mustard oil would attract ants, and the turmeric powder acts as an antibiotic for healing the cracks. Suggestion includes the use of *aak* leaf-filled gunny bags in irrigation channels to kill termites. In a document of the early nineteenth century from the Mewar region of Rajasthan, the use of sesame oil for soil and foliar application to trees to protect from frost and termites is mentioned. In olden days Indian farmers used to keep asafoetida (an oleoresin dry gum obtained from *Ferula* spp.) in a pack of cotton cloth at two or three points in 10–15 m irrigation channels, for controlling termites in affected crops.

The use of chemical insecticides is roughly 100 years old, and their soil application for termite control in crops was developed relatively late. The earliest reference to soil-applied termiticides (called at the time "soil poisons" or, less commonly, "chemical insulation") was a 1928 test in California for protection of utility poles (St. George 1952). Although adopted rapidly for use in structures, soil applications were thought to be less desirable than the good building practices (which minimize susceptibility) and wood treatments (which protect the wood directly). In the early 1950s, chlorinated hydrocarbons, viz., aldrin, dieldrin, chlordane, and heptachlor, were used as soil termiticides and continued to dominate termite control scenario till the mid 1980s, when they were withdrawn from market due to public outcry for their long persistence in soil and other health hazards. In 1978 the US Environmental Protection Agency (EPA) canceled the use of chlordane on food crops and phased out other aboveground uses over the following 5 years. In 1988, all previously approved uses of chlordane in the USA were canceled by the EPA (http://www.epa.gov). With the advent of newer insecticides, the application methodology changed. Subsequently several organophosphates (isofenphos and chlorpyrifos) and pyrethroids (cypermethrin, permethrin, bifenthrin, and fenvalerate) were marketed for termite control in the USA.

In India, Roonwal and Chhotani (1961 and 1967) dealt with the Indian wood destroying termites, as described in the monograph by Sen-Sarma et al. (1975) on the Indian wood destroying termites. Other authors such as Beeson (1941), Harris (1961, 1971), Hickin (1971), Narayanan and Rattan (1952), Roonwal (1979), and Kapur and Bose (1972) provided some accounts on termites that are injurious to agriculture crops and their control. Thakur et al. (1956, 1957) suggested the use of aldrin and dieldrin for termite control in field crops. Gupta (1959) controlled termites with 5% BHC dust at 20 kg/ha, when applied in furrows in sugarcane.

Agarwala (1955) found that treatments of sugarcane setts with insecticides, viz., benzene hexachloride, aldrin, chlordane, dieldrin, etc., as dust or sprinkle were effective in termite control. Parihar (1985) reported that castor seed dressing with aldrin 30 E.C. at 10 ml/kg protected the crop against termites. He opined that pre-sowing soil application of 5% aldrin dust at 37.5 kg/ha also gave good results and the commonly used 10% HCH the least effective. Scheffrahn et al. (1997) suggested chemical toxicity, formulation, application method, drywood termite behavior, and gallery system architecture influenced the performance of local chemical treatments.

Singh and Singh (2001) reported sugarcane sett treatment with a 0.2% solution of imidacloprid 70 WS and soil treatment with phorate 10 G 2.50 kg a.i./ha, chlorpyrifos 15 G 2.50 kg a.i./ha, and chlorpyrifos 20 EC at 1 kg a.i./ha were the most performing in controlling the termite infestations. Rana et al. (2001) observed that plots where wheat seeds were treated with chlorpyrifos and endosulfan at 0.9 and 2.4 g a.i./kg respectively were least infested by termites. Delgarde and Rouland (2002) found the effective dose of thiamethoxam for *Trinervitermes trinervius* Rambur, *Odontotermes smeathmani* Fuller, and *Amitermes evuncifer* Silvestri to be 0.3 ppm, which resulted in 100% mortality within 2–8 days, depending on the species. They further observed that *O. smeathmani* consumed the product, and thiamethoxam could be transmitted in the colony from contaminated individuals to healthy individuals. Thiamethoxam acted as an antifeedant for *T. trinervius* and *A. evuncifer* and not as repellent. Sekamatte et al. (2003) observed that soybean and groundnut were more effective in suppressing termite attack than common beans in maize-legume cropping system and suggested an integrated management strategy for termites in smallholder cropping systems in East Africa.

Santos et al. (2004) observed the phenomenon of social facilitation among termites when poisoned with endosulfan and chlorpyrifos. They found that group size significantly affected the median time for death. This observation has important practical implications because density of individuals (population density) within the colony has significant role to play in time required to eliminate the colony and the doses are based on colony size and not on the population density reached inside the nest.

The sand termite, *Psammotermes hypostoma* Desneux, prefers to infest the places with a high moisture content and warm temperature (Hafez 1980; Moharram et al. 1992). Ahmed et al. (2015) found chlorpyrifos (48 EC) as the most potent, acetamiprid (20 SP) most toxic, and thiamethoxam (40 WG) most powerful insecticide for management of *P. hypostoma.* They further observed that the reduction percentages on palm fronds damage had a linear relationship with increase in exposure period (from 15–60 days).

Rust and Saran (2008) found acetamiprid very active against *R. hesperus* in topical applications. Moreover, they demonstrated that termites were quickly affected by short exposures to sand treated with acetamiprid (1 ppm) as within 1 h their locomotion was impaired. They also observed that acetamiprid was transferred from donors to recipients only when donors were held on deposits ≥50 ppm for 1 h. Deposits even as low as 1 ppm were repellent with termites failing to tunnel into the treated sand, without any significant mortality.

Ahmed et al. (2007) suggested the use of thiamethoxam and imidacloprid either as sett treatment or soil application in connection with irrigation could be a good alternative to chlorpyrifos and bifenthrin, in sugarcane crops in Pakistan, where insecticides are the usual approach to manage termites in this crop (Sattar and Salihah 2001; Ahmed et al. 2007). Singh and Singh (2002) suggested application of chlorpyrifos, imidacloprid, and fipronil as sett treatment in furrows before the first irrigation.

Iqbal and Saeed (2013) evaluated toxicities of insecticides against *Microtermes mycophagus* Desneux collected from four locations (tree plantation, untreated building, treated building, agriculture area) of Multan, Pakistan. They found that the population collected from agricultural area was more tolerant to all insecticides compared to those of other three locations. They ranked the insecticide order of average toxicity as follows: chlorfenapyr > spinosad > thiamethoxam > fipronil > indoxacarb > imidacloprid. Bhagawati et al. (2014) reported under field conditions sugarcane setts treated with clothianidin 50 WDG at concentration of 1 ml/liter registered the lowest infestation of termites and showed statistical parity with the combined application of acephate 50% + imidacloprid 1.8% at same concentration.

Gao et al. (1985) reported successful control of termite infestations with dechlorane bait in fields. Actually, workers all around the globe are actively involved in developing new integrated termite management program which would be effective and sustainable.

11.8 Factors Deciding Management Methodology

It is very difficult to decide the best method for termite management because every situation is different. The major factors that determine the methodology are as follows:

(a) Whether termite management is to be done in field, wooden structures, or buildings.
(b) Termite species in question, whether mound forming or non-mound forming or subterranean or arboreal.
(c) Soil characteristics, viz., clay content, organic matter content, pH, moisture content, bulk density, cation exchange capacity, etc.
(d) The presence of previous crop residues, current crop present in the field, and cropping system followed by the farmer.
(e) Type of insecticide/termiticide available and availability of particular formulations in the local market.
(f) Proper knowledge of cultural practices and pest control measures being followed.
(g) Cost of protection, a very important factor. In less developed countries, small and marginal farmers dominate the farming community. The newer termiticides are expensive so they prefer to use cheaper organophosphate-based products.

(h) Whether a new structure is being constructed or termite management is to be done for an existing structure/building.
(i) Soil type and structure, viz., clay, loamy, sandy loam, or sandy soils.
(j) Depth of foundation and perimeter area.
(k) Availability of proper barrier materials, especially for physical barriers.
(l) Availability of termite management professionals, bait stations, and monitoring professionals.

11.9 Conclusion

Management of termites varies widely, depending on the situation and cost incurred in the procedure. Today numerous termiticides are available in the market, including organophosphates and pyrethroids, which are more economic to use and expensive non-repellent termiticides. Termites are highly organized and "intelligent" social insects; they reproduce at an alarming rate and are able to replenish any shortage in number of workers or soldiers quickly. Termite management science has progressed tremendously in the last two decades. Detailed studies related to their behavior and biology need to be carried out to achieve success in management.

Termite workers have generated voluminous amount of literature which has helped in developing new management technologies. Biological control is still in its infancy. However, some natural enemies have been reported, but literature indicating their successful use is limited. A number of plant products have been reported to have deleterious effect on termites, but economically and viable application technologies need to be developed. A thorough perusal of the literature shows that the use of synthetic chemical insecticides is the preferred way to manage termites. We believe, however, that an integrated approach represents the right path to follow. Insects have survived for more than 1.5 million years, and it is their adaptability that makes them stand against humans more firmly than any other species. Application of bait technology and liquid termiticides is used in different situations. These technologies have their own merits and demerits. We have to weigh the benefits against the demerits and decide which strategy to follow. No matter whatever strategy we may adopt, synthetic chemical insecticides still remain the backbone of termite management, and more experimental studies need to be carried out, to work out strategies to avoid their damage in a more pragmatic way.

References

Agarwala, S. B. D. (1955). Control of termites (1946–1953). *Journal of Economic Entomology, 48*, 533–537.

Ahmed, M. A. I., & Saba, R. M. (2014). Comparative toxicological evaluation between different formulations of two selected neonicotinoid pesticides on *Culex pipiens* (Diptera: Culicidae) mosquito. *Advances in Environmental Biology, 8*, 1169–1174.

Ahmed Shiday, B. M., & French, J. R. J. (2013). The efficacy of flufenoxuron as potential bait toxicant against termites. *African Journal of Agricultural Research, 8*, 3688–3694.

Ahmed, S., Riaz, M. A., & Hussain, A. (2007). Assessment of the damaged and population of termites (*Odontotermes* & *Unicolor*) under various methods of insecticide application. *International Journal of Agriculture and Biology, 9*, 125–128.

Ahmed, M. A. I., Eraky, S. A., Fakeer, M., & Soliman, A. S. (2014). Toxicity assessment of selected neonicotinoid pesticides against the sand termite, *Psammotermes hypostoma* Desneux workers (Isoptera: Rhinotermitidae) under laboratory conditions. *Australian Journal of Basic and Applied Sciences, 8*, 238–240.

Ahmed, M. A. I., Eraky, S. A., Mahmoud, F. M., & Soliman, A. S. (2015). Potential toxicity assessment of novel selected pesticides against sand termite, *Psammotermes hypostoma* Desneux workers (Isoptera: Rhinotermitidae) under field conditions in Egypt. *Journal of Plant Protection Research, 55*, 193–197.

Altson, R. A. (1947). A fungus parasitic on *Coptotermes curvignathus*, Holmgr. *Nature, 160*, 120.

Anonymous. (2002). *Pest Control, 70*, S1–S23.

Austin, J. W., Glenn, G. J., & Gold, R. E. (2008). Protecting urban infrastructure from Formosan termite (Isoptera: Rhinotermitidae) attack: A case study for United States Railroads. *Sociobiology, 51*, 231–247.

Bagnères, A. G., Pichon, A., Hope, J., Davis, R., & Clément, J. L. (2009). Contact versus feeding intoxication by fipronil in *Reticulitermes* termites (Isoptera: Rhinotermitidae): Laboratory evaluation of toxicity, uptake, clearance, and transfer among individuals. *Journal of Economic Entomology, 102*, 347–356.

Beal, R. H., & Carter, F. L. (1968). Initial soil penetration by insecticide emulsions used for subterranean termite control. *Journal of Economic Entomology, 61*, 380–383.

Beal, R. H., & Kais, V. (1962). Apparent infection of subterranean termites by *Aspergillus flavus*. *Insect Pathology, 4*, 488–489.

Beard, R. L. (1974). Termite biology and bait-block method of control. *Connecticut* Agr Esp Sta Bull 748.

Beeson, C. F. C. (1941). *The ecology and control of forest insects in India and the neighbouring countries*. Government of India, ii+ 1008 pp. Dehra Dun (Vasant Press).

Bhagawati, S., Bhattacharyya, B., Mishra, H., & Gogoi, D. (2014). Chemical management of termites (*Odontotermes obesus*) in preserved setts of sugarcane (*Saccharum officinarum*). *Journal of Entomology and Zoology Studies, 5*, 856–859.

Bokhtiar, S. M., & Sakurai, K. (2005). Integrated use of organic manure and chemical fertilizer on growth, yield and quality of sugarcane in high Ganges river floodplain soils of Bangladesh. *Communications in Soil Science and Plant Analysis, 36*, 1823–1837.

Boucias, D. G., Stokes, C., Storey, G., & Penland, J. C. (1996). The effects of imidacloprid on the termite *Reticulitermes flavipes* and its interactions with the mycopathogen *Beauveria bassiana*. *Pflanzenschutz-Nachr Bayer, 49*, 103–144.

Carson, R. (1962). *Silent Spring*. Boston: Houghton Mifflin.

Cleghorn, H. F. C. (1861). *The forests and gardens of South India*. London: WH Allen Co.

Cordova, D., Benner, E. A., Sacher, M. D., Rauh, J. J., Sopa, J. S., Lahm, G. P., Selby, T. P., Stevenson, T. M., Flexner, L., Gutteridge, S., Rhoades, D. F., Wu, L., Smith, R. M., & Tao, Y. (2006). Anthranilic diamides: A new class of insecticides with a novel mode of action, ryanodine receptor activation. *Pesticide Biochemistry and Physiology, 84*, 196–214.

Cropper, M. L., Evans, W. N., Berardi, S. J., Ducla-Soares, M. M., & Portney, P. R. (1992). The determinants of pesticide regulation: A statistical analysis of EPA decision making. *Journal of Political Economy, 100*, 175–197.

Davis, R. W., & Kamble, S. T. (1992). Distribution of subslab injected Dursban TC (chlorpyrifos) in a loamy sand soil when used for subterranean termite control. *Bulletin of Environmental Contamination and Toxicology, 48*, 585–591.

DeBach, P. H., & McOmie, W. A. (1939). New diseases of termites caused by bacteria. *Annals of the Entomological Society of America, 32*, 137–146.

Delgarde, S., & Rouland, L. C. (2002). Evaluation of the effects of thiamethoxam on three species of African termite (Isoptera: Termitidae) crop pests. *Journal of Economic Entomology, 95*, 531–536.

Doppelreiter, V. H., & Korioth, M. (1981). Entwicklungshemmung durch Diflubenzuron bei den Bodentermiten *Heterotermes indicola* und *Reticulitermes flavipes*. *Zeitschrift für Angewandte Entomologie, 91*, 131–137.

Edwards, R., & Mill, A. E. (1986). *Termites in buildings: Their biology and control*. East Grinstead: Rentokil Ltd.

Erhorn, E. M. (1934). The termites of Hawaii, their significance and control, and the distribution of termites by commerce. In C. A. Kofoid (Ed.), *Termites and termite control* (pp. 321–333). Berkeley: University of California Press.

Esenther, G. R., & Beal, R. H. (1974). Attractant-mirex bait suppresses activity of *Reticuliternes*. *Journal of Economic Entomology, 67*, 85–88.

Esenther, G. R., & Beal, R. H. (1978). Insecticidal baits on field plot perimeters suppress *Reticulitermes*. *Journal of Economic Entomology, 71*, 604–607.

Esenther, G. R., & Gray, D. E. (1968). Subterranean termite studies in southern Onrurio. *Canadian Entomology, 100*, 827–834.

Forschler, B. T. (1994). Survivorship and tunneling activity of *Reticulitermes flavipes* (Kollar) (Isoptera: Rhinotermitidae) in response to termiticide soil barriers with and without gaps of untreated soil. *Journal of Entomological Science, 29*, 43–54.

Forschler, B. T., & Townsend, M. L. (1996). Mortality of eastern subterranean termites (Isoptera: Rhinotermitidae) exposed to four soils treated with termiticides. *Journal of Economic Entomology, 89*, 678–681.

French, J. R. J. (1991). Baiting techniques for control of *Coptotermes* species within existing buildings in Australia. In M. I. Haverty, & W.W. Wilmx (Eds.), *Proceedings of symposium of current research on wood-destroying organisms* (pp. 46–50). USDA forest service' general technical report PSW-128, Berkeley.

Froggatt, W. W. (1905). Economic entomology white ants (Termitidae) with suggestions for dealing with them in houses and orchards. *The Agricultural Gazette of New South Wales, 16*(632–656), 753–774.

Fullaway, D. T. (1920). Termites, or white ants, in Hawaii. *Hawaiian Agriculture, 17*, 294–301.

Gahlhoff, J., & Koehler, P. (1999). To kill or not to kill? *Pest Control Technology, 27*, 22–28.

Gahlhoff, J. E., & Koehler, P. G. (2001). Penetration of the eastern subterranean termite into soil treated at various thicknesses and concentrations of Dursban TC and Premise 75. *Journal of Economic Entomology, 94*, 486–491.

Gao, D., Zhu, B., Gan, H. S., & Yuan, S. (1985). A new toxic bait for the control of forest-infesting termites. *Journal of Nanjing Institute, 3*, 128–131.

Glenn, G. J., & Gold, R. E. (2003). Evaluation of commercial termiticides and baiting systems for pest management of the Formosan subterranean termite, *Coptotermes formosanus* (Isoptera: Rhinotermitidae). *Sociobiology, 41*, 193–196.

Gold, R. E., Howell, H. N., Jr., Pawson, B. M., Wright, M. S., & Lutz, J. C. (1996). Persistence and bioavailability of termiticide to subterranean termites (Isoptera: Rhinotermitidae) from five soil types and locations in Texas. *Sociobiology, 28*, 337–363.

Gould, K. A., Herrick, J., & Lemma, H. (2001). Refuse to refuge: Dry season use and modification of cattle dung by subterranean termites in Guanacaste, Costa Rica. *Biotropica, 33*, 121–130.

Grace, J. K., & Yamamoto, R. T. (1994). Simulation of remedial borate treatments intended to reduce attack on Douglas-fir lumber by the Formosan subterranean termite (Isoptera: Rhinotermitidae). *Journal of Economic Entomology, 87*, 1547–1554.

Grace, J. K., Abdallay, A., & Farr, K. R. (1989). Eastern subterranean termite (Isoptera: Rhinotermitidae) foraging territories and populations in Toronto. *Canadian Entomologist, 121*, 551–556.

Grace, J. K., Yates, J. R., Tamashiro, M., & Yamamoto, R. T. (1993). Persistence of organochlorine insecticides for Formosan subterranean termite (Isoptera: Rhinotermitidae) control in Hawaii. *Journal of Economic Entomology, 86*, 761–765.

Grace, J. K., Tome, C. H. M., Shelton, T. G., Oshiro, R. J., & Yates, J. R. (1996). Baiting studies and considerations with *Coptotermes formosanus* (Isoptera: Rhinotermitidae) in Hawaii. *Sociobiology, 28*, 511–520.

Gunasekara, A. S., Truong, T., Goh, K. S., Spurlock, F., & Tjeerdema, R. S. (2007). Environmental fate and toxicology of fipronil. *Journal of Pesticide Science, 32*, 189–199.

Gupta, B. D. (1959). Studies on the control of termites in sugarcane. *Proceedings of International Society of Sugar Cane Technologists, 10*, 961–977.

Haagsma, K. A., & Rust, M. K. (2005). Effect of hexaflumuron on mortality of the western subterranean termite (Isoptera: Rhinotermitidae) during and following exposure and movement of hexaflumuron in termite groups. *Pest Management Science, 61*, 517–531.

Haagsma, K. A., & Rust, M. K. (2007). The effect of imidacloprid on mortality, activity, and horizontal transfer in the western subterranean termite (Isoptera: Rhinotermitidae). *Sociobiology, 50*, 1127–1148.

Hafez, M. (1980). Highlights of the termite problem in Egypt. *Sociology, 5*, 147–156.

Harris, W. V. (1961). *Termites, their recognition and control* (p. 187). London: Longmans Green and Co. Ltd.

Harris, W. V. (1971). *Termites: Their recognition and control* (2nd ed.p. 186). London: Longmans.

Harris, C. (1972). Factors influencing the effectiveness of soil insecticides. *Annual Review of Entomology, 17*, 177–198.

Haynes, K. F. (1988). Sublethal effects of neurotoxic insecticides on insect behavior. *Annual Review of Entomology, 33*, 149–168.

Henderson, G., & Forschler, B. T. (1997). Termite bait tests. *Louisiana Agriculture, 40*, 9–11.

Hickin, N. E. (1971). *Termites: A world problem* (1st ed.). London: Hutchinson and Co. Ltd.

Hu, X. P. (2005). Evaluation of efficacy and nonrepellency of indoxacarb and fipronil treated soil at various concentrations and thicknesses against two subterranean termites (Isoptera: Rhinotermitidae). *Journal of Economic Entomology, 98*, 509–517.

Hu, X. P., & Hickman, B. (2006). Exterior perimeter plus limited interior treatments with fipronil as an IPM option for subterranean termite management. *International Pest Control, 48*, 200–203.

Hu, X. P., Song, D., & Scheler, C. (2005). Transfer of indoxacarb among workers of *Coptotermes formosanus* Shiraki (Isoptera: Rhinotermitidae): Effects of dose, donor/recipient ratio, and post-exposure time. *Pest Management Science, 61*, 1209–1214.

Hu, X. P., Song, D., & Presley, W. (2006). Horizontal and vertical transfer of fipronil within functional colonies. In T. Sutphin, B. Cartwright, & R. Houseman (Eds.), *Proceedings of 2006 national conference on urban entomology, 21–24 May 2006* (pp. 39–44). North Carolina: Raleigh.

Ibrahim, S. A., Henderson, G., & Fei, H. (2003). Toxicity, repellency, and horizontal transmission of fipronil in the Formosan subterranean termite (Isoptera: Rhinotermitidae). *Journal of Economic Entomology, 96*, 461–467.

Iqbal, N., & Saeed, S. (2013). Toxicity of six new chemical insecticides against the termite, *Microtermes mycophagus* D. (Isoptera: Termitidae: Macrotermitinae). *Pakistan Journal of Zoology, 45*, 709–713.

Iwata, R., Ito, T., & Shinjo, G. (1989). Efficacy of fenitrothion microcapsule against termites, *Coptotermes formosanus* Shiraki (Isoptera: Rhinotermitidae). II Transmissibility of fenitrothion through grooming. *Applied Entomology and Zoology (Japanese), 24*, 213–221.
Jepson, F. P. (1930). The control of tea termites. *Tropical Agriculture, 75*, 191–195.
Jones, S. C. (1984). Evaluation of two insect growth regulators for the bait-block method of subterranean termite (Isoptera: Rhinotermitidae) control. *Journal of Economic Entomology, 77*, 1086–1091.
Kalshoven, L. G. E. (1955). The ability of *Coptotermes* to locate exposed timber. *Idea, 10*, 43–49.
Kapur, A. P., & Bose, G. (1972). Termite problems in tropical ecology in India. In *Termite problems in India* (pp. 31–37). New Delhi: Biological Research Committee.
Kard, B. M. (1998). Premise termiticide field test results. *Pest Control, 66*, 64–65,105.
Kard, B. (2001). Gulfport studies stay the course. *Pest Control, 69*, 30–33.
Kard, B. M. (2003). Integrated pest management of subterranean termites (Isoptera). *Journal of Entomological Science, 38*, 200–224.
Kard, B. M., Mauldin, J. K., & Jones, S. C. (1989). Evaluation of soil termiticides for control of subterranean termites (Isoptera). *Sociobiology, 15*, 285–297.
Keuchenius, P. E., Corporaal, J. B., & vijanden van, D. (1922). *Hevea brasiliensis* (animal enemies of *Hevea brasiliensis*) In N. L. Swart, A. A. L. Rutgers, & J. H. de Bussy (Eds), *Handboek voor de rubber cultuur in Nederlandsch-Indie (Handbook for rubber culture in Netherlands-India)* (pp. 216–229).
Kevorkian, A. G. (1937). Studies in the entomophthoraceae. I. Observations on the genus *Conidiobolus*. *Journal of Agriculture of the University of Puerto Rico, 21*, 191–200.
Khan, K. I., Fazal, Q., & Jafri, R. H. (1977). Pathogenicity of locally discovered *Bacillus thuringiensis* strain to the termites: *Heterotermes indicola* (Wassman) and *Microcerotermes championi* (Snyder). *Pakistan Journal of Scientific Research, 29*, 12–13.
King, E. G., & Spink, W. T. (1969). Foraging galleries of the Formosan termite, *Coptotermes formosanus,* in Louisiana. *Annals of the Entomological Society of America, 62*, 537–542.
Kofoid, C. A., Light, S. F., Horner, A. C., Randall, M., Herms, W. B., & Bowe, E. E. (1934). *Termites and termite control*. Berkeley: University of California Press.
Kramm, K. R., West, D. F., & Rockenbach, P. G. (1982). Termite pathogens: Transfer of the entomopathogen *Metarhizium anisopliae* between *Reticulitermes* sp. termites. *Journal of Invertebrate Pathology, 40*, 1–6.
Krejzová, R. (1971). Infektionsversuche mit einigen Entomophthora-Arten bei den termiten Coptotermes formosanus (Shiraki) und Reticulitermes lucifugus (Rossi). *Entomophaga, 16*, 221–231. (in German).
Lai, P. Y. (1977). Biology and ecology of the Formosan subterranean termite, *Coptotermes formosanus*, and its susceptibility to the entomogenous fungi, *Beauveria bassiana* and *Metarrhizium anisopliae*. PhD dissertation, University of Hawaii, Honolulu
Lax, A. R., & Osbrink, W. L. A. (2003). United States Department of Agriculture—Agriculture Research Service research on targeted management of the Formosan subterranean termite *Coptotermes formosanus* Shiraki (Isoptera: Rhinotermitidae). *Pest Management Science, 59*, 788–800.
Lee, C. Y., Vongkaluang, C., & Lenz, M. (2007). Challenges to subterranean termite management of multi-genera faunas in Southeast Asia and Australia. *Sociobiology, 50*, 213–221.
Lenz, M. (1976). The dependence of hormone effects in termite caste determination on external factors. In M. LuÈscher (Ed.), *Phase and caste determination in insects. Endocrine aspects* (pp. 73–89). Oxford: Pergamon Press.
Lenz, M. (2002). Termite problem species and management of termite problems in Australia. *Sociobiology, 40*, 11–12.
Lenz, M., & Evans, T. A. (2002). Termite bait technology: Perspectives from Australia. In S. C. Jones, J. Zhai, W. H. Robertson, & S. C. Charleston (Eds.), *Proceedings of the 4th international conference on urban pests* (pp. 27–36).
Lenz, M., & Runko, S. (1994). Protection of buildings, other structures and materials in ground contact from attack by subterranean termites (Isoptera) with a physical barrier – A fine mesh of high grade stainless steel. *Sociobiology, 24*, 1–16.

Logan, J. W. M., Cowie, R. H., & Wood, T. G. (1990). Termite (Isoptera) control in agriculture and forestry by non-chemical methods: A review. *Bulletin of Entomological Research, 80*, 309–330.
LuÈscher, M. (1958). Ueber die Entstehung der Soldaten bei Termiten. *Revue Suisse de Zoologie, 65*, 372–376.
Lund, A. E. (1962). Subterraneans and their environment – New concepts of termite ecology. *Pest Control, 30*, 60–61.
Lund, A. E. (1965). Piggyback bacteria can kill termite colony. *Pest Control, 33*, 22–24.
Lund, A. E. (1966). US Patents 3249,492; 3249,439; 3249,494; 3249,500; 3249,501
Lund, A. E. (1971). Microbial control of termites. In H. D. Burges & N. W. Hussey (Eds.), *Microbial control of insects and mites* (pp. 385–386). London: Academic.
Mando, A., & Stroosnijder, L. (1999). The biological and physical role of mulch in the system wide program on integrated pest management. *IPM Research Briefs, 2*, 11–26.
Mando, A., Brussaard, L., & Stroosnijder, L. (1999). Termites and mulch mediated rehabilitation of vegetation on crusted soil in West Africa. *Restoration Ecology, 7*, 33–41.
Manzoor, F., & Pervez, M. (2014). Efficacy of Biflex and Fipronil in controlling subterranean termites (Isoptera: Rhinotermitidae) in different soils types in Pakistan. In B. T. Forschler (Ed.), *10th Pacific-termite research group conference held at Kuala Lumpur, Malaysia*, February 26–28, 2014.
Manzoor, F., Sayyed, A. H., Rafique, T., & Malik, S. A. (2012). Toxicity and repellency of different insecticides against *Heterotermes indicola* (Isoptera: Rhinotermitidae). *The Journal of Animal and Plant Sciences, 22*, 65–71.
Mao, L., Henderson, G., & Scherer, C. W. (2011). Toxicity of seven termiticides on the Formosan and eastern subterranean termites. *Journal of Economic Entomology, 104*, 1002–1008.
McCall, P. J., Swann, R. L., Laskowski, D. A., Vrona, S. A., Unger, S. M., & Dishburger, H. J. (1979). Prediction of chemical mobility in soil from sorption coefficients. In *Proceedings of aquatic toxic hazard* (pp. 49–58). Philadelphia: ASTM.
McCann, S. F., Annis, G. D., Shapiro, R., Piotrowski, D. W., Lahm, G. P., Long, J. K., Lee, K. C., Hughes, M. M., Myers, B. J., & Griswold, S. M. (2001). The discovery of indoxacarb: Oxadiazines as a new class of pyrazoline-type insecticides. *Pest Management Science, 57*, 53–164.
Merrill, J. H., & Ford, A. L. (1916). Life history and habits of two new nematodes parasitic on insects. *Journal of Agricultural Research, 6*, 115–127.
Mix, J. (1988). King of the hill. *Pest Control, 56*, 34–35.
Moharram, A. M., Bagy, M. M. K., & Abdel-Galil, F. A. (1992). Fungi associated with the sand termite *Psammotermes hypostoma* in Assiut, Egypt. *Mycologia, 84*, 930–935.
Myles, T. G. (1996). Development and evaluation of a transmissible coating for control of subterranean termites. *Sociobiology, 28*, 373–457.
Narayanan, E. S., & Rattan, L. (1952). A short review of recorded information on the control of termites damaging crops in India, along with results of recent control experiments at Delhi. *Indian Journal of Entomology, 14*, 21–50.
Nauen, R., & Bretschneider, T. (2001). New modes of action of insecticides. *Pesticide Outlook, 13*, 241–245.
Ngee, P. S., Yoshimura, T., & Lee, C. Y. (2004). Foraging populations and control strategies of subterranean termites in the urban environment, with special reference to baiting. *Japanese Journal of Environmental Entomology and Zoology, 15*, 197–215.
NPCA (National Pest Control Association). (1985). *Approved reference procedures for subterranean termite control*. Dunn Loring: NPCA Inc.
Osbrink, W. L. A., & Lax, A. R. (2002). Effect of tolerance to insecticides on substrate penetration by Formosan subterranean termites (Isoptera: Rhinotermitidae). *Journal of Economic Entomology, 95*, 989–1000.
Osbrink, W. L. A., Lax, A. R., & Brenner, R. J. (2001). Insecticide susceptibility in *Coptotermes formosanus* and *Reticulitermes virginicus* (Isoptera: Rhinotermitidae). *Journal of Economic Entomology, 94*, 1217–1228.

Page, R. Z. (1966). Affidavit submitted to the US Navy, Hawaii (p. 46)

Page, R. Z. (1967). Termite control by induced epizootics of entomophagous microorganisms. US patent #3337,395.

PANNA. (2009). PAN pesticides database – Pesticide products. Pesticide action network, North America. http://www.pesticideinfo.org/Detail_Product.jsp?REG_NR=00796900210&DIST_NR=007969

Parihar, D. R. (1985). Crop termite damage in Indian desert and its control in castor. *Zeitschrift furAngewandte Zoologie, 72*, 309–315.

Paton, R., & Miller, L. R. (1980). Control of *Mastotermes darwiniensis* Froggatt (Isoptera: Mastrotermitidae) with mirex baits. *Australian Forestry Research, 10*, 249–258.

Pawson, B. M., & Gold, R. E. (1996). Evaluation of baits for termites (Isoptera: Rhinotermitidae) in Texas. *Sociobiology, 28*, 485–510.

Pemberton, C. E. (1928). Nematodes associated with termites in Hawaii, Borneo and Celebes. *Proceedings of the Hawaiian Entomological Society, 7*, 148–150.

Peterson, C. (2010). Considerations of soil-applied insecticides for termite control. *Outlooks on Pest Management, 21*, 89–93.

Peterson, C. J., Gerard, P. D., & Ems-Wilson, J. (2004). Recommendations for treated-area choice assays with termites (Isoptera). *Sociobiology, 44*, 171–185.

Potter, M. F. (1997). Termites. In D. Moreland (Ed.), *Handbook of pest control* (pp. 233–332). Cleveland: Mallis Handbook and Technical Training Co.

Potter, M. F., & Hillery, A. E. (2002). Exterior-targeted liquid termiticides: an alternative approach to managing subterranean termites (Isoptera: Rhinotermitidae) in buildings. *Sociobiology, 39*, 373–405.

Potter, M. F., & Hillery, A. E. (2003). Trench warfare. *Pest Control Technology, 31*, 28–32.

Prestwich, G. D., Mauldin, J. K., Engstrom, J. B., Carvalho, J. F., & Cupo, D. Y. (1983). Comparative toxicity of fluorinated lipids and their evaluation as bait-block toxicants for the control of *Reticulitermes* spp. (Isoptera: Rhinotermitidae). *Journal of Economic Entomology, 76*, 690–695.

Quarcoo, F. Y., Appel, A. G., & XP, H. (2010). Effects of indoxacarb concentration and exposure time on onset of abnormal behaviors, morbundity, and death in eastern subterranean termite (Isoptera: Rhinotermitidae). *Journal of Economic Entomology, 103*, 762–769.

Ramakrishnan, R., Suiter, D. R., Nakatsu, C. H., & Bennett, G. W. (2000). Feeding inhibition and mortality in *Reticulitermes flavipes* (Isoptera: Rhinotermitidae) after exposure to imidacloprid-treated soils. *Journal of Economic Entomology, 93*, 422–428.

Rana, J. S., Ombir, & Dahiya, K. K. (2001). Management of termite, *Microtermes obesi* (Holmgren) in wheat, *Triticum aestivum* through seed treatment. *Annales Biologiques, 17*, 207–209.

Randall, M., & Doody, T. C. (1934). Poison dusts. I. Treatments with poisonous dusts. In C. A. Kofoid (Ed.), *Termites and termite control* (pp. 463–476). Berkeley: University of California Press.

Rath, A. C. (2000). The use of entomopathogenic fungi for control of termites. *Biocontrol Science and Technology, 10*, 563–581.

Reid, B. L., Brinkmann, R., Smith, G., Ishizaka, K., Palis, B., & DeVilliers, V. (2002). Imidacloprid use in termite control operations globally and changing use patterns in the United States. In S. C. Jones, J. Zhai, & W. H. Robertson (Eds.), *Proceedings of the 4th international conference on urban pests, 7–10 July 2002, Charleston, South Carolina* (pp. 355–368). Blacksburg: Pocahontas Press.

Remmen, L. N., & Su, N. Y. (2005). Tunneling and mortality of eastern and Formosan subterranean termites (Isoptera: Rhinotermitidae) in sand treated with thiamethoxam or fipronil. *Journal of Economic Entomology, 98*, 906–910.

Ripa, R., Paola Luppichini, S. N. Y., & Michael, K. R. (2007). Field evaluation of potential control strategies against the invasive eastern subterranean termite (Isoptera: Rhinotermitidae) in Chile. *Journal of Economic Entomology, 100*, 1391–1399.

Roonwal, M. L. (1979). *Termite life and termite control in tropical South Asia* (p. 177). Jodhpur: Scientific Publishers.

Roonwal, M. L., & Chhotani, O. B. (1961). *Termite fauna of Assam region, Eastern India* (p. 406). Monograph, Zoological Survey of India.

Roonwal, M. L., & Chhotani, O. B. (1967). Indian wood-destroying termites. *Journal of the Bombay Natural History Society, 63*, 354–364.

Rust, M. K. (1986). Managing household pests. In G. W. Bennett & J. M. Owens (Eds.), *Advances in urban pest management* (pp. 335–368). New York: Van Nostrand Reihald Co.

Rust, M. K., & Saran, R. K. (2006). Toxicity, repellency, and transfer of chlorfenapyr against western subterranean termites (Isoptera: Rhinotermitidae). *Journal of Economic Entomology, 99*, 864–872.

Rust, M. K., & Saran, R. K. (2008). Toxicity, repellency, and effects of acetamiprid on western subterranean termite (Isoptera: Rhinotermitidae). *Journal of Economic Entomology, 101*, 1360–1366.

Saltzman, S., & Yaron, B. (1986). *Pesticides in soil*. New York: Van Nostrand Reinhold Company.

Santos, C. A., DeSouza, O., & Guedes, R. N. C. (2004). Social facilitation attenuating insecticide-driven stress in termites (Isoptera: Nasutitermitinae). *Sociobiology, 44*, 1–7.

Saran, R. K., & Rust, M. K. (2007). Toxicity, uptake, and transfer efficiency of fipronil in western subterranean termite (Isoptera: Rhinotermitidae). *Journal of Economic Entomology, 100*, 495–508.

Saran, R. K., Ziegler, M., Kudlie, S., Harrison, D., Leva, D. M., Scherer, C., & Mark, A. C. (2014). Behavioral effects and tunneling responses of eastern subterranean termites (Isoptera: Rhinotermitidae) exposed to chlorantraniliprole-treated soils. *Journal of Economic Entomology, 107*, 1878–1889.

Sattar, A., & Salihah, Z. (2001, September 24–26). Detection and control of subterranean termites. In *Technologies for sustainable agriculture. Proceedings of national workshop* (pp. 195–198). NIAB, Faisalabad.

Scheffrahn, R. H., Su, N. Y., & Busey, A. (1997). Laboratory and field evaluations of selected chemical treatments for control of drywood termites (Isoptera: Kalotermitidae). *Journal of Economic Entomology, 90*, 492–502.

Schoknecht, U., Rudolph, D., & Hertel, H. (1994). Termite control with microencapsulated permethrin. *Pesticide Science, 40*, 49–55.

Sekamatte, M. B., Ogenga-Latigo, M., & Russell-Smith, A. (2003). Effects of maize-legume intercrops on termite damage to maize, activity of predatory ants and maize yields in Uganda. *Crop Protection, 22*, 87–93.

Sen-Sarma, P. K., Thakur, M. L., Mishra, S. C., & Gupta, B. K. (1975). *Wood destroying termites in India* (p. 187). Dehradun: Forest Research Institute.

Shelton, T. G., Bell, C. D., & Wagner, T. L. (2005). Lack of transfer of permethrin among nestmates of *Reticulitermes flavipes* in laboratory trials (Isoptera: Rhinotermitidae). *Sociobiology, 45*, 69–75.

Shelton, T. G., Mulrooney, J. E., & Wagner, T. L. (2006). Transfer of chlorfenapyr among workers of *Reticulitermes flavipes* (Isoptera: Rhinotermitidae) in the laboratory. *Journal of Economic Entomology, 99*, 886–892.

Shelton, T. G., Fye, D., & Ulyshen, M. (2014). U.S. Department of Agriculture-Forest Service Termiticide Report for 2013. *Pest Management Professional, 82*, 42–52.

Silver, K. S., & Soderlund, D. M. (2005). Action of pyrazoline-type insecticides at neuronal target sites. *Pesticide Biochemistry and Physiology, 81*, 136–143.

Singh, M., & Singh, N. B. (2001). Application of insecticide for termite control and its effect on yield contributing characters in sugarcane. *Sugar Tech, 3*, 146–153.

Singh, M., & Singh, N. B. (2002). Bud damage due to termites in sugarcane. *Cooperative Sugar, 33*, 655–658.

Smith, J. L., & Rust, M. K. (1990). Tunneling response and mortality of the western subterranean termite (Isoptera: Rhinotermitidae) to soil treated with termiticides. *Journal of Economic Entomology, 83*, 1395–1401.

Smith, J. L., & Rust, M. K. (1993). Cellulose and clay in sand affects termiticide treatments. *Journal of Economic Entomology, 86*, 53–60.

Smith, J. A., Pereira, R. M., & Koehler, P. G. (2008). Relative repellency and lethality of the neonicotinoids thiamethoxam and acetamiprid and an acetamiprid/bifenthrin combination to *Reticulitermes flavipes* termites. *Journal of Economic Entomology, 101*, 1881–1887.

Smythe, R. V., & Coppel, H. C. (1965). The susceptibility of *Reticulitermes flavipes* (Kollar) and other termite species to an experimental preparation of *Bacillus thuringiensis* Berliner. *Journal of Invertebrate Pathology, 7*, 423–426.

Smythe, R. V., & Coppel, H. C. (1966). Pathogenicity of externally occurring fungi to *Reticulitermes flavipes*. *Journal of Invertebrate Pathology, 8*, 266–267.

Song, D., & Hu, X. P. (2006). Effects of dose, donor–recipient interaction time and ratio on fipronil transmission among the Formosan subterranean termite nestmates (Isoptera: Rhinotermitidae). *Sociobiology, 48*, 237–246.

Spomer, N. A., & Kamble, S. T. (2011). Temporal changes in chlorantraniliprole and indoxacarb in four midwestern soils and bioefficacy against the eastern subterranean termite (Isoptera: Rhinotermitidae). *Journal of Economic Entomology, 104*, 990–1001.

Spomer, N. A., Kamble, S. T., Warriner, R. A., & Davis, R. W. (2008). Influence of temperature on rate of uptake and subsequent horizontal transfer of [14C] fipronil by eastern subterranean termites (Isoptera: Rhinotermitidae). *Journal of Economic Entomology, 101*, 902–908.

Spomer, N. A., Kamble, S. T., & Siegfried, B. D. (2009). Bioavailability of chlorantraniliprole and indoxacarb to eastern subterranean termites (Isoptera: Rhinotermitidae) in various soils. *Journal of Economic Entomology, 102*, 1922–1927.

St. George, R. A. (1952). Tests with new insecticides for termite control. *Pest Control, 20*, 20.

Stadykov, S. I. (1970). The use of entomopathogenic bacteria for the control of termites. *Journal of Economic Entomology, 49*, 1144–1149.

Su, N. Y. (1994). Field evaluation of hexaflumuron bait for population suppression of subterranean termites (Isoptera: Rhinotermitidae). *Journal of Economic Entomology, 87*, 389–397.

Su, N. Y. (2005). Response of the Formosan subterranean termites (Isoptera: Rhinotermitidae) to baits or nonrepellent termiticides in extended foraging arenas. *Journal of Economic Entomology, 98*, 2143–2152.

Su, N. Y. (2007). Hermetically sealed baits for subterranean termites (Isoptera: Rhinotermitidae). *Journal of Economic Entomology, 100*, 475–482.

Su, N. Y., & Scheffrahn, R. H. (1986). A method to access, trap, and monitor field populations of the Formosan subterranean termite (Isoptera: Rhinotermitidae) in the urban environment. *Sociobiology, 12*, 299–304.

Su, N. Y., & Scheffrahn, R. H. (1988). Foraging population and territory of the Formosan subterranean termite (Isoptera: Rhinotermitidae) in an urban environment. *Sociobiology, 14*, 353–359.

Su, N. Y., & Scheffrahn, R. H. (1990). Comparison of eleven soil termiticides against the Formosan subterranean termite and eastern subterranean termite (Isoptera: Rhinotermitidae). *Journal of Economic Entomology, 83*, 1918–1924.

Su, N. Y., & Scheffrahn, R. H. (1996). A review of the evaluation criteria for bait-toxicant efficacy against field colonies of subterranean termites (Isoptera). *Sociobiology, 28*, 521–530.

Su, N. Y., & Scheffrahn, R. H. (1998). A review of subterranean termite control practices and prospects for integrated pest management programmes. *Integrated Pest Management Reviews, 3*, 1–13.

Su, N. Y., Tamashiro, M., Yates, J. R., & Haverty, H. I. (1982). Effect of behaviour on the evaluation of insecticides for prevention of or remedial control of the formosan subterranean termite. *Journal of Economic Entomology, 75*, 188–193.

Su, N. Y., Tamashiro, M., & Haverty, M. I. (1985). Effects of three insect growth regulators, feeding substrates, and colony origin on survival and presoldier production of the Formosan subterranean termite (Isoptera: Rhinotermitidae). *Journal of Economic Entomology, 78*, 1259–1263.

Su, N. Y., Tamashiro, M., & Haverty, M. I. (1987). Characterization of slow-acting insecticides for the remedial control of the Formosan subterranean termite (Isoptera: Rhinotermitidae). *Journal of Economic Entomology, 80*, 1–4.

Su, N. Y., Scheffrahn, R. H., & Ban, P. M. (1989). Method to monitor initiation of aerial infestations by alates of the Formosan subterranean termite (Isoptera: Rhinotermitidae) in high-rise buildings. *Journal of Economic Entomology, 82*, 1643–1645.

Su, N. Y., Thoms, E. M., Ban, P. M., & Scheffrahn, R. H. (1995). Monitoring/baiting station to detect and eliminate foraging populations of subterranean termites (Isoptera: Rhinotermitidae) near structures. *Journal of Economic Entomology, 88*, 932–936.

Su, N. Y., Ban, P. M., & Scheffrahn, R. H. (1997a). Remedial baiting with hexaflumuron in above-ground stations to control structure-infesting population of Formosan subterranean termite (Isoptera: Rhinotermitidae). *Journal of Economic Entomology, 90*, 809–817.

Su, N. Y., Chew, V., Wheeler, G., & Scheffrahn, R. H. (1997b). Comparison of tunneling responses into insecticide treated soil by field populations and laboratory groups of subterranean termites (Isoptera: Rhinotermitidae). *Journal of Economic Entomology, 90*, 503–509.

Su, N. Y., Ban, P. M., Chew, V., & Scheffrahn, R. H. (1999a). Size and edge effects of concrete plots on chlorpyrifos degradation in sub-slab sand. *Journal of Economic Entomology, 92*, 409–415.

Su, N. Y., Ban, P. M., & Scheffrahn, R. H. (1999b). Longevity and efficacy of pyrethroid and organophosphate termiticides in field degradation studies using miniature slabs. *Journal of Economic Entomology, 92*, 890–898.

Su, N. Y., Ban, P. M., & Scheffrahn, R. H. (2001). Control of subterranean termites (Isoptera: Rhinotermitidae) using commercial prototype above ground stations and hexaflumuron baits. *Sociobiology, 37*, 111–120.

Su, N. Y., Ban, P., & Scheffrahn, R. H. (2004). Polyethylene barrier impregnated with Lambda-cyhalothrin for exclusion of subterranean termites (Isoptera: Rhinotermitidae) from structures. *Journal of Economic Entomology, 97*, 570–574.

Swoboda, L. E., Miller, D. M., Fell, R. J., & Mullins, D. E. (2004). The effects of nutrient compounds (sugars andamino-acids) on bait consumption by *Reticulitermes* spp. (Isoptera: Rhinotermitidae). *Sociobiology, 44*, 547–563.

Tamashiro, M. (1968). *Susceptibility of termites to microbes* (p. 7). A research proposal submitted to the office of Naval Research, Department of entomology, University of Hawaii, Honolulu.

Tamashiro, M., Yates, J. R., & Ebesu, R. H. (1987). The Formosan subterranean termite in Hawaii: Problems and control. In M. Tamashiro & N. Y. Su (Eds.), *Biology and control of the Formosan subterranean termite* (pp. 15–22). Honolulu: College of Tropical Agriculture and Human Resources, University of Hawaii.

Thakur, C., Prasad, A. R., & Singh, R. P. (1956). The use of different doses and formulations of aldrin and dieldrin against termites and their effect on soil fertility. *Indian Journal of Sugarcane Research and Development, 1*, 50–55.

Thakur, C., Butani Dhamo, K., & Prasad, A. R. (1957). Pests of sugarcane. *Eekh Samachar, 4*, 8.

Thorne, B. L., & Breisch, N. L. (2001). Effects of sublethal exposure to imidacloprid on subsequent behavior of subterranean termite *Reticulitermes virginicus* (Isoptera: Rhinotermitidae). *Journal of Economic Entomology, 94*, 492–498.

Thorne, B. L., & Forschler, B. T. (2000). Criteria for assessing efficacy of stand-alone termite bait treatments at structures. *Sociobiology, 36*, 245–255.

Tillman, P. G., Hammes, G. G., Sacher, M., Connair, M., Brady, E. A., & Wing, K. D. (2002). Toxicity of a formulation of the insecticide indoxacarb to the tarnished plant bug, *Lygus lineolaris* (Hemiptera: Miridae) and the big-eyed bug, *Geocoris punctipes* (Hemiptera: Lygaeidae). *Pest Management Science, 58*, 92–100.

Tomalski, M. D., & Vargo, E. L. (2004). Chain reaction. *Pest Control, 72*, 51–53.

Toumanoff, C., & Rombau, J. (1965). Action de certains champignons entomophages, cultivés sur les milieux appropriés attractifs, sur le termite de Saintonge Reticulitermes santonensis (de Feytaud). *Annales Parasitologie Humaine et Comparee, 40*, 605–609. (in French).

Toumanoff, C., & Toumanoff, C. (1959). Les épizooties dues à Serratia marcescens Bizio chez un termite (*Reticulitermes santonensis* de Feytaud). *Comptes Rendus Hébdomadaires de l'Academie Agricole Française, 45*, 216–218. (in French).

Tsunoda, K. (2006). Transfer of fipronil, a nonrepellent termiticide, from exposed workers of *Coptotermes formosanus* (Isoptera: Rhinotermitidae) to unexposed workers. *Sociobiology, 47*, 563–575.

Tsunoda, K., Matsuoka, H., & Yoshimura, T. (1998). Colony elimination of *Reticulitermes speratus* (Isoptera: Rhinotermitidae) by bait application and the effect on foraging territory. *Journal of Economic Entomology, 91*, 1383–1386.

USDA (US Department of Agriculture). (1951). *Decay and termite damage in houses*. US Department of Agriculture Farmer's Bull 1993.

Valles, S. M., & Woodson, W. D. (2002). Group effects on insecticide toxicity in workers of Formosan subterranean termite, *Coptotermes formosanus* Shiraki. *Pest Management Science, 58*, 769–774.

Valles, S. M., Oi, F. M., Wagner, T., & Brenner, R. J. (2000). Toxicity and in vitro metabolism of t-permethrin in eastern subterranean termite (Isoptera: Rhinotermitidae). *Journal of Economic Entomology, 93*, 1259–1264.

Van Zwaluwenberg, R. H. (1916). *Report of entomologist*. Annuul Report Porto Rico Agr Esp Sra 1915: 4245.

Vargo, E. L., & Husseneder, C. (2009). Biology of subterranean termite: Insights from molecular studies of *Reticulitermes* and *Coptotermes*. *Annual Review of Entomology, 54*, 379–403.

Verma, M., Sharma, S., & Prasad, R. (2009). Biological alternative for termite control: A review. *International Biodeterioration and Biodegradation, 63*, 959–972.

Wagner, T., Peterson, C., & Shelton, T. (2011). Termiticide efficacy reports 2010. *Pest Management Professional, 2011*, 28–34.

Watson, J. A. L. (1988). *Termites in the Canberra region* (2nd ed.). Canberra: CSIRO Publishing.

Wing, K. D., Sacher, M., Kagaya, Y., Tsuruvuchi, Y., Mulderig, L., Connair, M., & Schnee, M. (2000). Bioactivation and mode of action of the oxadiazine indoxacarb in insects. *Crop Protection, 19*, 537–545.

Wolcott, G. N. (1924). The comparative resistance of woods to the attack of the termite *Cryprotermes brevis* Walker. *Porro Rico lnsulur Exp Stu BUN, 33*, 3–15.

Yeoh, B. H., & Lee, C. Y. (2007). Tunneling responses of the Asian subterranean termite, *Coptotermes gestroi* in termiticide-treated sand (Isoptera: Rhinotermitidae). *Sociobiology, 50*, 457–478.

Chapter 12
Termites and Standard Norms in Wood Protection: A Proposal Targeting Drywood Termites

Lara Maistrello

Contents

L. Maistrello (✉)
Dipartimento di Scienze della Vita, Centro Interdipartimentale BIOGEST-SITEIA,
Università di Modena e Reggio Emilia, Reggio Emilia, Italy
e-mail: lara.maistrello@unimore.it

M.A. Khan, W. Ahmad (eds.), *Termites and Sustainable Management*,
Sustainability in Plant and Crop Protection,
https://doi.org/10.1007/978-3-319-68726-1_12

Abstract A standard is a technical document approved by a recognized certification body at national or international level, which defines and unifies the characteristics and specifications of a process, product or service, to ensure quality and safe, reliable performances in respect to environment. The use of international standards allows to remove barriers to world trade, and their importance is particularly recognized in the field of wood technology, to guarantee that preservatives are effective in protecting wood from biotic degradation agents, such as termites. In the European Union, the USA, Australia and Japan, the existing standard norms to test efficacy against termites are exclusively related to subterranean species (Rhinotermitidae). Due to the great differences in biological features, these standards are not suitable for drywood termites (Kalotermitidae) that, on the other hand, are increasingly indicated as serious wood pests, worldwide. This chapter aims at filling this gap by outlining the differences in biology, ecology and behaviour of the two types of termites and their importance as invasive pests, describing the importance and features of standard norms and reviewing the available standards for wood protection against termites. Finally, a proposal for a standard protocol is presented, specifically developed to determine the efficacy of preventive wood treatments on drywood termites.

Keywords Standard norms • Termite management • Wood preservatives • Drywood termites • Subterranean termites

12.1 Introduction

Termites have an extraordinary ecological impact on earth, playing a key role in global carbon cycle, decomposition processes and nutrients recycling, as a result of their multilevel co-evolution with microorganisms that allow a superefficient utilization of lignocellulosic materials (Kudo 2009; Hongoh 2010). However, for this ability to feed on any type of plant-derived matter, some termite species are considered as pests, when their activity interferes with human interests. In particular, of the 3000 known termite species, about 2.8% are reported to damage structural lumbers, furniture and other wooden artefacts. The estimate of the global economic impact of these pests in 2010 was estimated at 40 billion USD (Rust and Su 2012).

Invasive termites, i.e. species introduced by man's intervention in territories other than the native ones, are responsible for major economic losses to houses and wooden/paper materials when they become established in urban areas. Their overall importance is rising, both in terms of number of successfully introduced species and of increasing distribution ranges (Evans et al. 2013). According to Evans et al. (2013), invasive termite species (Evans 2010) belong mainly to the families Kalotermitidae and Rhinotermitidae. Overall, Kalotermitids, known with the common name of "drywood" termites, account for at least 20% of the estimated worldwide annual control and repair costs for termite damage (Rust and Su 2012). The Kalotermitids can thrive in xeric habitats with no need of direct contact with soil moisture and can create colonies in small wooden objects that can be easily transported everywhere, acting as propagules for further timber colonization (Evans et al. 2011). These specific biological features, combined with increased urbanization and worldwide movement of wood/other cellulose-based products, will allow drywood termites to raise their invasive pest status and global economic importance (Lewis and Forschler 2014).

A standard is a technical document approved by a recognized certification body at national or international level, which defines and unifies the specifications of a process/product/service to ensure quality, safety and efficiency. International standards are instrumental in facilitating international trade. Considering the technology and products for wood protection, in Europe and in the USA, the existing standard norms to test effectiveness against termites are exclusively related to subterranean species and are completely inadequate for testing Kalotermitids. The increasing importance of drywood termites as serious wood pests and therefore the need to develop tailored control measures and interventions aimed at preventing their infestation imply the need of specific standard norms that take into account the great differences in bio-ethology of the two types of termites. This chapter outlines the available information necessary to develop a proposal for a standard protocol to determine the efficacy of preventive wood treatments on drywood termites.

12.2 Drywood Versus Subterranean Termites as Invasive Pests

Wood and other cellulose-based materials (paper, cardboard, etc.) are highly traded and transported commodities: if these materials are infested with termites, the infestations may easily go undetected because of the extremely cryptic lifestyle of these insects. Upon arrival at destination, if no effective measures are taken to kill the living organisms inside the commodities, this could lead to the propagation of the infestation to other cellulose-based materials and to the introduction of new species that could establish, becoming invasive. Increasing global trade is highly facilitating this process. However, not all termite-infested commodities can become propagules. At the same time, not all termite species are likely to become invasive species.

Among the features of invasive termite species that render them pests of growing economic importance, Evans (2010) indicated the use of sound heartwood as food, the habit to nest inside the wood and the high tendency to generate neotenic reproductives. These traits are shared by Kalotermitidae and Rhinotermitidae. However, important differences between these two families must be recognized in terms of habitat preferences, feeding patterns, behaviour and colony size/structure. Both families are typical lower termites, classified in the "Type 1" feeding group, based on gut morphology and content analysis (Donovan et al. 2001), because they feed on non-humified plant matter digested with the aid of both gut flagellates and bacteria. Both families share a hemimetabolous pathway, characterized by high developmental plasticity (Korb and Hartfelder 2008) and a typical colony that includes (a) imagoes (developed gradually via several nymphal instars) acting as founders and primary reproductives of the new colonies after departing the natal nests, performing the nuptial flight and shedding their wings; (b) dependent larvae that are young immatures (up to the II–III instar) dependent on brood care by nestmates; and (c) older immatures that can feed and care themselves – they are larval instars (without visible wing buds) or nymphal instars (with visible wing buds) and may undergo different types of moults (progressive, regressive, stationary), including the specific moult to neotenic or the one that leads to presoldier-soldier. These individuals, which represent the majority of the colony, and not necessarily work or assist nestmates (Korb and Hartfelder 2008), can be considered "workers" in the broad sense, as "any kind of non reproductive individual which feeds itself and can provide some assistance to other individuals" (Roisin and Korb 2011); (d) soldiers that are sterile individuals developed from immature instars by means of specific moulting (first as presoldier and then as soldicr), morphologically and behaviourally specialized to defend the colony, dependent on nestmates for food; and (e) neotenic reproductives that develop from older immature instars, either nymphs or larvae, after a specific moult that allows reproductive maturity, while maintaining an immature appearance. Neotenics are also called secondary reproductives (Evans 2010) to indicate reproductive individuals that remain in the native colony or bud from it, retaining assistance from helper nestmates; they are sometimes called replacement or supplementary reproductives to indicate the fact that they can develop in absence of or in addition to primary reproductives.

The preference for dry wood is shared by the vast majority of the 456 Kalotermitid species, although there are a few species with damp wood or subterranean habits (Weesner 1970; Krishna et al. 2013). Hence, besides the taxonomic features (Krishna et al. 2013), a more reliable characterization of this family is given by the life type, based on nesting and feeding habits (Eggleton and Tayasu 2001; Korb and Hartfelder 2008). According to this categorization, Kalotermitidae are "single-piece nesters" that feed on and nest in the same single piece of substrate, which is always wood; they never leave the nest (except the winged adults), and colony life is limited by food availability (Korb and Hartfelder 2008). In these termites, workers are not a true differentiated caste, and the older immatures are often referred to as "pseudergates" (Roisin and Korb 2011). New infestations on wood objects are initiated exclusively by the alate imagoes, flying out from mature colonies. Drywood termites can tolerate dry conditions for prolonged periods (Rust et al. 1979) and have a low moisture requirement, so that they can live inside pieces of wood with no need to connect their nests to the soil. They produce dry six-sided faecal pellets that are usually expelled from the inhabited galleries (an example is given in Fig. 12.4). Piles of these pellets are a recognizable sign of drywood termite infestation.

This ability to exploit cellulose materials within xeric habitats is one of the key features of the success of the most invasive pest species of this family, such as *Cryptotermes* sp. (Evans et al. 2013; Lewis and Forschler 2014). However, the ability to retain metabolic water, that allows these species to thrive in dry substrates/environments, implies the evolutionary trade-off of small colony size, ranging between few hundred and few thousand individuals (Lewis et al. 2013; Lewis and Forschler 2014).

Rhinotermitidae exhibit features intermediate between lower and higher termites (Vargo and Husseneder 2009). In particular, in their colonies there is a true worker caste, a trait typical of higher termites (Korb and Hartfelder 2008), but they have "Type 1" feeding habits, typical of lower termites (Donovan et al. 2001). Considering the nesting habits, rhinotermitids are intermediate between the one-piece nesting of more basal lower termites and the multiple-pieces nesting, typical of the higher termites (Korb and Hartfelder 2008). As a result, they can have a nest in one piece of wood, but the workers will exploit new food resources outside the nest (foraging), and therefore the colony life is not limited by food availability.

The name "subterranean termites" comes from the strong need of moisture in their environment that is satisfied by nesting inside or in close contact with soil (Thorne 1998). While foraging, they usually reach the food sources from the underlying soil by performing subterranean galleries (tunnelling), and they always keep some connection with the soil through tunnels in wood or by building typical shelter tubes using "mud", a mix of soil, faecal material, bits of wood and saliva (Thorne 1998). Tunnel distribution is optimized for food searching and transport efficiency (Campora and Grace 2001; Su 2005; Lee et al. 2007), and tunnelling activity is affected by moisture (Su and Puche 2003) and decayed wood extracts (Su 2005).

Rhinotermitidae is the most widely distributed family of termites, especially abundant in temperate areas (Bignell and Eggleton 2000). They have very large colonies, ranging from a few thousands (often 100,000) to million individuals, and foraging galleries can extend up to 100 m (Su and Scheffrahn 1988). When attacking wood, galleries of these termites are typically recognizable because they are covered with moist faecal matter. The mud is also used to build the "carton" nest in which they live that is also known as "replacement wood". Globally, subterranean termites are the most important pests of structural wood, accounting for 80% of estimated economic impact of termites (Rust and Su 2012), with some species in the genus *Coptotermes* among the most invasive in the world (Evans et al. 2013).

12.2.1 Growing Economic Importance of Drywood Termites

Due to their biological features, drywood termites have higher chances to become invasive pests. In their colonies all individuals, except the soldiers, are potentially able to reproduce, either as primary or as neotenic reproductives; therefore, any piece of wood containing some of these termites is likely to become an infestation propagule (Evans et al. 2011). Besides, taking into account the size of the potential cellulosic propagule, invasive *Cryptotermes* are capable to accept pieces of wood smaller (8 cm^3) than those accepted by non-invasive species (Evans et al. 2011). It was indeed shown that *Cryptotermes* colonies can occupy objects as small as a book (Gulmahamad 1997). A detailed list of invasive Kalotermitids is reviewed in Evans (2010) and Evans et al. (2013), and the genera *Incisitermes* and *Cryptotermes* appear as the most problematic, worldwide. A study that analyzed the termite infestations detected onboard ships in Australia and Florida (Scheffrahn and Crowe 2011) showed that these two drywood termites represented more than 40% of the interceptions and pointed out that vessels can be colonized by alates of all termite taxa flying on board during dockage and that colonization by Kalotermitids can occur also during construction.

The so-called West Indian drywood termite, *Cryptotermes brevis*, a common pest of structural lumber and sheltered wood, is the most widespread invasive termite species (Evans et al. 2013), occurring in the New World tropics and tropical oceanic islands, in parts of Africa, Madagascar and Reunion Island, in Central and South America. The recognition of the native areas of this species remained unresolved for a long time. After an initial hypothesis that indicated it as native to the circum-Caribbean region (Emerson 1936) and many unsuccessful survey expeditions, the problem was solved only recently (Scheffrahn et al. 2008), when outdoor populations were detected in coastal deserts of Peru and Chile. Although the arrival dates are unknown for most locations, the authors suggest a post-Colombian release of *C. brevis* by shipboard infestations and the movement of infested wood during the early Spanish Empire to the present time (Scheffrahn et al. 2008). In Europe isolated infestations by *C. brevis* were reported in England (Gay 1969), Germany (Becker and Kny 1977), continental Portugal and Spain (Nunes et al. 2010). However, established populations are known to occur for years in Canary Islands (Martinez 1957), in Azores and other Atlantic Ocean Islands (Borges and Myles 2007) and in different cities in Italy (Raineri et al. 2001; Fontana and Buzzetti 2003; Liotta 2005). Presently, in the Azores archipelago, this termite is reported to infest a significant proportion of buildings, causing serious economic and patrimonial losses and threatening to spread to yet unaffected towns and islands (Guerreiro et al. 2014). It is likely that the wide distribution of this species is due to its capability to create viable propagules (Evans et al. 2013), as it was observed during the dissection of hardwood shipping pallets, when up to eight separate colonies were detected on a single board (Grace et al. 2009). Increased trade of wooden commodities/shipping materials is therefore likely to favour the further spread of species already known as invasive, as well as to increase the number of new invasive species characterized by features similar to *C. brevis*.

12.3 Standard Norms: Definitions and Importance

A standard is a technical document which defines and unifies the characteristics, requirements, specifications and guidelines that can be used consistently to ensure that materials, products, processes and services are fit for their purpose, guaranteeing safety, good quality, reliable performance and respect for the environment.

Primary activities of standards organization (or standards body, standards developing organization) include developing, coordinating, promulgating, revising, amending, reissuing, interpreting or otherwise producing technical standards addressing the needs of a relatively wide base of affected adopters. Certification demonstrates that a product or service meets minimum standards so to ensure the quality expected by customers. For some industries, certification is a legal or contractual requirement. Usually, the standards organizations do not enforce, regulate or certify compliance with the standards, the tasks being performed by specific, external certification bodies.

The procedures of standards organizations can be accredited at different geographic levels: national, regional and international. For example, ANSI (American National Standards Institute 2016) is the administrator and coordinator of the United States private sector voluntary standardization system; Standards Australia (Standards Australia 2016) is devoted to the development and adoption of standards in Australia, facilitating Australian participation in international standards development; CEN (European Committee for Standardization 2016), the association that gathers the National Standardization Bodies of 33 European countries, is responsible for developing and defining voluntary standards at European level; ISO (International Organization for Standardization 2016), which is the world's largest developer of voluntary international standards, is an independent, non-governmental organization, whose members are the standards organizations of the 162 member countries. The use of national/regional/international standards allows to overcome technical barriers in interlocal or interregional commerce, caused by differences among technical regulations and standards developed independently and separately by each local body.

Standards are designated using a format that starts with the acronym of the issuing body/bodies, the number of the standard, the year it was published and a title that describes the subject. Standards are regularly reviewed and revised, and usually updates are listed on the website of the originator or standard writing body. Some standards are public documents freely available on the Internet, libraries, etc. Others can be purchased directly from the issuing body, are available under intellectual property or are controlled/closed documents containing classified information or trade secrets.

12.3.1 Standard Norms in Wood Protection

The importance of standards is crucial in all aspects and products related to wood preservation technology, and the stakeholders include engineers, architects, producers of wood products (both treated and untreated), producers of wood preservatives and/or their components, end users of treated wood, government bodies, universities and other research/technological institutions and any other entity with a general interest in wood protection. Some countries/regions have standards organizations specifically dedicated to wood protection. For example, the AWPA (American Wood Protection Association 2016) is a non-profit organization accredited at ANSI, whose technical committees develop voluntary standards that are universally specified for wood preservation in the USA and are recognized worldwide. IEO-WEI (IEO-WEI 2016) that represents the wood preservation industries within the European Union, among its different activities, develops quality standards for wood preservatives and treated timber and is heavily involved in the work of CEN. In Australia the TPAA (Timber Preservers Association of Australia 2016) may help with establishing and adhering to standards for the treatment of timber and promote best practice in the production of treated wood.

12.3.2 Standard Norms Targeting Termites

Because of the importance of termites as agents of wood degradation and destruction, specific norms have been developed to establish procedures to assess the efficacy of wood treatments against these insects, in terms of remedial control interventions and/or preventative measures or related to the durability of wood. Table 12.1 lists the main features of the most widely used standard norms related to termites, and they include both laboratory and field methods. Almost all these standards are targeting subterranean termite species, and the laboratory methods standards are characterized by similar test conditions: a variable amount of termites (usually more than 150 individuals) is placed inside a container (or a glass tube) with a ventilated lid or a piece of aluminium foil, together with a substrate (sand or similar materials) that must be regularly moisturized with water; the wood test specimen is either placed inside the substrate or where the glass tube is used and it represents the basement of the apparatus; the test duration is between 4 and 8 weeks; the evaluation parameters consist mainly of (a) visual assessments of termite attack, based on scales that are reported in the standard text, (b) termite mortality rates and (c), occasionally, also wood mass loss rate, calculated from the difference between the initial and the final dry wood mass (obtained weighting the wood after placing it in an oven at a temperature usually higher than 100 °C for a certain number of hours). Only the Indonesian standard considers both subterranean and drywood termites and describes separate trial methods for each type. Considering standards for field tests, they usually imply the use of wood stakes, sometimes confined inside containers turned upside-down that are placed in areas with recognized high presence of termites (often in proximity of known termite colonies); the evaluations usually occur after 4–24 months and are based on visual ratings of termite damage.

Table 12.1 Main features of the most widely used standard norms in wood protection, targeting termites

Standard norm acronym (year)	Issuing body	Country/region	Title	Notes	Test type[1] (L/F)	Termite species and type[2] (S/D)	Wood specimen species	Apparatus	Substrate	Need for substrate moisture	Test duration	Evaluation parameters
ASTM D3345 (2008)	ASTM	The USA	Standard test method for laboratory evaluation of wood and other cellulosic materials for resistance to termites		L	*Reticulitermes* sp., *Coptotermes* sp., (S)	*Pinus* sp. sapwood	Glass jar	Sand	Yes	4 weeks	Termite damage visual rating
E1-16 (2016)	AWPA	The USA	Laboratory methods for evaluating the termite resistance of wood-based materials: choice and no-choice tests		L	*Reticulitermes* sp., *Coptotermes* sp., (S)	*Pinus* sp. sapwood	Glass jar	Sand	Yes	4 weeks	Termite damage visual rating, termite mortality rate
E7-15 (2015)	AWPA	The USA	Standard field test for evaluation of wood preservatives to be used in ground contact (UC4A, UC4B, UC4C); stake test		F	*Reticulitermes* sp., *Coptotermes* sp., (S)	*Pinus* sp. sapwood		Soil in habitats with long-time termite presence	Yes	6–20 months	Termite damage visual rating

(continued)

Table 12.1 (continued)

Standard norm acronym (year)	Issuing body	Country/ region	Title	Notes	Test type[1] (L/F)	Termite species and type[2] (S/D)	Wood specimen species	Apparatus	Substrate	Need for substrate moisture	Test duration	Evaluation parameters
E21-15 (2015)	AWPA	The USA	Standard field test for evaluation of wood preservatives to be used for interior applications (UC1 and UC2); full-size commodity termite test		F	*Reticulitermes* sp., *Coptotermes* sp., (S)	*Pinus* sp. sapwood		Soil in habitats with long-time termite presence	Yes		Termite damage visual rating
E26-15 (2015)	AWPA	The USA	Standard field test for evaluation of wood preservatives to be used for interior applications (UC1 and UC2); ground proximity termite test		F	*Reticulitermes* sp., *Coptotermes* sp., (S)	*Pinus* sp. sapwood		Soil in habitats with long-time termite presence	Yes		Termite damage visual rating
UNI EN 117 (2013)	CEN	The EU	Wood preservatives; determination of toxic values against *Reticulitermes* species (European termites) (laboratory method)	Vacuum impregnation treatment	L	*Reticulitermes* sp. (S)	*Pinus sylvestris* sapwood	Test containers of suitable size and material	White quartz sand or aluminium-iron-magnesium silicate	Yes	8 weeks	Termite damage visual rating, termite mortality rate

UNI EN 118 (2014)	CEN	The EU	Wood preservatives; determination of preventive action against *Reticulitermes* species (European termites) (laboratory method)	Surface treatment	L	*Reticulitermes* sp. (S)	*Pinus sylvestris sapwood*	A glass tube placed upon the wood specimen, topped with aluminium foil	White quartz sand or aluminium-iron-magnesium silicate	Yes	8 weeks	Termite damage visual rating, termite mortality rate
EN 350-2 (1996)	CEN	The EU	Durability of wood and wood-based products; natural durability of solid wood; guide to the natural durability and treatability of selected wood species of importance in Europe	Termites among target organisms	na	Unspecified	Various	na	na	na	na	Wood durability ratings
EN 460 (1996)	CEN	The EU	Durability of wood and wood-based products; natural durability of solid wood; guide to the durability requirements for wood to be used in hazard classes	Termites among target organisms	na	Unspecified	Various	na	na	na	na	Wood durability ratings

(continued)

Table 12.1 (continued)

Standard norm acronym (year)	Issuing body	Country/ region	Title	Notes	Test type[1] (L/F)	Termite species and type[2] (S/D)	Wood specimen species	Apparatus	Substrate	Need for substrate moisture	Test duration	Evaluation parameters
AWPC protocols (2015)	TPAA	Australia	Protocols for assessment of wood preservatives	As appropriate	L	(a) Subt. species, i.e. *Coptotermes acinaciformis*	Unspecified species, sapwood	Glass jars with ventilated lid	(a) Matrix of mound material	Yes	(a) 8 weeks	Wood mass loss rate, termite mortality rate
						(b) *Mastotermes darwiniensis*			(b) Vermiculite and sawdust of susceptible timber		(b) 6 weeks	
AWPC protocols (2015)	TPAA	Australia	Protocols for assessment of wood preservatives	As appropriate	F	(a) Subt. species, i.e. *Coptotermes acinaciformis*	Unspecified species, sapwood	Ventilated steel exposure containers	Soil in habitats with long-time termite presence	Yes	min = 16 weeks, up to 6–12 months	Wood mass loss rate
						b)*Mastotermes darwiniensis*						
JIS K 1571 (2004)	Japanese standards association	Japan	Test methods for determining the effectiveness of wood preservatives and their performance requirements	Injection and surface treatments	L, F	*Coptotermes formosanus* (S)	*Cryptomeria* sapwood	L: acrylic resin cylinder	L: moisturized cotton;	Yes	L: 21 days	Wood mass loss rate (l,f), termite damage visual rating (l,f), termite mortality rate (l)
								F: upside-down box container	F: soil in habitats with long-time termite presence		F: 2 years	

SNI 01-7207 (2006)	Standar Nasional Indonesia	Indonesia	Endurance test for timber and timber products against wood destroying organisms	Tests for: (a) drywood termites;	L	(a) *Cryptotermes cynocephalus* (D)	Unspecified	(a) Glass tube placed upon the wood specimen, topped with a cotton swab	(a) None	(a) no	(a) 8 weeks	Wood mass decrease rate, termite damage visual rating, termite mortality rate
				(b) Subterranean termites		(b) *Coptotermes curvignathus*		(b) Glass jar topped with aluminium foil	(b) Sand	(b) Yes	(b) 6 weeks	

[1]Test type: *L* laboratory test performed indoor, *F* field test performed outdoor
[2]Termite type: *S* subterranean, *D* drywood; *na* not applicable

12.4 Preventative Measures: Need for a Specific Standard for Drywood Termites

The above-mentioned differences between drywood and subterranean termites regarding the biology, ecology and behaviour necessarily imply radically different management approaches. Remedial interventions have been reviewed previously (Thorne 1998; Lewis 2003; Rust and Su 2012; Lewis and Forschler 2014). In general, preventive interventions consist in the application of treatments to the wood to stop or avoid infestation by biotic agents that can feed/develop on wooden substrates, i.e. bacteria, fungi or insects. The economic importance of wood preservatives is considerable: estimates indicate that they make about 75% of the total poundage of pesticides used in North America (Freeman et al. 2003). Treatments can be applied by totally dipping the wooden material inside tanks containing the wood preservative or by means of pressure treatments under vacuum in appropriate retorts. Considering termites, the use of preservatives aims at preventing wood colonization by subterranean termites during their foraging activities or, in the case of drywood termites, the initial colony formation (Lewis 2003).

Conventional chemical wood protection is based on a broad spectrum of biocide formulations such as copper/organic biocides, copper-organometallics and metal-free preservatives (Freeman et al. 2003). Among them, chromate copper arsenate (CCA) was a worldwide distributed wood preservative for more than 30 years (Eaton and Hale 1993). However, due to its high mammalian and environmental toxicity, it has been banned by EU legislation in 2003 and is being phased out also by other countries. Presently, anti-termite wood preservative treatments include those containing boron (Gentz and Grace 2006; Nami Kartal et al. 2007; Kose et al. 2011; Han et al. 2012; Li et al. 2012), copper and/or zinc (Tascioglu and Tsunoda 2010; Wu et al. 2012; Maistrello et al. 2012a; Akhtari and Nicholas 2013). More treatments are being discovered and tested, such as those derived from plants (Kartal et al. 2011, 2012, 2013).

The majority of trials to assess effectiveness of preventive treatments on termites are based on standards currently in use for subterranean termites. No standards are currently available in the USA, EU, Australia and Japan for evaluation on drywood termites, in spite of their worldwide increasing importance as pests (Evans et al. 2013; Lewis and Forschler 2014).

12.5 Guidelines for a Standard Protocol to Assess Efficacy of Preventive Wood Treatments on Drywood Termites

12.5.1 Background and Aim

Laboratory testing gives a basis to assess the effectiveness of a wood preservative, when applied as a surface or pressure treatment, against drywood termite species, members of the family Kalotermitidae, such as *Cryptotermes* sp. or *Kalotermes* sp.

This proposed method was developed taking in account the specific biological features of these termites and considering the available literature related to laboratory trials, testing different types of wood treatments or the durability of wood species against different drywood termites (Moein and Rust 1992; Hadi et al. 1998, 2005, 2010, 2012; Hwang et al. 2006; Sunarta et al. 2011; Suhasman et al. 2012; Maistrello et al. 2012a, b). The present method has been tested on *Kalotermes flavicollis*; however, it can be applied to different members of Kalotermitidae.

Regarding the products to be tested, the method is applicable to:

- Water-insoluble chemicals which are being studied as active ingredients
- Organic formulations, as supplied or as prepared in the laboratory by dilution of concentrates
- Organic water-dispersible formulations, supplied or prepared in the laboratory by dilution of concentrates
- Water-soluble materials, for example, salts

The method can be used also in conjunction with an ageing procedure.

This proposal specifically describes the procedure for the trial with termites and is not meant to describe or consider the wood treatment procedures that must be performed according to specific, appropriate protocols.

12.5.2 Principle

In a no-choice feeding test, drywood termites are forced to attack the wood they are exposed to. It relies on exposure of wood test specimens, surface or pressure treated (treatment procedures are not described here) to specified groups of Kalotermitidae species and on assessment of the attack suffered after exposure under fixed conditions and over a given time period. It includes comparison of results with those obtained from untreated control test specimens.

12.5.3 Test Materials

12.5.3.1 Biological Material

Older immature instars and soldiers of an identified termite species of the Kalotermitidae family (i.e. *Cryptotermes* sp., *Kalotermes* sp.). Older immature instars are the individuals acting as "workers" in the broad sense (see par. 2) that include both older larvae (instars with no wing buds) and nymphs (with wing buds). An ethogram-type study showed that in drywood termites there are no differences in feeding and other behavioural patterns performed by older larvae and nymphs (Crosland et al. 2004).

The termite species and the locality of origin should be stated in the test report, and their identification should be proved. In providing biological validation of individual species, it is essential that the locality of origin of each test termite species is given. The description of the locality should at least include the district name.

The termites should be obtained from colonies reared in the laboratory in optimal conditions or from outdoor colonies collected from infested wood materials.

12.5.3.2 Apparatus and Testing Chamber

The testing apparatus consists in glass tube domes (as many as the number of wood specimens to be tested) that are assembled to their respective wood specimens after termites are placed inside the domes. All assembled items are kept in a testing chamber for the whole duration of the trial.

Glass Tube Domes (the bottom part of glass tubes) (Fig. 12.1a, b), with a small hole (1 mm diameter) in the middle of the dome end, and open at the other end, which is the ground:

- Interior diameter at the ground end: 30 mm
- Height (from the ground end to the upper centre of the dome): 35 mm

Adhesive (Glue) which cannot be attacked by the termites and is non-toxic, for securing the glass domes to the wood specimens. This adhesive shall also not react with the preservative applied to the wood. It should be chosen among those that dissolve in alcohol or water, in order to ensure fast unsealing at the end of the test.

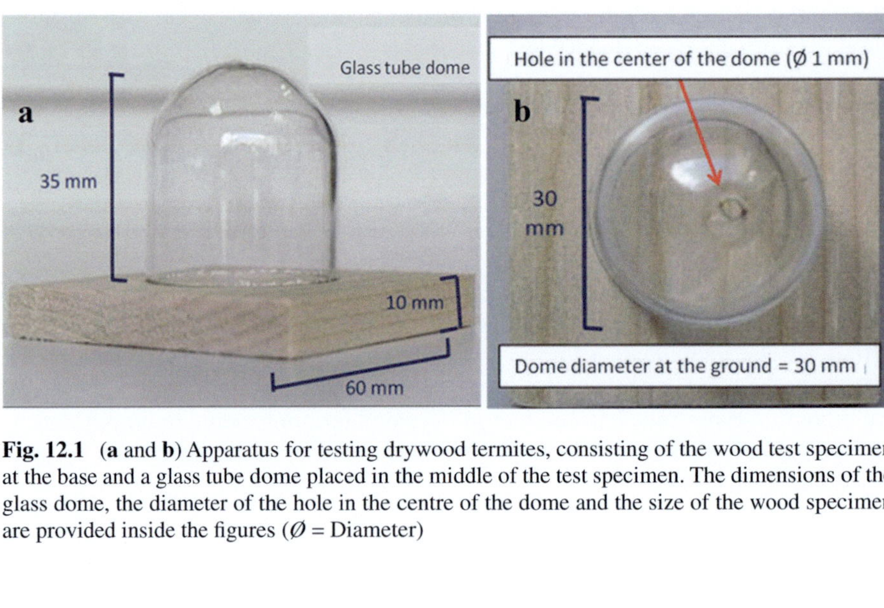

Fig. 12.1 (**a** and **b**) Apparatus for testing drywood termites, consisting of the wood test specimen at the base and a glass tube dome placed in the middle of the test specimen. The dimensions of the glass dome, the diameter of the hole in the centre of the dome and the size of the wood specimen are provided inside the figures (Ø = Diameter)

Testing Chamber protected from light and ventilated, controlled at temperature 26 ± 2 °C and relative humidity of 60 ± 5%.

12.5.4 Test Specimens

12.5.4.1 Species of Wood

The reference genus is pine (*Pinus* sp.). Additional tests could be made with other timber species that should be stated in the test report.

12.5.4.2 Wood Quality

The wood shall be free from visible cracks, stain, decay, insect damage and other defects. It should not be water stored, floated, chemically treated or steamed. Wood that has been kiln dried at temperatures below 60 °C may be used.

The wood must be exclusively sapwood containing little resin, between 2.5 and 8 annual rings per 10 mm. The proportion of latewood in the annual rings must not exceed 30% of the whole. It is recommended to use test specimens of similar growth rate within a single assay.

12.5.4.3 Provision of Test Specimens

Prepare planed strips with a fine-sawn finish and having a cross-section of 60 ± 0.5 mm × 10 ± 0.5 mm, removing a minimum of 2 mm from any surfaces exposed during drying. The longitudinal faces shall be parallel to the direction of the grain. The annual rings shall have a contact angle of 45 ± 15 mm to the broad faces. Make transverse cuts, neatly, to give sharp edges and a fine-sawn finish to the end-grain surfaces, to give test specimens 60 ± 0,5 mm long. The test specimens shall originate from a minimum of three trees or shall be taken at random from a stock originally of more than 500 test specimens.

12.5.4.4 Size of Test Specimens

The dimensions of each specimen shall be 60 ± 0.5 mm × 60 ± 0.5 mm × 10 ± 0.5 mm (Fig. 12.1a).

12.5.4.5 Number and Distribution of Test Specimens

The test specimens shall be divided as follows:

(a) Treated test specimens: these are subject to attack by the drywood termite species; their number is at least 6 specimens for each treatment type.
(b) Untreated control test specimens to check the virulence of the termites taken for the test: these untreated test specimens are subjected to attack by the drywood termite species; their number is at least 3 for each series of tests.

12.5.5 Procedure

12.5.5.1 Preparation of the Wood Test Specimens

Preparation of Test Specimens Prior to Treatment

Before treatment procedures, each test specimen should be marked with a pencil on at least three different sides, so that it can be identified throughout the test.

Test specimens should be oven dried at 103 ± 2 °C for 18 h and then weighed with an analytical balance with an accuracy of 0.001 g.

Treatment of Test Specimens

Treatment procedures of wood specimens should be performed according to specific, appropriate protocols for either surface or pressure treatments.

After the treatment has been performed, all test specimens are left 24 h to air dry at room temperature, oven dried at 103 ± 2 °C for 18 h and then reweighed to determine the treatment retention. Before using the specimens, these should be left at least 2 weeks in the conditions of the testing chamber.

12.5.5.2 Exposure of the Test Specimens to the Insects

Collecting and Selecting the Termites

Pick up the insects individually using forceps. Make groups of 50 late immature instars that can include both older larvae and early instars nymphs (N1–N2), rejecting those insects which are moulting (indicated by the dull white colour of the abdomen) and also those which appear wounded or remain motionless. To each group made up in this way, add a number of soldiers corresponding to the proportion found in the colony.

The number of groups to be prepared as indicated above is equal to the total number of test specimens to be subjected to the termites attack. If the required number of termites is more than that in a single culture, the control series and test series shall contain the same number of groups from each origin colony. Termites from different origin colonies shall not be mixed in a single group.

Installation of the Termites in the Apparatus

Distribute in each tube dome a group of 50 termites made up as described previously. Place one wood test specimen over a tube dome, so that the dome is in the centre of the test specimen, and then turn it upside down, so that the wood test specimen represents the base of the apparatus (Fig. 12.2).

Fig. 12.2 An example of assembled testing unit, with termites visible inside the glass dome, and reference numbers visible on two sides of the wood specimen

Securing the Glass Tube Domes
Attach with adhesive the ground glass end of one tube at the centre of the surface of each test specimen (Fig. 12.3), both for treated and untreated wood specimens.

Testing Unit
A testing unit is considered as the assembly of one glass tube dome + one group of termites (50 immature instars individuals + soldiers in proportion to original colony) + one secured (with glue) wood test specimen (Fig. 12.2). For each treatment series, consider at least six testing units and for the untreated series at least three testing units.

12.5.5.3 Conditions and Duration of the Tests

Place the testing units (as described above) on separate plastic trays (one tray for each treatment series and one for the untreated series), and keep them inside the testing chamber (Sect. 12.5.3.2) and leave them there for 12 weeks.

It is recommended that, throughout the duration of the test, each testing unit be inspected at regular intervals, and any necessary action is taken to maintain the termites in the best possible condition without disturbing their activity. The results of the inspections should be properly recorded.

These inspections cover in particular the presence, location and activity of the termites and the moults to neotenic (secondary) reproductives and those to imagoes (see Sect. 12.2) that could eventually occur. Action can be taken if the termites are escaping (i.e. by performing galleries inside the wood).

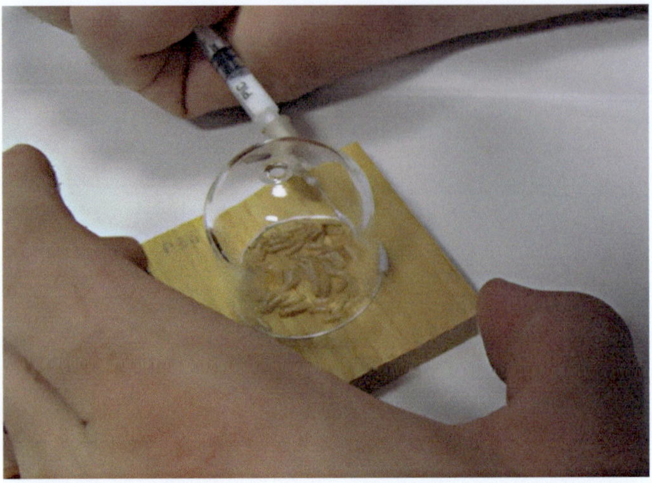

Fig. 12.3 Usage of the glue to secure with an adhesive the ground glass end of dome to the centre of the surface of the test specimen

12.5.5.4 Examination of the Test Specimens and the Colonies at the End of the Test

At the end of the test, the testing units are taken out from the testing chamber and are dismantled. Each testing unit is disassembled by unsealing the glass dome from the wood specimen (see Sect. 12.5.3.2). Fragments of dead termites and termite faecal pellets should be carefully removed also from the galleries eventually performed inside the wood specimens (Fig. 12.4).

Assessment of the efficacy of the preventive treatments is obtained using three types of parameters to be recorded at the end of the experiment: (a) survival of termites, (b) visual examination and rating of termite damage on the wood test specimens and (c) wood mass loss (in percentage).

Termite Survival
For each testing unit, count the number of live termites, considering (a) immature and immature-derived individuals: this include the total number of immature instars (larvae, nymphs) and of those individuals eventually derived from them after moulting to imagoes or neotenic reproductives and (b) soldiers. Determine the survival rate of (a).

Visual Examination and Rating of Termite Damage
Carry out a visual examination of each wood test specimen, and classify any evidence of attack by its locality, extent and depth. Rate the results of this examination in accordance with the following schedule:

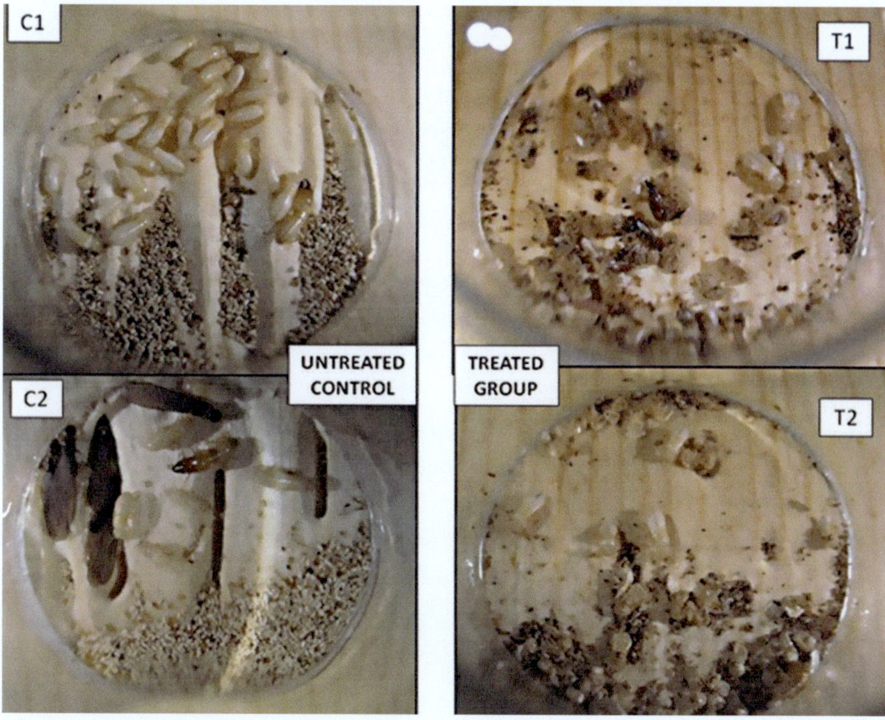

Fig. 12.4 Examples of testing units at the end of trial (after 12 weeks) from the untreated control (left) and from one of the treated groups (right), taken from the experiment where the proposed protocol was used (Bergamonti et al. 2016). Notice the high number of live termites and of faecal pellets, arranged in an orderly manner on the wood surface in the untreated control, in contrast with many corpses/fragments of dead termites and few faecal pellets scattered all over the surface in the treated group. Some imagoes (alates) are visible in one of the untreated group testing unit

- 0 no attack
- 1 attempted attack

  (i) Superficial erosion of insufficient depth to be measured on an unlimited area of the test specimen, or (ii) attack to a depth of 0,5 mm provided that this is restricted to an area or areas not more than 30 mm^2 in total or (iii) combination of (i) and (ii)

- 2 slight attack

  (i) Erosion of 1 mm in depth limited to not more than 1/10 of the surface area of the test specimen, or (ii) single tunnelling to a depth of up to 3 mm or (iii) combination of (i) and (ii)

- 3 average attack

  (i) Erosion of <1 mm in depth over more than 1/10 of the surface area of the test specimen, or (ii) erosion of >1 mm to <3 mm in depth limited to not more than 1/10 of the surface area of the test specimen, or (iii) isolated tunnelling of a depth > 3 mm not enlarging to form cavities or (iv) any combination of (i), (ii) or (iii)

- 4 strong attack:

  (i) Erosion of >1 mm to <3 mm in depth of more than 1/10 of the surface area of the test specimen, or (ii) tunnelling penetrating to a depth > 3 mm and enlarging to form a cavity in the body of the test specimen or (iii) combination of (i) and (ii)

Wood Mass Loss (In Percentage)
At the end of the test, the wood test specimens are oven-dried at 103 ± 2 °C for 18 h and finally weighed to assess wood consumption. The wood mass loss in percentage is therefore calculated.

Validity of the Tests
The test is valid if the three untreated virulence control test specimens correspond to level 4 rating, when visually examined and if the corresponding termite groups have at least 50% survivors.

12.5.6 Expression of Results in the Test Report

A final test report should include the following data:

(a) Number and date of the document.
(b) Data on wood preservative treatment (supplier, name and type of preservative tested, name and concentration of active ingredient, treatment type (surface or pressure treatment), density of the preservative, any solvent or diluent used and, if necessary, the dilution used for the test, date of application of the product, exact amount of the product absorbed by each test specimen, expressed in g/m^2 and correspondingly in ml/m^2, if applicable, any ageing procedure applied, detailing the nature, conditions and duration (if possible by reference to a standard).
(c) Termite species used in the test and locality of origin.
(d) Date of exposure of the test specimens to the termites (start of experiment).
(e) Location and date of final examination.
(f) For each testing unit report: (a) rate of immatures and immature-derived individuals' survival at the end of the test (see Sect. 12.5.5.4) and, if any, the presence of living soldiers, (b) degree of attack on each test specimen as determined by the visual rating and (c) wood mass loss in percentage.
(g) Name of the laboratory where the report is prepared.
(h) Name and signature of the officer(s) in charge.

(i) At the end of the report this note may be included:

> The interpretation and the practical conclusions that can be drawn from of this test report demand a specialized knowledge of the subject of wood preservation and, for this reason, this test report cannot by itself constitute an approval certificate.

12.5.7 An Example of Usage of the Developed Protocol

The developed protocol has been used to test innovative wood preservatives based on polyamidoamines (Bergamonti et al. 2016), and the trial was performed using *K. flavicollis* as representative of drywood termite species.

By using this protocol, the efficacy of the different types of tested treatments on this termite was demonstrated, both in terms of insect survival, wood consumption, number and arrangement of the faecal pellets, as it can be observed from the pictures taken at the end of the assay (Figs. 12.4 and 12.5). During the experiment, in some of the untreated control groups, some nymphs initially placed in the testing units had moulted into imagoes (alates) (Fig. 12.4).

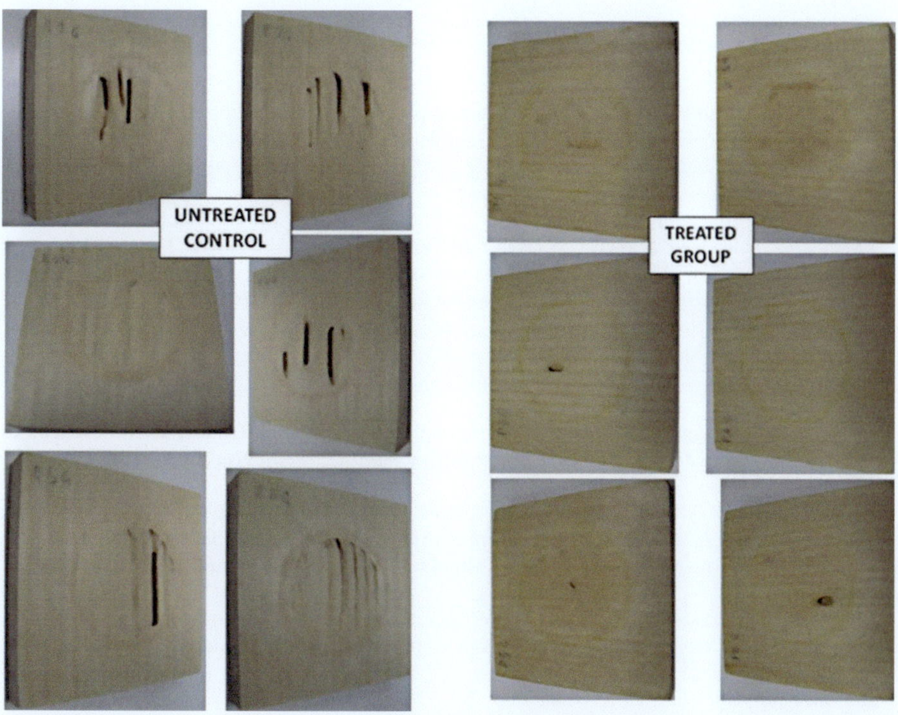

Fig. 12.5 Examples of appearance of the wood test specimens at the end of the trial in the untreated control (left) and in one of the treated groups (right), taken from the experiment where the proposed protocol was used (Bergamonti et al. 2016)

12.6 Conclusion

In a world of increasing global trade, the use of standard norms allows to overcome technical barriers in interregional commerce, and it is crucial to safeguard the consumers and the end users of products and services about their safety, good quality and reliable performance. The standards importance is highly recognized in the wood preservation technology field, also to guarantee that preservatives are effective in protecting wood from biotic degradation agents, i.e. xylophagous insects such as termites, in respect to environment. On the other hand, increased trade of wooden commodities favours, in particular, the spread of some species of drywood termites that become invasive pests, due to their peculiar biological features. In spite of the worldwide increasing economic importance of drywood termites, almost all standard norms in wood protection are focused on subterranean termites, and no standards targeting drywood termites are currently available in the USA, EU, Australia and Japan. This chapter aimed at filling this gap, by proposing a standard specifically developed to assess the efficacy of preventative wood treatments on drywood termites, based on their unique biological features.

References

Akhtari, M., & Nicholas, D. (2013). Evaluation of particulate zinc and copper as wood preservatives for termite control. *European Journal of Wood and Wood Products, 71*, 395–396.

American National Standards Institute (2016). ANSI-American National Standards Institute. https://www.ansi.org/. Accessed 2 Sept 2016.

American Wood Protection Association. (2016). AWPA – American Wood Protection Association – Standards for Treated Wood Since 1904. http://www.awpa.com/. Accessed 2 Sept 2016.

Becker, G., & Kny, U. (1977). Uberleben und Entwicklung der Trockenholz-Termite *Cryptotermes brevis* (Walker) in Berlin. *Anz Fur Schadlingskunde Pflanzenschutz Umweltschutz, 50*, 177–179.

Bergamonti, L., Berzolla, A., Chiappini, E., Feci, E., Maistrello, L., Palanti, S., Predieri, G., & Vaccari, G. (2016). Polyamidoamines (PAAs) functionalized with siloxanes as wood preservatives against fungi and insects. *Holzforschung, 71*(1), 65–75. https://doi.org/10.1515/hf-2016-0010.

Bignell, D. E., & Eggleton, P. (2000). Termites in ecosystems. In T. Abe, D. E. Bignell, & M. Higashi (Eds.), *Termites: Evolution, sociality, symbioses, ecology* (pp. 363–387). Dordrecht: Kluwer Academic.

Borges, P., & Myles, T. (2007). *Termitas dos Acores*. Lisbon: Principia Ed.

Campora, C. E., & Grace, J. K. (2001). Tunnel orientation and search pattern sequence of the Formosan subterranean termite (Isoptera: Rhinotermitidae). *Journal of Economic Entomology, 94*, 1193–1199.

Crosland, M. W., Traniello, J. F. A., & Scheffrahn, R. H. (2004). Social organization in the drywood termite, *Cryptotermes cavifrons*: Is there polyethism among instars? *Ethology Ecology and Evolution, 16*, 117–132.

Donovan, S. E., Eggleton, P., & Bignell, D. E. (2001). Gut content analysis and a new feeding group classification of termites (Isoptera). *Ecological Entomology, 26*, 356–366.

Eaton, R. A., & Hale, M. D. C. (1993). *Wood: Decay, pests, and protection*. London: Chapman and Hall.

Eggleton, P., & Tayasu, I. (2001). Feeding groups, lifetypes and the global ecology of termites. *Ecological Research, 16*, 941–960.

Emerson, A. E. (1936). Distribution of termites. *Science, 83*, 410–411.

European Committee for Standardization. (2016). CEN – European Committee for Standardization https://www.cen.eu/Pages/default.aspx. Accessed 9 Sept 2016.

Evans, T. A. (2010). Invasive termites. In D. E. Bignell, Y. Roisin, & N. Lo (Eds.), *Biology of termites: A modern synthesis* (pp. 519–562). Dordrecht: Springer.

Evans, T. A., Inta, R., & Lai, J. C. S. (2011). Foraging choice and replacement reproductives facilitate invasiveness in drywood termites. *Biological Invasions, 13*, 1579–1587.

Evans, T. A., Forschler, B. T., & Grace, J. K. (2013). Biology of invasive termites: A worldwide review. *Annual Review of Entomology, 58*, 455–474.

Fontana, P., & Buzzetti, F. M. (2003). New record of *Cryptotermes brevis* Walker, 1853 in northern Italy (Insecta Isoptera Kalotermitidae). *Boll Mus Civ Storia Nat Venezia, 54*, 35–44.

Freeman, M. H., Shupe, T. F., Vlosky, R. P., & Barnes, H. M. (2003). Past, present, and future of the wood preservation industry. *Forest Products Journal, 53*, 8–15.

Gay, F. J. (1969). Species introduced by man. In K. Khrishna & F. M. Weesner (Eds.), *Biology of termites* (Vol. 1, pp. 459–494). New York: Academic.

Gentz, M. C., & Grace, J. K. (2006). A review of boron toxicity in insects with an emphasis on termites. *Journal of Agricultural and Urban Entomology, 23*, 201–207.

Grace, J. K., Woodrow, R. J., & Oshiro, R. J. (2009). Expansive gallery systems of one-piece termites (Isoptera: Kalotermitidae). *Sociobiology, 54*, 37–44.

Guerreiro, O., Cardoso, P., Ferreira, J. M., Ferreira, M. T., & Borges, P. V. (2014). Potential distribution and cost estimation of the damage caused by *Cryptotermes brevis* (Isoptera: Kalotermitidae) in the Azores. *Journal of Economic Entomology, 107*, 1554–1562.

Gulmahamad, H. (1997). Naturally occurring infestations of drywood termites in books. *Pan-Pacific Entomologist, 73*, 245–247.

Hadi, Y. S., Nawawi, D. S., Herliyana, E. N., & Lawniczak, M. (1998). Termite attack resistance of four polystyrene-impregnated woods from Poland. *Forest Products Journal, 48*, 60–62.

Hadi, Y. S., Westin, M., & Rasyid, E. (2005). Resistance of furfurylated wood to termite attack. *Forest Products Journal, 55*, 85–88.

Hadi, Y. S., Nurhayati, T., Jasni, Y. H., & Kamiya, N. (2010). Smoked wood resistance against termite. *Journal of Tropical Forest Science, 22*, 127–132.

Hadi, Y. S., Nurhayati, T., Jasni, Y. H., & Kamiya, N. (2012). Resistance of smoked wood to subterranean and dry-wood termite attack. *International Biodeterioration and Biodegradation, 70*, 79–81.

Han, G., Cheng, W., Manning, M., & Eloy, P. (2012). Performance of zinc borate-treated oriented structural straw board against mold fungi, decay fungi, and termites – a preliminary trial. *BioResources, 7*, 2986–2995.

Hongoh, Y. (2010). Diversity and genomes of uncultured microbial symbionts in the termite gut. *Bioscience, Biotechnology, and Biochemistry, 74*, 1145–1151.

Hwang, W. J., Yoshimura, T., Tsunoda, K., & Imamura, Y. (2006). Efficacy of two alkylammonium compounds in controlling the attack on wood by western drywood termite, *Incisitermes minor* (Isoptera : Kalotermitidae). *Sociobiology, 48*, 471–478.

IEO – WEI. (2016). What is WEI? http://www.wei-ieo.org/aboutwei.html. Accessed 9 Sept 2016.

International Organization for Standardization. (2016). ISO Standards. In ISO. http://www.iso.org/iso/home/standards.htm. Accessed 2 Sept 2016.

Kartal, S. N., Terzi, E., Kose, C., Hofmeyr, J., & Imamura, Y. (2011). Efficacy of tar oil recovered during slow pyrolysis of macadamia nut shells. *International Biodeterioration and Biodegradation, 65*, 369–373.

Kartal, S. N., Terzi, E., Yoshimura, T., Arango, R., Clausen, C. A., & Green, F. (2012). Preliminary evaluation of storax and its constituents: Fungal decay, mold and termite resistance. *International Biodeterioration and Biodegradation, 70*, 47–54.

Kartal, S. N., Aysal, S., Terzi, E., Yilgor, N., Yoshimura, T., & Tsunoda, K. (2013). Wood and bamboo-PP composites: Fungal and termite resistance, water absorption, and FT-IR analyses. *BioResources, 8*, 1222–1244.

Korb, J., & Hartfelder, K. (2008). Life history and development – a framework for understanding developmental plasticity in lower termites. *Biological Reviews, 83*, 295–313.

Kose, C., Terzi, E., Kartal, S. N., Erilkun, B., & Imamura, Y. (2011). Preliminary evaluation of boron release and biological resistance of wood treated with disodium octoborate tetrahydrate (DOT) and a water-repellent compound. *African Journal of Biotechnology, 10*, 1833–1839.

Krishna, K., Grimaldi, D. A., Krishna, V., & Engel, M. S. (2013). Treatise on the Isoptera of the world. *Bulletin of the American Museum of Natural History, 377*, 307–408.

Kudo, T. (2009). Termite-microbe symbiotic system and its efficient degradation of lignocellulose. *Bioscience, Biotechnology, and Biochemistry, 73*, 2561–2567.

Lee, S. H., Bardunias, P., & Su, N. Y. (2007). Optimal length distribution of termite tunnel branches for efficient food search and resource transportation. *Biosystems Engineering, 90*, 802–807.

Lewis, V. R. (2003). IPM for drywood termites (Isoptera: Kalotermitidae). *Journal of Entomological Science, 38*, 181–199.

Lewis, V., & Forschler, B. (2014). Management of drywood termites: Past practices, present situation, and future prospects. In P. Dhang (Ed.), *Urban insect pests: Sustainable management strategies* (pp. 130–153). Boston: CABI.

Lewis, V. R., Leighton, S., Tabuchi, R., Baldwin, J. A., & Haverty, M. I. (2013). Influence of environmental factors on activity patterns of *Incisitermes minor* (Isoptera: Kalotermitidae) in naturally infested logs. *Journal of Economic Entomology, 106*, 338–346.

Li, Z., Jiang, M., Ma, X., Jiang, Z., & Fei, B. (2012). A laboratory test of the leachability and decay resistance of some synthesized borate compounds. *Forest Products Journal, 62*, 134–138.

Liotta, G. (2005, February 22–27). Le termiti agenti di degrado delle strutture lignee dei beni culturali. In *Proceedings of the international conference "Conservation of Historic Wooden Structures"* (pp. 3–10), Florence.

Maistrello, L., Berzolla, A., Macias-Pavon, I., Vignali, F., Predieri, G., & Chiappini, E. (2012a). Wood impregnated with metal chelates dissolved in organic media tested for termite resistance. *Journal of Entomological and Acarological Research, 43*, 277–285.

Maistrello, L., Martini, L., Macias-Pavon, I., Bortolini, S., & Marchettini, N. (2012b). Evaluation of polyphenols-rich natural compounds as treatments to prevent attacks by subterranean and drywood termites: Preliminary results. *Journal of Entomological and Acarological Research, 43*, 261–267.

Martinez, J. (1957). El termes de Madera seca (*Cryptotermes brevis*) en las islas Canarias. *Montes*, 147–161.

Moein, S., & Rust, M. (1992). The effect of wood degradation by fungi on the feeding and survival of the west-Indian drywood termite, *Cryptotermes brevis* (Isoptera, Kalotermitidae). *Sociobiology, 20*, 29–39.

Nami Kartal, S., Hwang, W.-J., Yamamoto, A., Tanaka, M., Matsumura, K., & Imamura, Y. (2007). Wood modification with a commercial silicon emulsion: Effects on boron release and decay and termite resistance. *International Biodeterioration and Biodegradation, 60*, 189–196.

Nunes, L., Gaju, M., Krecek, J., Molero, R., Teresa Ferreira, M., & De Roca, C. B. (2010). First records of urban invasive *Cryptotermes brevis* (Isoptera:Kalotermitidae) in continental Spain and Portugal. *Journal of Applied Entomology, 134*, 637–640.

Raineri, V., Rey, A., Marini, M., & Zaffagnini, V. (2001). A new discovery of *Cryptotermes brevis* in genoa, Italy (Isoptera). *Bollettino della Società Entomologica I taliana, 133*, 99–102.

Roisin, Y., & Korb, J. (2011). Social organisation and the status of workers in termites. In D. E. Bignell, Y. Roisin, & N. Lo (Eds.), *Biology of termites: A modern synthesis* (pp. 133–164). Heidelberg: Springer.

Rust, M. K., & Su, N. Y. (2012). Managing social insects of urban importance. *Annual Review of Entomology, 57*, 355–375.

Rust, M. K., Reierson, D. A., & Scheffrahn, R. H. (1979). Comparative habits, host utilization and xeric adaptations of the southwestern drywood termites, *Incisitermes fruticavus* rust and *Incisitermes minor* (Hagen). *Sociobiology, 4*, 239–255.

Scheffrahn, R. H., & Crowe, W. (2011). Ship-borne termite (Isoptera) border interceptions in Australia and onboard infestations in Florida, 1986–2009. *Florida Entomologist, 94*, 57–63.

Scheffrahn, R. H., Krecek, J., Ripa, R., & Luppichini, P. (2008). Endemic origin and vast anthropogenic dispersal of the west Indian drywood termite. *Biological Invasions, 11*, 787–799.

Standards Australia. (2016). Standards Australia. http://www.standards.org.au/Pages/default.aspx. Accessed 9 Sept 2016.

Su, N. Y. (2005). Directional change in tunneling of subterranean termites (Isoptera: Rhinotermitidae) in response to decayed wood attractants. *Journal of Economic Entomology, 98*, 471–475.

Su, N. Y., & Puche, H. (2003). Tunneling activity of subterranean termites (Isoptera: Rhinotermitidae) in sand with moisture gradients. *Journal of Economic Entomology, 96*, 88–93.

Su, N. Y., & Scheffrahn, R. H. (1988). Foraging population and territory of the Formosan subterranean termite (Isoptera: Rhinotermitidae) in an urban environment. *Sociobiology, 14*, 353–359.

Suhasman, S., Nadi, Y. S., Massijaya, M. Y., & Santoso, A. (2012). Binder less particleboard resistance to termite attack. *Forest Products Journal, 62*, 412–415.

Sunarta, S., Darmadji, P., Uehara, T., & Katoh, S. (2011). Production and characterization of palm fruit shell bio-oil for wood preservation. *Forest Products Journal, 61*, 180–184.

Tascioglu, C., & Tsunoda, K. (2010). Laboratory evaluation of wood-based composites treated with alkaline copper quat against fungal and termite attacks. *International Biodeterioration and Biodegradation, 64*, 683–687.

Thorne, B. (1998). Biology of subterranean termites of the genus *Reticulitermes*. In *NPCA research report on subterranean termites* (pp. 1–30). Dunn Loring: National Pest Management Association.

Timber Preservers Association of Australia. (2016). TPAA – Timber Preservers Association of Australia. http://www.tpaa.com.au/. Accessed 9 Sept 2016.

Vargo, E. L., & Husseneder, C. (2009). Biology of subterranean termites: Insights from molecular studies of *Reticulitermes* and *Coptotermes*. *Annual Review of Entomology, 54*, 379–403.

Weesner, F. M. (1970). Termites of the Nearactic region. In K. Krishna & F. M. Weesner (Eds.), *Biology of termites* (Vol. 2, pp. 477–525). New York: Academic.

Wu, Q., Lei, Y., Lian, K., & Qi, Y. (2012). Copper/carbon core shell nanoparticles as additive for natural fiber/wood plastic blends. *BioResources, 7*, 3213–3222.

Chapter 13
Biotechnology: A Tool in Termite Management

Tariq Ahmad, Shabnum Nabi, and Qazi Humera

Contents

T. Ahmad (✉)
Entomology Research Laboratory, Department of Zoology, University of Kashmir, Srinagar, J&K 190006, India
e-mail: drtariqento@kashmiruniversity.ac.in

S. Nabi
J.N. Medical College, Aligarh Muslim University, Aligarh 202002, India

Q. Humera
Department of Commerce & Management, University of Kashmir, Srinagar, J&K 190006, India

M.A. Khan, W. Ahmad (eds.), *Termites and Sustainable Management*, Sustainability in Plant and Crop Protection,
https://doi.org/10.1007/978-3-319-68726-1_13

Abstract Termites are the silent invaders, which affect life and property. Being regarded as one of the important agricultural and urban pests, they are of national and international concern to scientists and farmers in particular and to masses in general. It is being estimated that the annual cost of termite damage to the buildings in USA is greater than that of combined cost of fires, storms, and floods, as such hinting an urgent need for termite management. Control strategies have shifted focus on biotechnological approaches for all-inclusive termite management. Biotechnology, globally recognized as a rapidly emerging and far-reaching field, is the "technology of hope" for its promising role in food, health, and environmental sustainability. Latest and enduring advances in life sciences offer a promising scenario, with a large number of agri- and industrial biotech products that have enormously helped mankind. Biotechnology is necessary to sustain an agriculture competitive and remunerative and to achieve nutrition security in the face of major present challenges. Investment in agricultural-related biotechnology has resulted in significantly enhanced research and development capability and institutional building over the years. However, progress has been rather slow in converting the research leads into usable product. In this chapter, therefore, we examine the potential of biotechnology as a tool in termite management.

Keywords Termite • Biotechnological approaches • Sustainable management

13.1 Introduction

Owing to increasing problems taking place in the pesticide applications, the use of biotechnological tools has become inevitable to reduce insect pest losses. Biotechnological approaches for pest management and ecological sustainability may offer alternative strategies to alleviate biotic stresses, by means of tools that can be used in pest management and sustainable crop production. Present focus concerns the biosafety of transgenic crops and environment conservation, covering issues ranging from host plant resistance to insect pests and applied molecular approaches. Also, data are required on phenotyping transgenic plants, mapping populations, and identifying physiochemical and molecular markers associated to insect resistance, checking the real potential of insect-resistant transgenic crops for pest management, or the use of biotechnological tools for diagnosing insects and monitoring their resistance to insecticides.

Biotechnology is necessary to maintain agriculture competitive and remunerative, while on the other hand nutritional security is of major challenges in present times. Biotechnology can play a decisive role in dealing with problems such as declining per capita availability of arable land, lower productivity of crops, livestock and fisheries, heavy production losses due to biotic and abiotic stresses, postharvest crop damage, and water scarcity problems. Investment in agricultural-related biotechnology has resulted in significantly enhanced research and development capability and institutional building over the years. However, progress has been rather slow in converting the research leads into usable products. Further, the spectrum of biotechnological applications in agriculture is wide, comprising of improved crops, animals, plants of agroforestry importance, microorganisms, molecular markers to tag genes of interest, accelerated breeding through marker-assisted selection, cultivars fingerprinting, and germplasm stocks. Further resources include DNA-based diagnostics for pests/pathogens of plants, farm animals, and fish, the assessment and monitoring of biodiversity, in vitro mass multiplication of elite planting material, embryo transfer technology for animal breeding, and finally food and feed biotechnology (NBDS 2015). Plants and animals are also used for the production of therapeutic or industrial products, where the emphasis is on improving efficiency and lowering production costs. Biotechnology plays an important role in developing and processing value-added products with improved nutritive quality, by checking food quality and safety. For example, studies in Indian subcontinent on bio-fertilizers promulgated them as best alternatives to chemical fertilizers. Bio-pesticides show approx. 2.5% share of the total pesticide market, with an annual growth rate of 10–15% (NBDS 2015). Regardless of their plentiful advantages, a number of constraints have limited their wider usage like inconsistent quality, short shelf life, and sensitivity to drought, temperature, and agronomic conditions. This situation appears related to the limited number of organisms that have been studied and tested, as well as to the focus restricted mainly to agronomic parameters. So, biotechnology tools can be used not only for pest management but also for isolating beneficial components of the insect pests' environment. At this regard, the termite management would aim, in a broader prospective, at using eco-friendly approaches for population control, exploiting, at the same time, termites for medical, industrial, and other health benefits of humans.

Termites are among the most successful groups of insects on earth, colonizing most landmasses except Antarctica. Their colonies range in size from a couple of hundred individuals to enormous societies with several million individuals. Due to their wood-eating habits, many termite species can produce great damage to unprotected buildings and other wooden structures (Tokuda et al. 1997) (Table 13.1). Their

Table 13.1 Worldwide crop losses, building damage, and economic losses due to termites

Country	Crop losses (%)	Building damage (%)	Economic losses/annum (million US $)
Australia	–	–	>95.24
Brazil	–	42.7	–
China	–	80–90	248.68–292.79
Europe	–	–	313
India	15–25 (maize crop)	–	35.12
Japan	–	–	800
Malaysia	–	70 – residential 20 – industrial 10 – commercial	8–10
Southern Africa	3–100	–	–
Spain	–	53.2	–
USA	–	–	>1000

Source: Verma et al. (2009)

concealed habit often results in their presence being undetected until the timbers are severely damaged, leaving a thin layer of a wall that protects them from the environment. Of all known species, only 183 cause damage, while 83 may lead to significant damage in wooden structures (Culliney and Grace 2000; Costa-Leonardo 2006). In North America, nine subterranean species are pests, while in Australia, it has been reported that 16 species cause economic damage. In the Indian subcontinent, 26 species are considered as pests, while 24 in tropical Africa and 17 in Central America and West Indies (Culliney and Grace 2000; Costa-Leonardo 2006). Among their genera, *Coptotermes* has the highest number of species of any genus, with 28 known to cause damage. Less than 10% of drywood termites are pests, infecting wooden structures and furniture in tropical, subtropical, and other regions. As such, the need is to use modern pest management tools through an intelligent pest management requiring the integration of multiple approaches for sustainability. Further, the integrated pest management (IPM) should be a socially acceptable, an environmentally responsible and an economically practical method of controlling pest populations. Finally, termites are a delicacy in the diet of some human cultures and are used in traditional medicines or as sacred creatures for worship, in countries such as Malaysia, Singapore, and Thailand.

The complex mounds built by termites have inspired scientists to develop autonomous robots which build castle-like structures and pyramids and other intricate structures without human assistance. Termite-inspired robots, also called as TERMES robots, were developed by researchers of Harvard University in 2014. Roboticists were interested to design robots which may have limited sensory and motor capabilities like termites but still have the ability to build up elaborate and complex structures. These robots travel along a grid, move and climb a step, as well as lift and put down bricks. They can also use sensors to detect other robots and existing bricks and while reacting to these stimuli, they can perceive different tasks such as when to lay a brick or climb a step higher. Such robots are considered useful for future projects on Mars, building castles and for preventing floods (Gibney 2014; Werfel et al. 2014). Termites possess a range of similar evolutionary adaptations which are attributed to their eusocial lifestyles. Recent research focus has been on different aspects of their biology such as caste polyphenism, lignocellulose digestion, and microbial symbiosis, with applications in diverse biotechnological niches. Termite biotechnology can be studied under following headings:

(a) Termite-targeted biotechnology for management of pests
(b) Termite-modeled biotechnology for use in industry

The first category includes some of the promising candidates such as RNA interference, digestive inhibition, pathogen enhancement, endocrine disruption, and primer pheromone mimicry. The second category includes termite digestomes which can be exploited in biomass, industrial, and processing applications (Scharf 2015).

13.2 Termite-Targeted Biotechnology

13.2.1 RNA Interference

RNA interference (RNAi) appears as a successful techniques for termite management. Its prime focus is on development of RNAi insecticides for use against numerous insect pests of agricultural importance (Baum et al. 2007; Scharf 2008). RNAi is a method by which exogenous double-stranded (ds) RNA molecules induce the destruction of endogenous RNAs of identical sequences. Various workers have deliberated on termites and their management under the heading "sociogenomics" (Robinson 1999; Robinson et al. 2005) focusing on gene expression and functional RNAi. These studies revealed caste differentiation as well as identification of genes involved in the termite sociality (Miura and Scharf 2011). It has been experimentally proven that RNAi disrupts protein synthesis. If this disruption is exploited as a pesticidal activity of RNAi, the insect pest management could become a targeted reality. RNAi pesticides have been observed to act via environmental RNAi, whereby sequence-specific gene silencing occurs in response to environmentally acquired dsRNA (Scharf 2015). Therefore, by proper gene targeting, the dsRNA, while inside the organism, can induce a systemic gene silencing response, with lethal effects (Baum et al. 2007; Huvenne and Smagghe 2010).

To date, 16 genes have already been successfully silenced by RNAi in termites and symbiotic protozoa (Scharf 2015). For the first time, the silencing of endogenous hexamerin and cellulose genes was achieved in *Reticulitermes flavipes* through both injection and feeding routes (Zhou et al. 2006, 2008a). These hexamerins are reported to be part of an environmentally responsive caste-regulatory mechanism that attenuates juvenile hormone (JH) potency and limits JH-dependent worker-to-presoldier differentiation (Scharf et al. 2007; Terrapon et al. 2014). Studies showed that hexamerin dsRNA fed, in combination with ectopic JH exposure, result in abnormally high levels of presoldier formation. This in turn induces long-term mortality in termites escalating colony stress, while other symptoms include morphogenic defects and rapid mortality in affected individuals (Zhou et al. 2008b). Although, to a greater extent, the focus is on developing dsRNA termiticides based on nanotechnology-regulatory climate in the USA (USEPA 2011), pesticidal properties of RNAi still remains untested and unverified in the field.

Other economically important insect pests, such as lepidopterans and coleopterans, principally depend on the expression of insecticidal proteins of *Bacillus thuringiensis* (James 2003; Vaughn et al. 2005), which pass the membranes of gut epithelial cells of susceptible insects (Rajamohan et al. 1998). Notwithstanding, the novel insect control strategies could be of significance for managing ever growing insect resistance. Baum et al. (2007) experimentally proved RNAi in quite a few coleopteran species, by supplying an artificial diet meshed with dsRNAs, observing larval stunting and mortality. The results were indicative of the fact that the RNAi pathway can be exploited to manage insect pests and associated diseases by means of common machinery for sequence-specific gene silencing, triggered by the presence of dsRNA (Hannon 2002). This method is referred to as RNAi in animals (Hannon 2002) and posttranscriptional gene silencing in plants (Baulcombe 2004). Similar studies have also been carried out in many other organisms such as the nematode *Caenorhabditis elegans* (Fire et al. 1998; Timmons and Fire 1998), planarian flatworms (Newmark et al. 2003), and ticks (Soares et al. 2005).

13.2.2 Digestive Inhibition

Digestion is the process of breaking down large insoluble food particles into smaller, water-soluble food molecules for their absorption into the blood plasma. In certain organisms, these substances are absorbed through the small intestine into the bloodstream albeit, termites with the aid of enzymes like cellulases, hemicellulases, amylases, and proteases recycle nutrients by means of grinding, decomposition, humification, and mineralization of cellulosic resources and their variants. They have suitable digestive mechanisms capable of metabolizing different biopolymers found in wood, fruits, tubers, crops, and soil components (Tayasu et al. 1997; Hartke and Baer 2011). Termites degrade lignocellulosic materials by the secretion of cellulases and hemicellulases. Cellulases have been used in the production of pharmaceuticals and detergents, wastewater treatment, and fruits or vegetables processing (Mamma et al. 2009). Moreover, by the action of amylases, termites degrade starch and glycogen, hydrolyzing α-1,4-glycosidic bonds in amylose chains and producing glucose, maltose, and maltotriose (Waller and LaFage 1986). These enzymes are biotechnologically applied in food, paper, detergent, and pharmaceutical industries, among others (Hmidet et al. 2009; Souza and Magalhaes 2010). Similarly, soil-feeding termites use peptidic components of humic acids as a source of carbon and energy, through the action of hindgut alkalinity, auto-oxidative processes, and proteolytic activity (Ji and Brune 2005). As such, the gut of soil-feeding termites is an important source of proteases which are used in processing of leather, milk clotting, meat maturation, enzymatic synthesis of sweeteners, and production of biodegradable plastics (Naveena et al. 2004; Ogino et al. 2008; Haddar et al. 2009; Merheb-Dini et al. 2009). The digestive enzymes of higher termites are encoded by the termite genome or produced by gut symbiotic bacteria, such as spirochetes and fibrobacteres (Warnecke et al. 2007). By sequencing, cloning, and expression of enzymes in heterologous systems, the termite enzymes can be produced on a large scale through genetic engineering techniques (Olempska-Beer et al. 2006). Apart from the biotechnological potential, the characterization of digestive enzymes from termites can contribute to the development of new insecticides.

Undoubtedly, the digestive apparatus of termites have a number of biotechnological applications and are a target for pest control. The high efficiency of the lignocellulolytic systems found in their gut makes their cellulases and hemicellulases vital models to be studied for use in the processing of lignocellulosic biomass for biofuel production (Sun and Scharf 2010; Scharf et al. 2011; Mathew et al. 2013). Apart from the biotechnological potential, the characterization of digestive enzymes from termites can also contribute to the development of new insecticides, as for the natural insecticides derived from plants. These applications may proceed from enzyme inhibitors, lectins, and secondary metabolites, known to interfere in digestive enzyme activities. For example, the *Microgramma vaccinifolia* rhizome lectin shows a termiticidal activity leading to imbalances in trypsin-like protease, acid phosphatase, and cellulase from the gut of termites (Albuquerque et al. 2012).

Lignocellulose digestion is of vital importance for termite nutrition. For that reason, an insecticidal inhibiting termite cellulases can be a key factor for management, as shown in in vitro studies on cellulose reduction by ethyl malonate (Misra and Vijayaraghavan 1956). Recent work focused on cellulase inhibitors, viz., β-glucosidase inhibition, an enzyme essential for liberating free glucose from cellulosic substrates. However, enzyme inhibitors which act upstream of β-glucosidase, like endoglucanases and cellobiohydrolases, have not been reported yet, due to the high cost as well as challenges of producing the large carbohydrate molecules necessary to block these enzymes. For this reason, Zhou et al. (2008a) supported the concept of RNAi targeting enzyme-coding RNAs, which can be an alternative approach for termite digestive inhibition, as well as for identifying the most susceptible target enzymes.

13.2.3 Pathogen Enhancement

Pathogen enhancement is a promising technology for insect pest management due to its low or lack of environmental risks. In termite management, the literature also indicates pathogens as effective biological control agents. However, the successful management of insects and termites ultimately relies on amalgamation of some or more techniques, for long lasting results. It has been projected that the market value of microbials for agricultural pest management in developed countries is only about 1% of overall crop protection strategies, while the prime sale is for products based on *B. thuringiensis* (Lisansky 1997; Lacey et al. 2001). Grace (1997) reported pathogens as prospective candidates for termite control, due to their environment being conducive to entomopathogens (Kramm et al. 1982; Rath 2000). Regardless of the huge available literature, however, there is no tangible progress in termite control (Logan et al. 1990, 1992; Culliney and Grace 2000), suggesting that the products developed possibly need a revalidation (Grace 2003).

Termite gut harbors a diverse community of microbes, and their survival is practically dependent on them in various ways. These include the supplement of nitrogen, carbon, and energy requirements (Breznak 2000). The gut flora also protects termites from foreign bacterial invasion by providing a "colonization resistance" (Veivers et al. 1982; Dillon and Dillon 2004). Gut microbes are naturally exchanged among colony members through grooming, food exchange (trophallaxis), and coprophagy (LaFage and Nutting 1978). The termite gut bacteria and protozoa are used as tools and targets for control by means of paratransgenesis, i.e., the use of genetically engineered microorganisms that reside within the host for exploitation as "shuttles" or "Trojan horses," so as to express in them some foreign genes. On the other hand, since termites live in moist subterranean environments conducive for microbial growth, their perceptive physiological and behavioral resistance mechanisms do not allow microbial outbreaks (Rosengaus et al. 2011).

There are also other reasons that enable pathogen tolerance, such as the cellulase enzymes which act as a preadaptation for disease resistance and the evolutionary convergence of digestive and immune functions (Hussain et al. 2013; Scharf et al. 2010). Further, wide-ranging microbial control methods have been tried and tested, but only few have proven successful under field conditions (Chouvenc et al. 2011). However, pathogen efficacy can be enhanced through targeted suppression of the insect resistance factors, by using both pharmacological agents and RNAi.

While applying pharmacological agents, the small glycomimetic molecule D-δ-gluconolactone inhibits gram-negative bacteria-binding proteins (GNBPs), which are commonly known as effector proteins (Bulmer et al. 2009; Hamilton et al. 2011). The second approach could be the use of RNAi for suppressing immune gene expression. Expression of GNBPs and termicin effector proteins has also been attenuated by RNAi, leading to enhanced fungal pathogenicity (Hamilton and Bulmer 2012). Based on these findings, it appears that the effector proteins are viable targets for enhancing pathogens efficacy, while the discovery of new immunosuppressive targets can also prove fruitful in termite control.

Eusocial behavior of termites is also contributing to reduce pathogen resistance due to disruption of hygienic behaviors such as allogrooming, sanitation, and corpse removal, which can have harmful effect on the termite colonies (Sun and Zhou 2013). Such behaviors were experimentally demonstrated by applying sublethal doses of nicotinoid insecticides and fungal pathogens, such as *Beauveria* and *Metarhizium* sp., which resulted in high-level mortality (Boucias et al. 1996; Ramakrishnan et al. 1999). Likewise, another option for successful termite microbial control is to combine pathogens with sublethal doses of CNS-active neurotoxins and drugs which also disturb their eusocial behavior (Scharf 2015).

13.2.4 Endocrine Disruption

Juvenile hormone (JH) plays an important role in termite development and caste differentiation, and its disruption can be exploited as an alternative option for control. It has been reported that soldier caste ranges from 5 to 10% in colonies, a density beyond which their numbers are detrimental to the colonies, as soldiers do not feed and groom themselves, but are cared by workers (Howard and Haverty 1979a, b). Increase in JH titers causes workers to differentiate into presoldiers and soldiers in 3 to 4 weeks (Miura and Scharf 2011). Nonetheless, this developmental change is also inducible by ectopic exposure to JH homologs or synthetic juvenoids (Scharf et al. 2003; Hrd'y et al. 2004; Hrd'y et al. 2006; Wimmer et al. 2007; Toga et al. 2009). Studies have been carried out on termiticidal potential of JH and juvenoids and on their effects in presoldier differentiation, in a number of termite species. However, not all termite species are susceptible to JH or juvenoids, while those susceptible do not show effective responses. The reasons were identified in temperature, seasonality, nutrition, JH stability, caste regulation, and determination of genetic caste. Ectopic JH has been observed most active at 27-33 °C, which is a key limiting factor for any JH-based control technology (Liu et al. 2005; Tarver et al. 2012). Other potential factors limiting JH and juvenoid efficacy are rapid JH metabolism and genetic control of soldier differentiation (Cornette et al. 2006; Hrd'y et al. 2006; Cornette et al. 2008), a phenomenon attributed to reproductive caste formation (Hayashi et al. 2007; Matsuura et al. 2009; Vargo and Parman 2012).

There are several other targets available for termite control, such as the bacteria and protozoa present in the gut and responsible for digestion and other biological processes. These organisms were reported for the first time in *Nasutitermes exitiosus* when antibiotics targeting bacteria and spirochetes reduced the life span of higher termites (Eutick et al. 1978). Subsequent work in lower termites like *Reticulitermes* demonstrated that tetracycline killed protists, which led to the eventual death of the host termites (Mauldin and Rich 1980). Further works on antibiotic applications led to the augmented aggression and mortality via elimination of bacteria-mediated chemical signatures (Matsuura 2001), which in turn can lead to termite mortality via physiological disruption and intercolony aggression. Further, gut bacteria have been used as a Trojan horse for delivery of antiprotozoal agents. This concept was at first tested using recombinant bacteria, *Escherichia coli*, expressing recombinant green fluorescent protein (GFP) as well as with indigenous gut bacteria, *Enterobacter cloacae* (Husseneder and Grace 2005). These studies revealed that exogenously acquired bacteria persist in the gut and can be shifted from donor to recipient termites.

JH-dependent reductions were also recognized in gut-symbiont gene expression (Sen et al. 2013), while further enlightening the role of symbiont population decline in the events preceding caste morphogenesis (Howard and Haverty 1979a, b; Brugerolle and Radek 2006). These methods suggest that disruption of symbiont populations by various routes may promote efficacy of endocrine and caste-disrupting control strategies.

13.2.5 Primer Pheromone Mimicry

In social insects, pheromones are hormone-like chemical messengers that are passed among individuals in colonies, triggering physiological responses in recipients. Primer pheromones are defined as hormone-like semiochemicals that can modulate caste differentiation, either positively or negatively. It has been proved and documented that termite soldier head extracts (SHEs) can be exploited in chemical defense and for chemosystematics purposes, as they also show primer pheromone-like activity (Lefeuve and Bordereau 1984; Okot-Kotber et al. 1991). The termite SHEs contain terpenes derived from the mevalonate pathway, the same biosynthetic pathway that gives rise to JH and other pheromones (Belles et al. 2005). However, its efficacy is also influenced by seasonality, colony caste composition, and other factors. Regardless of their chemical classification, primer pheromones are candidates for new classes of juvenoid and endocrine-disrupting termiticides.

Endocrine disruption can also be achieved by affecting JH or primer pheromone signaling pathways. This can be done through RNAi-based silencing of genes from JH-responsive networks (Tarver et al. 2010; Sen et al. 2013) or, at the protein level, with pharmacological agents or peptide hormones that disrupt signaling. Several peptide hormones and related signaling pathways have now been verified in termites, including neurologically active NPF peptides (Nuss et al. 2010), an allatostatin that regulates JH biosynthesis and responds to JH in the gut (Elliott et al. 2009; Sen et al. 2013), and JH-dependent insulin signaling (Hattori et al. 2013; Sen et al. 2013).

13.3 Termite-Modeled Biotechnology

13.3.1 Termite Digestomes

The term digestome is used to describe the collective pool of host and symbiont genes that collaborate to achieve high-efficiency lignocellulose digestion in the termite gut, as well as simple sugar fermentation, nutrient transport, and assimilation (Scharf and Tartar 2008). The digestive apparatus of termites has several biotechnological applications, while it is also a target for pest control. A lot of research has been carried out on lignocellulosic biomass as it is a source of simple sugars and aromatics, which can be utilized for fermentation to biofuels and biomaterials (Girio et al. 2010; Douglas et al. 2012). This has led to the establishment of some technologies to use this conversion process on an industrial level (Ragauskas et al. 2006; Menon and Rao 2012). However, these technologies need to be further developed and improved, to become commercial and cost-effective.

Various components of lignocellulose, such as cellulose and hemicellulose, are digested in the termite gut, releasing glucose and pentose monosaccharides that flow into essential metabolic pathways in both host and symbionts (Brune and Ohkuma 2011; Ohkuma and Brune 2011). Nonetheless, many workers believe the digestion of termite lignocelluloses is carried out by microbial symbiota like bacteria and protists positioned in the hindgut paunch (Ohkuma 2003; Brugerolle and Radek 2006; Bignell 2011; Hongoh 2011). On the other hand, Tartar et al. (2009) reported that not only microbial symbionts assist in lignocellulose digestion, but host-derived activities also contribute to digestion as most of the relevant host enzymes are expressed at much higher levels than any individual microbial enzyme. Regardless of previous studies confirming the role of host termite enzymes in lignocellulose digestion (Cleveland 1928; Hungate 1938), this view was accepted many decades later, when endogenous termite cellulase was identified (Watanabe et al. 1998; Lo et al. 2011). Consequently, extensive research was conducted in integrative enzymology confirming that not only cellulases but also other enzymes, such as phenol oxidases and other related lignases, contribute in termite digestion (Coy et al. 2010; Ke et al. 2013). Studies on digestomes of a range of termite species were carried out by different workers, as in *R. flavipes*, in which around 200 candidate lignocellulose genes from host and protist symbionts were reported (Sethi et al. 2013).

13.3.2 Biomass Digestion

Lignocellulose is one of the sustainable global resources that are receiving much attention as a renewable energy supply (Ohkuma 2003; Ni and Tokuda 2013). It is the most abundant, widespread, and renewable bioenergy resource available on earth, due to its plant origin (Scharf 2015). The use of plant biomass for energy production has its advantage when keeping in view its carbon neutral properties, if compared to fossil fuels (Ragauskas et al. 2006); nonetheless it needs to be produced strictly in sustainable ways (Johnson 2009). This is warranted by the fact that a lot of research has been focused on lignocellulosic biomass for the purpose of producing biofuels and biomaterials. Especially in the last decade, the use of feedstocks for biofuel production increased, particularly for second-generation feedstocks, as a source of alcohol fuel (Naik et al. 2010; Alonso et al. 2011).

13.3.3 Translational Digestomics

Scharf and Tartar (2008) advocated a reverse genetics approach for a gene sequencing project in termites due to the complexity of the enzyme systems essential for lignocellulose digestion, followed by translated gene sequencing and experimental characterization. It is reported that around 110 recombinant termite and symbiont digestive enzymes have been produced to date, while 13 termites have been studied for digestome enzymes. The majority of the enzymes account from a very few species, viz., *R. speratus* (24%), *Coptotermes formosanus* (17%), *R. flavipes* (15%), *Nasutitermes costalis* (11%), *N. takasagoensis* (9%), and *Neotermes koshunensis* (6%) (Scharf 2015). Further, 60% of these different recombinant digestive enzymes are derived from bacteria, 29% from host termite and 11% from protist symbionts. The different recombinant expression platforms employed during the whole process comprised of bacteria (47%), yeast (34%), baculovirus/insect (11%), and fungus (8%). Therefore at the moment, fungi, yeast, and bacteria are treated as the most sought-after expression systems from the industrial point of view (Scharf 2015).

13.3.4 Renewable Energy

The energy gathered from resources that are naturally stocked up on a human time scale such as sunlight, wind, rain, tides, waves, and geothermal heat is defined as a renewable energy. It is another key termite-modeled biotechnological technique, with research efforts performed on termite gut to replace fossil fuels with a cleaner renewable energy sources (Gibney 2014; Werfel et al. 2014). It is believed that termites are efficient bioreactors which are capable of producing two liters of hydrogen from a single sheet of paper (Zhang et al. 2010), while their hindgut harbors around 200 species of microbes. During digestion, they release hydrogen which is trapped inside wood and plants (Heather 1971; Machida et al. 2001). The enzymes in the termite gut break down lignocellulose polymers into sugars and hydrogen, while the gut bacteria convert sugar and hydrogen into cellulose acetate, an acetate ester of cellulose on which termites rely for energy (Breznak and Brune 1993). To better understand the metabolic pathway, community DNA sequencing of the microbes in the termite hindgut is of importance, while genetic engineering may facilitate us to generate hydrogen in bioreactors from any woody biomass (Breznak and Brune 1993).

Termites use sophisticated means to control the temperatures of their mounds. The shape and orientation of the mounds of the Australian compass termite stabilize their internal temperatures during the day, and as the towers heat up, the solar chimney effect (stack effect) creates an updraft of air within the mound (Heather 1971; Tan and Wong 2013). Wind blowing across the tops of the towers enhances the air circulation through the mounds, which also include side vents in their construction. The solar chimney effect has been in use for centuries in human societies for passive cooling (Heather 1971; Tan and Wong 2013). It is only relatively recently, however, that climate responsive construction techniques have become incorporated into modern architecture especially in Africa, as the stack effect has become a popular means to achieve natural ventilation and passive cooling in modern buildings.

For sustainable economy and clean environment, cellulose-based biofuels have recently received tremendous attention both in industry and academic communities. Lignocellulose biomass is a potential alternative to reduce dependence on fossil fuels and alleviate global climatic change, notwithstanding the cost-effective processes for converting biomass to biofuels is yet to be fully realized (Sun and Scharf 2010). The last two decades have mainly been devoted to the production of industrial bioethanol by means of biocatalysis and fermentation technologies from bacterial and fungal cellulolytic systems, in combination with breakthroughs in molecular genetics, enzyme engineering, and metabolic engineering. However, it is not out of place to mention here that the technology for biomass conversion for large-scale application is still in its infancy, due to issues concerning efficiency and processing economics. To progress in this direction, we need a sound strategy, learning, and state-of-the-art technologies to exploit organisms such as wood-feeding termites and other insects, which can process lignocellulosic biomass much more efficiently.

The cellulose-digesting insects which feed on agricultural crops and forest plantations include termites, wood-feeding roaches, beetles, wood wasps, aquatic insects, and silverfish. Cellulose digestion has been reported in more than 20 insect families belonging to 10 different insect orders, viz., Thysanura, Plecoptera, Dictyoptera, Orthoptera, Isoptera, Coleoptera, Trichoptera, Hymenoptera, Phasmida, and Diptera. These insects have been extensively studied for understanding the mechanism of lignocellulose digestion and its potential for biofuel production, with the aid of molecular approaches such as metagenomics, proteomics, and transcriptomics (Jian-Zhong and Scharf 2010).

Some of the objectives for investigating insect cellulolytic systems include screening out for genes and enzymes suitable for industrial applications, as well as the study of their physiochemical microhabitats. This can be done by pretreatment technologies and removal of substrate-specific barriers to cellulases, to improve cellulose digestion. The studies on host termite transcriptomes attribute ligninases and phenolic acid esterases as candidate enzymes which can modify lignin components from lignocellulosic substrate (Scharf and Boucias 2010).

Studies on in situ oxygen profiling within the fore-, mid-, and hindgut of two wood-feeding lower termite species revealed that the lignin disruption occurs in fore- and midgut regions, from where the wood particles move to the hindgut for further depolymerization by hindgut protozoa (Ke et al. 2010). Similarly, Zhang et al. (2010) described two recombinant endogenous glycosyl hydrolases from a lower termite species, which were expressed in *E. coli* to convert cellulose to glucose. On the other hand, examination of hydrogen and methane emissions as by-products from three lower termite species during cellulose degradation showed a characteristic method for producing bio-hydrogen as a by-product during cellulose conversion through gut cellulolytic and metabolic systems (Cao et al. 2010).

Apart from termites, enzyme biochemistry and lignocellulose degradation has been studied in detail in wood-feeding beetles such as the Asian longhorn beetle (Geib et al. 2010), through zymogram analysis to identify and characterize cellulases and hemicellulases as well as their prospect in the production of cellulosic biofuels. Another example of cellulose-consuming beetle species is the scarab beetle, whose larval gut microflora associated with digestive processes has a potential in the development of artificial bioreactors (Huang et al. 2010). Cook and Doran-Peterson (2010) described the crane fly gut to serve as a natural bio-refinery, a model which can be applied in improving and developing biomass-to-biofuel technology. In recent years, studies have been done in North American forests with an aim of dealing with the issues of pest management and landscape ecology as well as its prospect in biomass harvesting and production of biofuels (Landis and Werling 2010). Such studies are not only encouraging but are emerging as an important area in the field of industrial biotechnology.

13.4 Termites in Medicine

Since a long time, insects have been used in medicine and as food in different parts of the world. They are still a basic component of human diets and other animals (Raubenheimer and Rothman 2012), representing a significant biomass (Meyer-Rochow and Chakravorty 2013), that also played a role as a medicinal resource (Costa-Neto 2005; Dossey 2010). In Asia, America, and Africa, entomophagy (the practice of consuming insects) has long been included in human diet (Meyer-Rochow 2010). Besides, insects have been used by humans for medicinal purposes, a practice known as entomotherapy (Costa-Neto 2005).

Termites in human food have been negatively projected as mediators of plant organic matter decomposition and as means of soil formation; energy and nutrient flow principally in tropical forests (Lee and Wood 1971; Vasconcellos and Moura 2010). Nonetheless termites are used in traditional popular medicine (Solavan et al.

2006; Alves and Alves 2011) by treating a range of diseases like asthma, tonsillitis, bronchitis, influenza, whooping cough, sinusitis, and hoarseness (Alves 2009). Furthermore, termites have been an essential source of food in the diets of people who suffer from protein deficiency and malnutrition. They have being consumed for generations in different parts of the world, a practice that has increased their popularity in recent years (Van-Huis 2003; Shockley and Dossey 2014).

African tribes have termites as totems, and tribe members are prohibited to consume their reproductive alates (Meyer 1999; De Visse et al. 2008). In Nigeria *Macrotermes nigeriensis* is not only used for spiritual protection but also to treat wounds and sick pregnant women, while in Southeast Asian countries they are involved in traditional rituals. In other countries like Malaysia, Singapore, and Thailand, people worship termite mounds (Gibney 2014; Werfel et al. 2014), on the basis of a spiritual beliefs.

It is noteworthy to mention here that there is enough data available regarding antimicrobial activity of products isolated from termites such as espinigerine and termicine, which are isolated from *Pseudocanthotermes spiniger*. These products were reported to have antifungal and antibacterial activities (Ramos-Elorduy 2005). Similarly another termite species, *Nasutitermes macrocephalus*, has been used to treat various diseases, while the molecular biological studies on members of the genus *Nasutitermes* confirmed their potential as antimicrobial peptide producers (Lamberty et al. 2001; Bulmer and Crozier 2004). On the other hand, Bulmer and Crozier (2006) also recommended that *Nasutitermes corniger* and its nest are promising natural products for use in antimicrobial therapy. Consequently, keeping pace with the past research work, termites need to be studied thoroughly from the biotechnological and pharmacological point of view.

13.5 Termites as Food

As discussed above, termites are not only used in many traditional medicines but also as a delicacy in the diet of a few human cultures. They play a key role in energy transfer via trophic interactions at various scales. Many studies threw light on various species being exploited as a food for humans and a feed for livestock. They especially detailed out their beneficial role in under-developed countries where malnutrition is prevalent, as termite proteins substantially improve human health.

Termite consumption is now global regardless of their recent popularity in developed nations (Table 13.2). In Africa, the alates comprise an essential part in the diets of native populations. Tribals use different ways and means for collecting and cultivating insects and may gather soldiers from several species (Meyer 1999). On the other hand queens, which are difficult to collect, are also regarded as a delicacy, while termite alates are high in nutrition with adequate levels of fats and proteins. The alates are very pleasant in taste with nutlike flavor after cooking (Meyer 1999). Apart from humans, termites also act as an important energy source and food for many insects, mammals, birds, and reptiles (Longhurst et al. 1978; Kob and Hewitt 1990).

Table 13.2 Termite species and their use in different parts of the world

Termite species	Usage	Country(ies)
Hodotermes mossambicus	Child malnutrition and as food	Botswana, Zambia
Microhodotermes viator	Food	South Africa
Kalotermes flavicollis	Food and feed	Brazil, Thailand
Coptotermes formosanus	Food	China
Reticulitermes flavipes	Food	Thailand
Reticulitermes tibialis	Food	Mexico
Cubitermes atrox	Food	Indonesia
Labiotermes labralis	Food	Colombia
Macrotermes acrocephalus	Food	China
Macrotermes annandalei	Food	China
Macrotermes barneyi	Food	China
Macrotermes bellicosus	Treatment of suture wounds and as food and feed	Somalia, Central Africa, Congo, Nigeria, Angola, Zambia, Kenya, Guinea, Senegal, Tanzania, Uganda
Macrotermes falciger	Food	Zimbabwe, Congo, Benin, Zambia, South Africa
Macrotermes gabonensis	Food	Congo
Macrotermes gilvus	Food	Thailand, Malaysia
Macrotermes herus	Feed	Tanzania
Macrotermes lilljeborgi	Feed	Cameroon, Guinea
Macrotermes michaelseni	Feed	Malawi
Macrotermes muelleri	Feed	Congo, Cameroon, Guinea
Macrotermes natalensis	Feed	Africa, Zimbabwe, Congo, Nigeria
Macrotermes nobilis	Feed	Congo, Gabon, Cameroon
Macrotermes renouxi	Feed	Cameroon
Macrotermes subhyalinus	Food and feed	Angola, Zambia, Kenya, Senegal, Tanzania, Uganda
Macrotermes vitrialatus	Food	Zambia
Microcerotermes dubius	Food	Malaysia
Microcerotermes serrula	Food	Malaysia
Nasutitermes corniger	Food and to treat asthma, cough, flu, and sore throat	Venezuela, Brazil
Nasutitermes ephratae	Food	Venezuela
Nasutitermes macrocephalus	Food and to treat asthma, bronchitis, coughs, influenza, sore throat, sinusitis, tonsillitis, and hoarseness	Venezuela, Brazil
Nasutitermes surinamensis	Food	Venezuela
Odontotermes badius	Food	South Africa, Zambia, Kenya
Odontotermes capensis	Food	South Africa
Odontotermes feae	Food and to treat asthma	India

(continued)

Table 13.2 (continued)

Termite species	Usage	Country(ies)
Odontotermes formosanus	Food and to treat ulcer, body pain, rheumatics, anemia, etc.	India, China
Odontotermes kibarensis	Food	Uganda
Odontotermes yunnanensis	Food	China
Pseudacanthotermes militaris	Food and feed	Angola, Kenya, Tanzania, Uganda
Pseudacanthotermes spiniger	Food and feed as well as having antifungal and antibacterial properties	Congo, Zambia, Tanzania, Kenya, Uganda, Brazil
Syntermes aculeosus	Food	Venezuela, Brazil
Syntermes parallelus	Food	Colombia
Syntermes spinosus	Food	Brazil, Colombia, Venezuela
Syntermes tanygnathus	Food	Colombia
Termes fatalis	Food	Guyana, Indonesia
Macrotermes nigeriensis	Sickness in pregnant women as well as wounds	Nigeria
Microcerotermes exiguous	Asthma, bronchitis, influenza, whooping cough, flu	Brazil

Source: Figueiredo et al. (2015)

In other countries such as in Asia and America, termites are consumed in local or tribal areas. In Australia, however, local people are quite aware that termites are edible, but they do not consume them even in times of scarcity (Heather 1971). On the contrary, in countries like Kenya, Tanzania, Zambia, Zimbabwe, and South Africa, termite mounds are used as a main food source, a practice called "geophagy" (Heather 1971; Katayama et al. 2008). Since termites are high in protein content as well as can be used to convert inedible waste to consumable products for humans, some researchers considered them as suitable candidates for human consumption and space agriculture (Katayama et al. 2008). To sum up, a number of termite species are used as food by people from all over the world, substantiating the studies carried out by previous works on termite nutrients and benefits (Van-Huis 2013; Banjo et al. 2006). Most of the studies also unequivocally put them next to grasshoppers, for consumption around the globe (Anankware et al. 2015).

13.6 Termites in Agriculture

Termite advantages and disadvantages in agricultural sector run in tandem as they can be utilized in enhancement of agricultural produce but may also damage the quality and market value of crops. In any case, the question of yield losses due to termites in any agroecosystem needs to be thoroughly studied and established (Wood and Cowie 1988). Even in some cases, termite attack has not resulted in yield loss especially in the areas with high planting densities, because the plants which were lost through termite attack resulted in an increase in yield due to lower competition for water, nutrients, and space (Logan 1992). Termites play an important role in energy transfer via trophic interactions at different levels and with the aid of ambient environmental conditions, they breakdown organic matter which ultimately affects soil microbial activity (Keya et al. 1982; Moorhead and Reynolds 1991). They utilize soil from a variety of depths for nest and gallery construction, while the same soil is ultimately redistributed by erosion and weathering processes (Lee and Wood 1971), with noticeable effects on its structure and properties.

In Africa and India, farmers plough large mounds which in some areas have become an integral part of the cropping systems (Mielke and Mielke 1982). In Australia and other semiarid ecosystems, research efforts revealed a role of termites in soil enrichment and crop productivity (Schaefer and Whitford 1981; Park et al. 1994). The importance of termites in carbon and nitrogen fluxes in the humid environment has also been studied in West Africa, Malaysia, and South America (Martius 1994; Lawton et al. 1996).

Studies have also highlighted the role of termites in decomposition of plant matter in natural ecosystems (Holt 1987; Martius 1994). In Africa, farmers spread termite mound soil in their fields so as to perk up soil conditions and plant nutrients (Watson 1977; Nyamapfene 1986). Because of such colossal benefits, farmers make the most of termite mounds as a substitute to chemical fertilizers (Oliviera and Paiva, 1985; Bishosha and Boloy 1995). Likewise, Holt and Coventry (1988) in their extensive studies concluded that the diversity and population of termites often decline if their seminatural or natural habitats are disturbed. The termites influence surrounding vegetation, which is ascribed to soil partitioning by the colonies (Jones 1989). This process is a typical case of ecosystem engineering (Jones et al. 1994) which results in spatial gradients in soil nutrient and water availabilities (Salick et al. 1983; Noble et al. 1989). Conversely, this spatiality is transferred into agricultural systems which ultimately influence farmers' practices and agricultural productivity.

Keeping huge benefits of termites in mind, economically or socially essential crops like cotton, tobacco, and millet are often grown exclusively on termite mounds (Logan 1992). The Chitemene in Tanzania uses termite mounds in their agricultural practices (Mielke and Mielke 1982). At the same time in South and West Africa, low-income farmers utilize termite activities in least-risk farming strategies (Brinn et al. 1994; Carter and Murwira 1995).

Therefore, the diversity of termites and the biological stability of natural ecosystems are mutually of principal importance, indicating that reduction in termite diversity has negative impacts on the ecosystem functioning as well as on agroecosystems. Farming practices can also have a feedback effect on termite diversity and activity, linked to alterations in their ecological processes, in soil nutrient cycling and water conductivity. Increased tillage, pesticides, monoculture, and reduction of fallow length also decrease termite diversity. This further effects result in water scarcity to plants, influencing nutrient cycling by decomposers and increasing pest losses (Black and Okwakol 1997).

13.7 Conclusion

Biotechnology is an indispensable prerequisite for a systematic development in agriculture, medicine, food, or any other industries with a global demand. Biotechnological approaches are worth appreciation as they sustain human needs directly and indirectly. An example is given by termites which initially were used to be treated as damaging pests but now are mostly sought as useful insects for various industrial purposes. The related biotechnology has led to new vistas, concepts, ideas, and methods for their exploitation in various industries. Termite-targeted biotechnology has led to the management of pests by applying novel potential candidates such as RNA interference, digestive inhibition, pathogen enhancement, endocrine disruption, and primer pheromone mimicry. The termite-modeled biotechnology for use in industries focused on their digestomes for utilization in biomass, industrial, and processing applications (Scharf 2015). Undoubtedly, a lot of scientific interest has been focused on termites especially on defensive, biomaterial, engineering, and other exclusive evolutionary innovations with an aim of serving people. Therefore, these important aspects of termites biology and ecology demonstrate that these tiny creatures are human friends and not as foes, with a huge potential in various industrial sectors.

References

Albuquerque, L. P., Santana, G. M. S., Pontual, E. V., Napoleao, T. H., Coelho, L. C. B. B., & Paiva, P. M. G. (2012). Effect of *Microgramma vaccinifolia* rhizome lectin on survival and digestive enzymes of *Nasutitermes corniger* (Isoptera, Termitidae). *International Biodeterioration and Biodegradation, 75*, 158–166.

Alonso, D. M., Wettstein, S. G., Bond, J. Q., Root, T. W., & Dumesic, J. A. (2011). Production of biofuels from cellulose and corn stover using alkylphenol solvents. *ChemSusChem, 4*, 1078–1081.

Alves, R. R. N. (2009). Fauna used in popular medicine in Northeast Brazil. *Journal of Ethnobiology and Ethnomedicine, 5*, 1–30.

Alves, R. R. N., & Alves, H. N. (2011). The faunal drugstore: Animal-based remedies used in traditional medicines in Latin America. *Journal of Ethnobiology and Ethnomedicine, 7*, 1–43.

Anankware, P. J., Fening, K. O., Osekre, E., & Obeng-Ofori, D. (2015). Insects as food and feed: A review. *International Journal of Agricultural Research Review, 3*, 143–151.

Banjo, A. D., Lawal, O. A., & Songonuga, E. A. (2006). The nutritional value of fourteen species of edible insects in Southwestern Nigeria. *African Journal of Biotechnology, 5*, 298–301.

Baulcombe, D. (2004). RNA silencing in plants. *Nature, 431*, 356–363.

Baum, J. A., Bogaert, T., Clinton, W., Heck, G. R., & Feldmann, P. (2007). Control of Coleopteran insect pests through RNA interference. *Nature Biotechnology, 25*, 132–226.

Belles, X., Martın, D., & Piulachs, M. D. (2005). The mevalonate pathway and the synthesis of juvenile hormone in insects. *Annual Review of Entomology, 50*, 181–199.

Bignell, D. E. (2011). Morphology, physiology, biochemistry and functional design of the termite gut: An evolutionary wonderland. In D. E. Bignell DE, Y. Roisin, & N. Lo (Eds.), *Biology of termites: A modern synthesis* (pp. 375–412). Dordrecht: Springer.

Bishosha, M. K., & Boloy, N. (1995). Termitary soil and dried peanut straw as market-garden fertilizers in Yangambi (Zaire). *Cahiers Agricultures, 4*, 125–128.

Black, H. I. J., & Okwakol, M. J. N. (1997). Agricultural intensification, soil biodiversity and agroecosystem function in the tropics: The role of termites. *Applied Soil Ecology, 6*, 37–53.

Boucias, D. G., Stokes, C., Storey, G., & Pendland, J. C. (1996). The effects of imidacloprid on the termite *Reticulitermes flavipes* and its interaction with the mycopathogen *Beauveria bassiana*. *Pflanzenschutz-Nachrichten Bayer, 49*, 103–144.

Breznak, J. A. (2000). Ecology of prokaryotic microbes in the guts of wood- and litter-feeding termites. In T. Abe, D. E. Bignell, & M. Higashi (Eds.), *Termites: Evolution, sociality, symbioses, ecology* (pp. 209–231). Dordrecht: Kluwer Academic.

Breznak, J. A., & Brune, A. (1993). Role of microorganisms in the digestion of lignocellulose by termites. *Annual Review of Entomology, 39*, 453–487.

Brinn, P. J., Black, H. I. J., Spurway, J. K. R., & Mzezewa, J. (1994) Land microvariability in Southwest Zimbabwe; photointerpre- tive identifications of land microvariability. In *SADC, proceedings of the 4th annual scientific conference of the SADC-LW and MP research programme*, October 1993, Windhoek, Namibia, SADC, Gabarone.

Brugerolle, G., & Radek, R. (2006). Symbiotic protozoa of termites. In H. Konig & A. Varma (Eds.), *Intestinal microorganisms of termites and other invertebrates* (pp. 243–269). Berlin/Heidelberg: Springer.

Brune, A., & Ohkuma, M. (2011). Role of the termite gut microbiota in symbiotic digestion. In D. E. Bignell, Y. Roisin, & N. Lo (Eds.), *Biology of termites: A modern synthesis* (pp. 439–475). Dordrecht: Springer.

Bulmer, M. S., & Crozier, R. H. (2004). Duplication and diversifying selection among termite antifungal peptides. *Molecular Biology and Evolution, 21*, 2256–2264.

Bulmer, M. S., & Crozier, R. H. (2006). Variation in positive selection in termite GNBPs and Relish. *Molecular Biology and Evolution, 23*, 317–326.

Bulmer, M. S., Bachelet, I., Raman, R., Rosengaus, R. B., & Sasisekharan, R. (2009). Targeting an antimicrobial effect or function in insect immunity as a pest control strategy. *Proceedings of the National Academy of Sciences of the United States of America, 106*, 12652–12657.

Cao, Y., Sun, J. Z., Rodriguez, J. M., & Lee, K. C. (2010). Hydrogen emission by three wood-feeding subterranean termite species (Isoptera: Rhinotermitidae): Production and characteristics. *Insect Science, 17*, 237–244.

Carter, S. E., & Murwira, H. K. (1995). Spatial variability in soil fertility management and communal farming in Mutoko Communal Area, Zimbabwe. *Ambio, 24*, 77–84.

Chouvenc, T., NY, S., & Grace, J. K. (2011). Fifty years of attempted biological control of termites–analysis of a failure. *Biological Control, 59*, 69–82.

Cleveland, L. R. (1928). Further observations and experiments on the symbiosis between termites and their intestinal protozoa. *The Biological Bulletin, 54*, 231–237.

Cook, D. M., & Doran-Peterson, J. (2010). Mining diversity of the natural biorefinery housed within *Tipula abdominalis* larvae for use in an industrial biorefinery for production of lignocellulosic ethanol. *Insect Science, 17*, 303–312.

Cornette, R., Koshikawa, S., Hojo, M., Matsumoto, T., & Miura, T. (2006). Caste-specific cytochrome P450 in the damp-wood termite *Hodotermopsis sjostedti*. *Insect Molecular Biology, 15*, 235–244.

Cornette, R., Gotoh, H., Koshikawa, S., & Miura, T. (2008). Juvenile hormone titers and caste differentiation in the damp-wood termite *Hodotermopsis sjostedti*. *Journal of Insect Physiology, 54*, 922–930.

Costa-Leonardo, A. M. (2006). Morphology of the sternal gland in workers of *Coptotermes gestroi* (Isoptera, Rhinotermitidae). *Micron, 37*, 551–556.

Costa-Neto, E. M. (2005). Entomotherapy, or the Medicinal Use of Insects. *Journal of Ethnobiology, 25*, 93–114.

Coy, M. R., Salem, T. Z., Denton, J. S., Kovaleva, E. S., & Liu, Z. (2010). Phenol-oxidizing laccases from the termite gut. *Insect Biochemistry and Molecular Biology, 40*, 723–732.

Culliney, T. W., & Grace, J. K. (2000). Prospects for the biological control of subterranean termites (Isoptera: Rhinotermitidae), with special reference to *Coptotermes formosanus*. *Bulletin of Entomological Research, 90*, 9–21.

De Visse, S. N., Freymann, B. P., & Schnyder, H. (2008). Trophic interactions among invertebrates in termitaria in the African savanna: A stable isotope approach. *Ecological Entomology, 33*, 758–764.

Dillon, R. J., & Dillon, V. M. (2004). The gut bacteria of insects: Nonpathogenic interactions. *Annual Review of Entomology, 49*, 71–92.

Dossey, A. T. (2010). Insects and their chemical weaponry: New potential for drug discovery. *Natural Product Reports, 27*, 1737–1757.

Douglas, B. J., Michael, J. B., Jay, D. B., Bruce, S., Ronald, E. H., Charles, C. L., & Kurt, W. (2012). Plant cell walls to ethanol. *The Biochemical Journal, 442*, 241–252.

Elliott, K. L., Hehman, G. L., & Stay, B. (2009). Isolation of the gene for the precursor of Phe-Gly-Leu-amide allatostatins in the termite *Reticulitermes flavipes*. *Peptides, 30*, 85–560.

Eutick, M. L., Veivers, P., O Brien, R. W., & Slaytor, M. (1978). Dependence of the higher termite, *Nasutitermes exitiosus* and the lower termite, *Coptotermes lacteus* on their gut flora. *Journal of Insect Physiology, 24*, 363–368.

Figueirêdo, R. E. C., Vasconcellos, A., Policarpo, I. S., & Alves, R. R. N. (2015). Edible and medicinal termites: A global overview. *Journal of Ethnobiology and Ethnomedicine, 11*, 29.

Fire, A., Xu, S., Mary, K., Montgomery, M. K., Kostas, S. A., Driver, S. E., & Mello, C. C. (1998). Potent and specific genetic interference by double-stranded RNA in *Caenorhabditis elegans*. *Nature, 391*, 806–811.

Geib, S. M., Tien, M., & Hoover, K. (2010). Identification of proteins involved in lignocellulose degradation using in gel zymogram analysis combined with mass spectroscopy-based peptide analysis of gut proteins from larval Asian longhorned beetles, Anoplophora glabripennis. *Insect Science, 17*, 253–264.

Gibney, E. (2014). Termite-inspired robots build castles. *Nature*. https://doi.org/10.1038/nature.2014.14713.

Girio, F. M., Fonseca, C., Carvalheiro, F., Duarte, L. C., Marques, S., & Bogel-Lukasik, R. (2010). Hemicelluloses for fuel ethanol: A review. *Bioresource Technology, 101*, 4775–4800.

Grace, J. K. (1997). Biological control strategies for suppression of termites. *Journal of Agricultural Entomology, 14*, 281–289.

Grace, J. K. (2003). Approaches to biological control of termites. *Sociobiology, 41*, 115–121.

Haddar, A., Agrebi, R., Bougatef, A., Hmidet, N., Sellami-Kamoun, A., & Nasri, M. (2009). Two detergent stable alkaline serine-proteases from *Bacillus mojavensis* A21: Purification, characterization and potential application as a laundry detergent additive. *Bioresource Technology, 100*, 3366–3373.

Hamilton, C., & Bulmer, M. S. (2012). Molecular antifungal defenses in subterranean termites: RNA interference reveals *in vivo* roles of termicins and GNBPs against a naturally encountered pathogen. *Developmental and Comparative Immunology, 36*, 372–377.

Hamilton, C., Lay, F., & Bulmer, M. S. (2011). Subterranean termite prophylactic secretions and external antifungal defenses. *Journal of Insect Physiology, 57*, 259–1266.

Hannon, G. J. (2002). RNA interference. *Nature, 418*, 244–251.

Hartke, T. R., & Baer, B. (2011). The mating biology of termites: A comparative review. *Animal Behaviour, 82*, 927–936.

Hattori, A., Sugime, Y., Sasa, C., Miyakawa, H., & Ishikawa, Y. (2013). Soldier morphogenesis in the dampwood termite is regulated by the insulin signaling pathway. *The Journal of Experimental Zoology B, 320*, 295–306.

Hayashi, Y., Lo, N., Miyata, H., & Kitade, O. (2007). Sex-linked genetic influence on caste determination in a termite. *Science, 318*, 985–987.

Heather, N. W. (1971). The exotic drywood termite *Cryptotermes brevis* (Walker) (Isoptera: Kalotermitidae) and endemic Australian drywood termites in Queensland. *Australian Journal of Entomology, 10*, 134–141.

Hmidet, N., Ali, N. E., Haddar, A., Kanoun, S., Alya, S., & Nasri, M. (2009). Alkaline proteases and thermostable α-amylase co-produced by *Bacillus licheniformis* NH1: Characterization and potential application as detergent additive. *Biochemical Engineering Journal, 47*, 71–79.

Holt, J. A. (1987). Carbon mineralization in semi-arid northeastern Australia: The role of termites. *Journal of Tropical Ecology, 3*, 255–263.

Holt, J. A., & Coventry, R. J. (1988). The effects of tree clearing and pasture establishment on a population of mound-building termites (Isoptera) in north Queensland, Australia. *Australian Jorunal of Ecology, 13*, 321–326.

Hongoh, Y. (2011). Toward the functional analysis of uncultivable, symbiotic microorganisms in the termite gut. *Cellular and Molecular Life Sciences, 68*, 1311–1325.

Howard, R. W., & Haverty, M. I. (1979a). Comparison of feeding substrates for evaluating effects of insect growth regulators on subterranean termites. *Journal of Georgia Entomological Society, 14*, 3–7.

Howard, R. W., & Haverty, M. I. (1979b). Termites and juvenile hormone analogs: A review of methodology and observed effects. *Sociobiology, 4*, 269–278.

Hrd'y, I., Kuldova, J., & Wimmer, Z. (2004). Juvenogens as potential agents in termite control: Laboratory screening. *Pest Management Science, 60*, 1035–1042.

Hrd'y, I., Kuldov'a, J., Hanus, R., & Wimmer, Z. (2006). Juvenile hormone III, hydroprene and a juvenogen as soldier caste differentiation regulators in three *Reticulitermes* species: Potential of juvenile hormone analogues in termite control. *Pest Management Science, 62*, 848–854.

Huang, S. W., Zhang, H. Y., Marshall, S., & Jackson, T. A. (2010). The scarab gut: A potential bioreactor for bio-fuel production. *Insect Science, 17*, 175–183.

Hungate, R. E. (1938). Studies on the nutrition of *Zootermopsis* II: The relative importance of the termite and the protozoa in wood digestion. *Ecology, 19*, 1–25.

Hussain, A., Li, Y. F., Cheng, Y., Liu, Y., Chen, C. C., & Wen, S. Y. (2013). Immune-related transcriptome of *Coptotermes formosanus* Shiraki workers: The defense mechanism. *PloS One, 8*, e69543.

Husseneder, C., & Grace, J. K. (2005). Genetically engineered termite gut bacteria (*Enterobacter cloacae*) deliver and spread foreign genes in termite colonies. *Applied Microbiology and Biotechnology, 68*, 360–367.

Huvenne, H., & Smagghe, G. (2010). Mechanisms of dsRNA uptake in insects and potential of RNAi for pest control: A review. *Journal of Insect Physiology, 56*, 227–235.

James, C. (2003). Global review of commercialized transgenic crops. *Current Science, 84*, 303–309.

Ji, R., & Brune, A. (2005). Digestion of peptidic residues in humic substances by an alkali stable and humic-acid-tolerant proteolytic activity in the gut of soil-feeding termites. *Soil Biology and Biochemistry, 37*, 1648–1655.

Jian-Zhong, S. & Scharf, M. E. (2010). Exploring and integrating cellulolytic systems of insects to advance biofuel technology. *Insect Science*, *17*(3), 163–165.

Johnson, E. (2009). Goodbye to carbon neutral: Getting biomass footprints right. *Environmental Impact Assessment Review, 29*, 165–168.

Jones, J. A. (1989). Environmental influences on soil chemistry in central semiarid Tanzania. *Soil Science Society of America Journal, 53*, 1748–1758.

Jones, C. G., Lawton, J. H., & Shachak, M. (1994). Organisms as ecosystem engineers. *Oikos, 69*, 373–386.

Katayama, N., Ishikawa, Y., Takaoki, M., Yamashita, M., Nakayama, S., Kiguchi, K., Kok, R., Wada, H., & Mitsuhashi, J. (2008). Entomophagy: A key to space agriculture. *Advances in Space Research, 41*, 701–705.

Ke, J., Sun, J. Z., Nguyen, H. D., Singh, D., Lee, K. C., Beyenal, H., & Chen, S. L. (2010). In-situ oxygen profiling and lignin modification in guts of wood-feeding termites. *Insect Science, 17*, 277–290.

Ke, J., Laskar, D. D., & Chen, S. (2013). Tetramethyl ammonium hydroxide (TMAH) thermochemolysis for probing in situ softwood lignin modification in each gut segment of the termite. *Journal of Agricultural and Food Chemistry, 61*, 1299–1308.

Keya, S. O., Mureria, N. K., & Arshad, M. A. (1982). Population dynamics of soil microorganisms in relation to proximity of termite mounds in Kenya. *Journal of Arid Environments, 5*, 353–359.

Kob, O. B., & Hewitt, P. H. (1990). Bird and mammal predators of the harvester termite *Hodotermes mossambicus* (Hagen) in semi-add regions of South Africa. *South African Journal of Science, 86*, 34–37.

Kramm, K. R., West, D. F., & Rockenbach, P. G. (1982). Termite pathogens: Transfer of the entomopathogen *Metarhizium anisopliae* between *Reticulitermes* sp. termites. *Journal of Invertebrate Pathology, 40*, 1–6.

Lacey, L. A., Frutos, R., Kaya, H. K., & Vail, P. (2001). Insect pathogens as biological control agents: Do they have a future? *Biological Control, 21*, 230–248.

LaFage, J. P., & Nutting, W. L. (1978). Nutrient dynamics of termites. In M. V. Brian (Ed.), *Production ecology of ants and termites* (pp. 165–232). London: Cambridge University Press.

Lamberty, M., Zachary, D., Lanot, R., Bordereau, C., Robert, A., & Hoffmann, J. A. (2001). Insect immunity constitutive expression of a cysteine-rich antifungal and a linear antibacterial peptide in a termite insect. *The Journal of the Biological Chemistry, 276*, 4085–4092.

Landis, D. A., & Werling, B. P. (2010). Arthropods and biofuel production systems in North America. *Insect Science, 17*, 220–236.

Lawton, J. H., Bignell, D. E., Bloemers, G. F., Eggleton, P., & Hodda, M. E. (1996). Carbon flux and diversity of nematodes and termites in Cameroon forest soils forest soils. *Biodiversity and Conservation, 5*, 261–273.

Lee, K. E., & Wood, T. G. (1971). *Termites and soils* (p. 251). London: Academic.

Lefeuve, P., & Bordereau, C. (1984). Soldier formation regulated by a primer pheromone from the soldier frontal gland in a higher termite, *Nasutitermes lujae*. *Proceedings of the National Academy of Sciences of the United States of America, 81*, 7665–7668.

Lisansky, S. (1997). Microbial pesticides. In H. F. Evans (Ed.), *Microbial insecticides: Novelty or necessity, British crop protection council monograph symposium proceedings* (vol. 68, pp. 3–10).

Liu, Y., Henderson, G., Mao, L., & Laine, R. A. (2005). Effects of temperature and nutrition on juvenile hormone titers of *Coptotermes formosanus*. *Annals of the Entomological Society of America, 98*, 732–737.

Lo, N., Tokuda, G., & Watanabe, H. (2011). Evolution and function of endogenous termite cellulases. In D. E. Bignell, Y. Roisin, & N. Lo (Eds.), *Biology of Termites: A modern synthesis* (pp. 51–67). Dordrecht: Springer.

Logan, J. W. M. (1992). Termites (Isoptera) – a pest or resource for small farmers in Africa. *Tropical Science, 32*, 71–79.

Logan, J. W. M., Cowie, R. H., & Wood, T. G. (1990). Termite (Isoptera) control in agriculture and forestry by non-chemical methods: A review. *Bulletin of Entomological Research, 80*, 309–330.

Logan, J. W. M., Rajagopal, D., Wightman, J. A., & Pearce, M. J. (1992). Control of termites and other soil pests of groundnuts with special reference to controlled released formulations of non-persistent insecticides in India and Sudan. *Bulletin of Entomological Research, 82*, 57–66.

Longhurst, C., Johnson, J. A., & Wood, T. G. (1978). Predation by *Megaponera foetans* (Fabr) (Hymenoptera; Formicidae) on termites in the Nigerian southern guinea savanna. *Oecologia, 32*, 101–107.

Machida, M., Kitade, O., Miura, T., & Matsumoto, T. (2001). Nitrogen recycling through proctodeal trophallaxis in the Japanese damp-wood termite *Hodotermopsis japonica* (Isoptera, Termopsidae). *Insectes Sociaux, 48*, 52–56.

Mamma, D., Hatzinikolaou, D., Kekos, D., Stamatis, H., & Kalogeris, E. (2009). Adsorption of major endoglucanase from Thermoascus aurantiacus on cellulosic substrates. *World Journal of Microbiology and Biotechnology, 25*, 781–788.

Martius, C. (1994). Diversity and ecology of termites in Amazo- nian forests. *Pedobiologia, 38*, 407–428.

Mathew, G. M., Mathew, D. C., Lo, S. C., Alexios, G. M., Yang, J. C., Sashikumar, J. M., Shaikh, T. M., & Huang, C. C. (2013). Synergistic collaboration of gut symbionts in *Odontotermes formosanus* for lignocellulosic degradation and bio-hydrogen production. *Bioresource Technology, 145*, 337–344.

Matsuura, K. (2001). Nestmate recognition mediated by intestinal bacteria in a termite, *Reticulitermes speratus*. *Oikos, 92*, 20–26.

Matsuura, K., Vargo, E. L., Kawatsu, K., Labadie, P. E., & Nakano, H. (2009). Queen succession through asexual reproduction in termites. *Science, 323*, 1687.

Mauldin, J. K., & Rich, N. M. (1980). Effect of chlortetracycline and other antibiotics on protozoan numbers in the eastern subterranean termite. *Journal of Economic Entomology, 73*, 123–128.

Menon, V., & Rao, M. (2012). Trends in bioconversion of lignocellulose: Biofuels, platform chemicals and biorefinery concept. *Progress in Energy and Combustion Science, 38*, 522–550.

Merheb-Dini, C., Gomes, E., Boscolo, M., & Silva, R. (2009). Production and characterization of a milk-clotting protease in the crude enzymatic extract from the newly isolated Thermomucor indicae-seudaticae N31 (milk-clotting protease from the newly isolated Thermomucor indicae-seudaticae N31). *Food Chemistry, 120*, 87–93.

Meyer, V. W. (1999). Distribution and density of termite mounds in the northern Kruger National Park with specific reference to those constructed by *Macrotermes Holmgren* (Isoptera: Termitidae). *African Entomology, 7*, 123–130.

Meyer-Rochow, V. B. (2010). Entomophagy and its impact on world cultures: The need for a multidisciplinary approach. In P. B. Durst, D. V. Johnson, R. N. Leslie, & K. Shono (Eds.), *Forest insects as food: Humans bite back* (pp. 23–36). Thailand: Food and Agriculture Organization of the United Nations.

Meyer-Rochow, V. B., & Chakravorty, J. (2013). Notes on entomophagy and entomotherapy generally and information on the situation in India in particular. *Applied Entomology and Zoology, 48*, 105–112.

Mielke, H. W., & Mielke, P. W. (1982). Termite mounds and chitemene agriculture: A statistical analysis of their association in southwestern Tanzania. *Journal of Biogeography, 9*, 499–504.

Misra, J. N., & Vijayaraghavan, P. K. (1956). Ethyl malonate – an inhibitor of termite cellulase. *Current Science, 25*, 229–230.

Miura, T., & Scharf, M. E. (2011). Molecular basis underlying caste differentiation in termites. In D. E. Bignell, Y. Roisin, & N. Lo (Eds.), *Biology of termites: A modern synthesis* (pp. 211–253). Dordrecht: Springer.

Moorhead, D. L., & Reynolds, J. F. (1991). A general model of litter decomposition in the northern Chihuahuan desert. *Ecological Modelling, 56*, 197–219.

Naik, S. N., Goud, V. V., Rout, P. K., & Dalai, A. K. (2010). Production of first and second generation biofuels: A comprehensive review. *Renewable and Sustainable Energy Reviews, 14*, 578–597.

Naveena, B. M., Mendiratta, S. K., & Anjaneyulu, A. S. R. (2004). Tenderization of buffalo meat using plant proteases from *Cucumis trigonus* Roxb (Kachri) and Zingiber officinale Roscoe (Ginger rhizome). *Meat Science, 68*, 363–369.

NBDS. (2015). *The national biotechnology development strategy (NBDS)-2015–2020*. Department of Biotechnology. Ministry of Science & Technology, Government of India. http://pib.nic.in/newsite/PrintRelease.aspx?relid=134035

Newmark, P. A., Reddien, P. W., Cebria, F., & Alvarado, A. S. (2003). Ingestion of bacterially expressed double-stranded RNA inhibits gene expression in planarians. *Proceedings of the National Academy of Sciences of the United States of America, 100*, 11861–11865.

Ni, J., & Tokuda, G. (2013). Ligno cellulose-degrading enzymes from termites and their symbiotic microbiota. *Biotechnology Advances, 31*, 838–850.

Noble, J. C., Diggle, P. J., & Whitford, W. G. (1989). The spatial distribution of termite pavements and hummock feeding sites on a semi-arid woodland in eastern Australia. *Acta Oecologica/Oecologia Generalis, 10*, 355–376.

Nuss, A. B., Forschler, B. T., Crim, J. W., TeBrugge, V., Pohl, J., & Brown, M. R. (2010). Molecular characterization of neuropeptide F from the eastern subterranean termite *Reticulitermes flavipes*. *Peptides, 31*, 419–428.

Nyamapfene, K. W. (1986). The use of termite mounds in Zimbabwe peasant agriculture. *Tropical Agriculture, 63*, 191–192.

Ogino, H., Otsubo, T., & Ishikawa, H. (2008). Screening, purification, and characterization of a leather-degrading protease. *Biochemical Engineering Journal, 38*, 234–240.

Ohkuma, M. (2003). Termite symbiotic systems: Efficient bio-recycling of lignocellulose. *Applied Microbiology and Biotechnology, 61*, 1–9.

Ohkuma, M., & Brune, A. (2011). Diversity, structure, and evolution of the termite gut microbial community. In D. E. Bignell, Y. Roisin, & N. Lo (Eds.), *Biology of termites: A modern synthesis* (pp. 413–438). Dordrecht: Springer.

Okot-Kotber, B. M., Ujvary, I., Mollaaghahaba, R., Szurdoki, F., Matolcsy, G., & Prestwich, G. D. (1991). Physiological influence on fenoxycarb pro-insecticides and soldier head extracts of various termite species on soldier differentiation in *Reticulitermes flavipes*. *Sociobiology, 19*, 77–89.

Olempska-Beer, Z. S., Merker, R. I., Ditto, M. D., & DiNovi, M. J. (2006). Food-processing enzymes from recombinant microorganisms –a review. *Regulatory Toxicology and Pharmacology, 45*, 144–158.

Oliviera, L. A., & Paiva, W. O. (1985). Use of termite nests and chicken manure as fertilizers for lettuce in red yellow podzolic soils of the Manaus region. *Acta Amazonica, 15*, 13–18.

Park, H. C., Majer, J. D., & Hobbs, R. J. (1994). Contribution of the Western Australian wheatbelt termite, *Drepanotermes tamminensis* (Hill), to the soil nutrient budget. *Ecological Research, 9*, 351–356.

Ragauskas, A. J., Williams, C. K., Davison, B. H., Britovsek, G., & Cairney, J. (2006). The path forward for biofuels and biomaterials. *Science, 311*, 484–489.

Rajamohan, F., Lee, M. K., & Dean, D. H. (1998). Bacillus thuringiensis insecticidal proteins: Molecular mode of action. *Progress in Nucleic Acid Research and Molecular Biology, 60*, 1–27.

Ramakrishnan, R., Suiter, D. R., Nakatsu, C. H., Humber, R. D., & Bennett, G. W. (1999). Imidacloprid-enhanced *Reticulitermes flavipes* susceptibility to the entomopathogen *Metarhizium anisopliae*. *Journal of Economic Entomology, 92*, 1125–1132.

Ramos-Elorduy, J. (2005). Insects: A hopeful food source. In M. G. Paoletti (Ed.), *Ecological implications of minilivestock* (pp. 263–291). Enfield: Science Publishers.

Rath, A. C. (2000). The use of entomopathogenic fungi for control of termites. *Biocontrol Science and Technology, 10*, 563–581.

Raubenheimer, D., & Rothman, J. (2012). Nutritional ecology of entomophagy in humans and other primates. *Annual Review of Entomology, 58*, 141–160.

Robinson, G. E. (1999). Integrative animal behavior and socio-genomics. *Trends in Ecology & Evolution, 5*, 202–205.

Robinson, G. E., Grozinger, C. M., & Whitfield, C. W. (2005). *Socio-genomics*: Social life in molecular terms. *Nature Reviews Genetics, 6*, 257–270.

Rosengaus, R. B., Traniello, J. F. A., & Bulmer, M. S. (2011). Ecology, behavior and evolution of disease resistance in termites. In D. E. Bignell, Y. Roisin, & N. Lo (Eds.), *Biology of termites: A modern synthesis* (pp. 165–191). Dordrecht: Springer.

Salick, J., Herrerra, R., & Jordan, C. F. (1983). Termitaria, nutrient patchiness in nutrient-deficient rain forests. *Biotropica, 15*, 1–7.

Schaefer, D. A., & Whitford, W. G. (1981). Nutrient cycling by the subterranean termite *Gnathamitermes tubiformans* in a Chihuahuan desert ecosystem. *Oecologia, 48*, 277–283.

Scharf, M. E. (2008). Silent pesticides. *Chemistry and Industry Managemenet, 11*, 20–23.

Scharf, M. E. (2015). Termites as targets and models for biotechnology. *Annual Review of Entomology, 60*, 77–102.

Scharf, M. E., & Boucias, D. G. (2010). Potential of termite based biomass pre-treatment strategies for use in bioethanol production. *Insect Science, 17*, 166–174.
Scharf, M. E., & Tartar, A. (2008). Termite digestomes as sources for novel lignocellulases. *Biofuels, Bioproducts and Biorefining, 2*, 540–552.
Scharf, M. E., Ratliff, C. R., Hoteling, J. T., & Bennett, G. W. (2003). Caste differentiation responses of two sympatric *Reticulitermes* termite species to juvenile hormone homologs and synthetic juvenoids in two laboratory assays. *Insectes Sociaux, 50*, 346–354.
Scharf, M. E., Buckspan, C. E., Grzymala, T. L., & Zhou, X. (2007). Regulation of polyphenic caste differentiation in the termite *Reticulitermes flavipes* by interaction of intrinsic and extrinsic factors. *The Journal of Experimental Biology, 210*, 4390–4398.
Scharf, M. E., Kovaleva, E. S., Jadhao, S., Campbell, J. H., Buchman, G. W., & Boucias, D. G. (2010). Functional and translational analyses of a beta-glucosidase gene (glycosyl hydrolase family 1) isolated from the gut of the lower termite *Reticulitermes flavipes*. *Insect Biochemistry and Molecular Biology, 40*, 611–620.
Scharf, M. E., Karl, Z. J., Sethi, A., & Boucias, D. G. (2011). Multiple levels of synergistic collaboration in termite lignocellulose digestion. *PloS One, 6*, 21709.
Sen, R., Raychoudhury, R., Cai, Y., Sun, Y., Lietze, V. U., Boucias, D. G., & Scharf, M. E. (2013). Differential impacts of juvenile hormone, soldier head extract and alternate caste phenotypes on host and symbiont transcriptome composition in the gut of the termite *Reticulitermes flavipes*. *BMC Genomics, 14*, 491.
Sethi, A., Slack, J. M., Kovaleva, E. S., Buchman, G. W., & Scharf, M. E. (2013). Lignin-associated metagene expression in a lignocellulose-digesting termite. *Insect Biochemistry and Molecular Biology, 43*, 91–101.
Shockley, M., & Dossey, A. T. (2014). Insects for human consumption. In J. Morales-Ramos, G. Rojas, & D. I. Shapiro-Ilan (Eds.), *Mass production of beneficial organisms* (pp. 617–652). New York: Academic.
Soares, C. A., Lima, C. M., Dolan, M. C., Piesman, J., Beard, C. B., & Zeidner, N. S. (2005). Capillary feeding of specific dsRNA induces silencing of the isac gene in nymphal *Ixodes scapularis* ticks. *Insect Molecular Biology, 14*, 443–452.
Solavan, A., Paulmurugan, R., & Wilsanand, V. (2006). Effect of the subterranean termite used in the South Indian folk medicine. *Indian Journal of Traditional Knowledge, 5*, 376–379.
Souza, P. M., & Magalhães, P. O. (2010). Application of microbial α-amylase in industry– a review. *Brazilian Journal of Microbiology, 41*, 850–861.
Sun, J. Z., & Scharf, M. E. (2010). Exploring and integrating cellulolytic systems of insects to advance biofuel technology. *Insect Science, 17*(3), 163–165.
Sun, Q., & Zhou, X. (2013). Corpse management in social insects. *International Journal of Biological Sciences, 9*, 313–321.
Tan, A., & Wong, N. (2013). Parameterization studies of solar chimneys in the tropics. *Energies, 6*, 145–163.
Tartar, A., Wheeler, M. M., Zhou, X., Coy, M. R., Boucias, D. G., & Scharf, M. E. (2009). Parallel meta transcriptome analyses of host and symbiont gene expression in the gut of the termite *R. flavipes*. *Biotechnology for Biofuels, 2*, 25.
Tarver, M. R., Zhou, X., & Scharf, M. E. (2010). Socio-environmental and endocrine influences on developmental and caste-regulatory gene expression in the eusocial termite *Reticulitermes flavipes*. *BMC Molecular Biology, 11*, 28.
Tarver, M. R., Florane, C. B., Zhang, D., Grimm, C., & Lax, A. R. (2012). Methoprene and temperature effects on caste differentiation and protein composition in the Formosan subterranean termite, *Coptotermes formosanus*. *Insect Sci., 12*, 18.
Tayasu, I., Abe, T., Eggleton, P., & Bignell, D. E. (1997). Nitrogen and carbon isotope ratios in termites: An indicator of trophic habit along the gradient from wood-feeding to soil-feeding. *Ecological Entomology, 22*, 343–351.
Terrapon, N., Li, C., Robertson, H. M., Ji, L., Meng, X., et al. (2014). Molecular traces of alternative social organization in a termite genome. *Nature Communications, 5*, 3636.

Timmons, L., & Fire, A. (1998). Specific interference by ingested dsRNA. *Nature, 395*, 854.

Toga, K., Hojo, M., Miura, T., & Maekawa, K. (2009). Presoldier induction by a juvenile hormone analog in the nasute termite *Nasutitermes takasagoensis*. *Zoological Science, 26*, 382–328.

Tokuda, G., Watanabe, H., Matsumoto, T., & Noda, H. (1997). Cellulose digestion in the wood-eating higher termite, *Nasutitermes takasagoensis* (Shiraki): Distribution of cellulases and properties of endo-beta-1,4-glucanase. *Zoological Science, 14*, 83–93.

USEPA. (2011). US Environmental Protection Agency: Policies concerning products containing nanoscale materials. *Federal Register, 76*, 2011–14943. EPA-HQ-OPP-2010-0197-0001.

Van-Huis, H. (2003). Insects as food in Sub-Saharan Africa. *Insect Science and its Application, 23*, 163–185.

Van-Huis, H. (2013). Potential of insects as food and feed in assuring food security. *Annual Review of Entomology, 58*, 563–583.

Vargo, E. L., & Parman, V. (2012). Effect of fipronil on subterranean termite colonies in the field. *Journal of Economic Entomology, 105*, 523–532.

Vasconcellos, A., & Moura, F. M. S. (2010). Wood litter consumption by three species of *Nasutitermes* termites in an area of the Atlantic Coastal Forest in northeastern Brazil. *Journal of Insect Science, 10*, 1–9.

Vaughn, T., Cavato, T., Brar, G., Coombe, T., Gooyer, D., Ford, S., Groth, M., Howe, A., Johnson, S., Kolacz, K., Pilcher, C., Purcell, J., Romano, C., English, L., & Pershing, J. (2005). A method of controlling corn rootworm feeding using a *Bacillus thuringiensis* protein expressed in transgenic maize. *Crop Science, 45*, 931–938.

Veivers, P. C., O'Brien, R. W., & Slaytor, M. (1982). Role of bacteria in maintaining the redox potential in the hindgut of termites and preventing the entry of foreign bacteria. *Journal of Insect Physiology, 28*, 947–951.

Verma, M., Sharma, S., & Prasad, R. (2009). Biological alternatives for termite control: A review. *International Biodeterioration and Biodegradation, 63*, 959–972.

Waller, D. A., & LaFage, J. P. (1986). Nutritional ecology of termites. In F. Slansky Jr. & J. G. Rodriguez (Eds.), *Nutritional ecology of insects, mites, spiders, and related invertebrates* (pp. 487–532). New York: Wiley.

Warnecke, F., Luginbühl, P., Ivanova, N., Ghassemian, M., Richardson, T. H., Stege, J. T., Cayouette, M., McHardy, A. C., Djordjevic, G., Aboushadi, N., Sorek, R., Tringe, S. G., Podar, M., Martin, G. H., Kunin, V., Dalevi, D., Madejska, J., Kirton, E., Platt, D., Szeto, E., Salamov, A., Barry, K., Mikhailova, N., Kyrpides, N. C., Matson, E. G., Ottesen, E. A., Zhang, X., Hernández, M., Murillo, C., Acosta, L. G., Rigoutsos, I., Tamayo, G., Green, B. D., Chang, C., Rubin, E. M., Mathur, E. J., Robertson, D. E., Hugenholtz, P., & Leadbetter, J. R. (2007). Metagenomic and functional analysis of hindgut microbiota of a woodfeeding higher termite. *Nature, 450*, 560–565.

Watanabe, H., Noda, H., & Lo, N. A. (1998). A cellulase gene of termite origin. *Nature, 394*, 330–331.

Watson, J. P. (1977). The use of mounds of the termite Macroter- rues falciger (Gerstacker) as a soil amendment. *Journal of Soil Science, 28*, 664–672.

Werfel, J., Petersen, K., & Nagpal, R. (2014). Designing collective behavior in a termite-inspired robot construction team. *Science, 343*, 754–758.

Wimmer, Z., Jurcek, O., Jedlicka, P., Hanus, R., & Kuldova, J. (2007). Insect pest management agents: Hormonogen esters (juvenogens). *Journal of Agricultural and Food Chemistry, 55*, 7387–7393.

Wood, T. G., & Cowie, R. H. (1988). Assessment of on-farm losses in cereals in Africa due to soil insects. *International Journal of Tropical Insect Science, 9*, 709–716.

Zhang, D., Lax, A. R., Bland, J. M., Yu, J., Fedorova, N., & Nierman, W. C. (2010). Hydrolysis of filter-paper cellulose to glucose by two recombinant endogenous glycosyl hydrolases of *Coptotermes formosanus*. *Insect Science, 17*, 245–252.

Zhou, X., Oi, F. M., & Scharf, M. E. (2006). Social exploitation of hexamerin, RNAi reveals a major caste regulatory factor in termites. *Proceedings of the National Academy of Sciences of the United States of America, 103*, 4499–4504.

Zhou, X., Wheeler, M. M., Oi, F. M., & Scharf, M. E. (2008a). Inhibition of termite cellulases by carbohydrate-based cellulase inhibitors: Evidence from *in vitro* biochemistry and *in vivo* feeding studies. *Pesticide Biochemistry and Physiology, 90*, 31–41.

Zhou, X., Wheeler, M. M., Oi, F. M., & Scharf, M. E. (2008b). RNA interference in the termite *Reticulitermes flavipes* by ingestion of double-stranded RNA. *Insect Biochemistry and Molecular Biology, 38*, 805–815.

Index

M.A. Khan, W. Ahmad (eds.), *Termites and Sustainable Management*, Sustainability in Plant and Crop Protection,
https://doi.org/10.1007/978-3-319-68726-1

MIX
Papier aus verantwortungsvollen Quellen
Paper from responsible sources
FSC® C105338

If you have any concerns about our products,
you can contact us on
ProductSafety@springernature.com

In case Publisher is established outside the EU,
the EU authorized representative is:
Springer Nature Customer Service Center GmbH
Europaplatz 3, 69115 Heidelberg, Germany

Printed by Libri Plureos GmbH
in Hamburg, Germany